Regional Development and Its Spatial Structure

Dadao Lu

Regional Development and Its Spatial Structure

Springer

Dadao Lu
Institute of Geographic Sciences
and Natural Resources Research
Chinese Academy of Sciences
Beijing, China

ISBN 978-981-99-7682-9 ISBN 978-981-99-7683-6 (eBook)
https://doi.org/10.1007/978-981-99-7683-6

This project was sponsored by the National Natural Science Foundation of China.

Translation from the Chinese Simplified language edition: “科学出版社” by Dadao Lu, © The author 1995. Published by Science Press. All Rights Reserved.

This Springer imprint is published by the registered company Springer Nature Singapore Pte Ltd.
The registered company address is: 152 Beach Road, #21-01/04 Gateway East, Singapore 189721, Singapore

Preface

Regional development issues have become major global socio-economic issues since the 1950s. New ideas and theories have appeared from time to time. The number of scholars engaged in research in this field in many countries is increasing, and much progress has been made in the process of solving practical problems in social and economic development. Accordingly, disciplines such as regional science, regional economics, and social and economic geography adapt to regional social and economic development needs and develop exceptionally actively. The research on the issues of regional development has become the frontier of the development of these disciplines.

Research on regional development issues before the 1980s in China was lifeless. It tended to be formulaic due to implementing a management system of authority centralization and planned economy plus the spiritual shackles. A few works and papers on the issues of China's regional development (regional exploitation) are limited to discussing the macro-policies proposed by the leaders. Those beyond this scope are carefully bracketed. In this regard, the author did some reviews in the *Theory and Practice of China's Industrial Distribution* published in 1989. The ideological opening-up has led to the gradual decentralization of the national economic management authority since 1978, and the local economic strength has increased rapidly. Meanwhile, the difference between the rich and the poor among regions has become a serious issue, particularly concern by the local and central governments. This is a great impetus to the research on regional exploitation and development issues. More than ten major works on the regional economy, territorial development, and industrial distribution have been published in recent years. Among them, the most recommended are as follows: *Theory and Practice of Economic Layout* by Chen Dongsheng (1989), *Regional Economics* by Zhou Qiye, Liu Zaisheng, et al. (1989), *China's Regional Industrial Structure* by Zhou Shulian, et al. (1991), and *A Study on China's Regional Development* by Yang Kaizhong (1989). In terms of purely theoretical research, there are the *Principles of Location Theory* by Yang Wuyang (1989), *Spatial Analysis of Economic Activities* by Chen Zongxing, etc.

In China, the research on issues of location and layout has not attracted the interest and attention of economists (mainly regional economists) for a long time. Therefore, it seems to be only a field of concern for geographers. Of course, this has been broken through since the 1980s. My German friend, P. Klemmer, the director of the Nordrhein-Westfalen Economic Research Institute and a regional and environmental economist, had been in touch with his Chinese colleagues several times and expressed his feeling (early 1988): He did not know why Chinese economists ignored the issues of social and economic location. When I entered Ruhr-University Bochum in Germany as a visiting scholar in 1980, I saw that the names, work, and theories of famous scholars such as economist A. Weber (Germany), regional economist W. Isard (USA), geographer W. Christaller (Germany), W. L. Garrison (USA), and T. Hagerstrand (Sweden) appeared widely in textbooks, dictionaries, and quotations of economics and geography. There were no boundaries of disciplines. This is the initial inspiration for connecting regional development to spatial structure in this book. In the research work for over ten years, especially through the summary and analysis of territorial development, regional development process, and experience for decades in China, I sincerely feel that spatial distribution and spatial structure are essential aspects and contents of regional development. The internal mechanism and state of regional development cannot be scientifically revealed and understood if the connection and integrity of the two are separated.

In the early and middle 1980s, academia summarized the gains and losses of China's territorial and regional development policies for over 30 years. The trend of "strategic shift" was also prevailing in society. I was writing a paper submitted to the National Economic Geography and Land Economics Symposium convened in Urumqi at the end of September of that year from July to August 1984, trying to conceive new spatial concepts and models for China's social and economic development in the future. Based on my long-term experience, and according to W. Christaller's central place theory and the development axis assumption proposed in the regional planning of Bavaria, Germany, I initially proposed the "point-axis system" theory and the T-shaped structure of China's territorial development and economic development, that is, the space strategy that the coastal zone and the coast along the Yangtze River will be taken as China's first-level development axis in the next few decades. I participated in the first compilation of the territorial planning outline from 1985 to 1987. The T-shaped structure and the concept of the "point-axis system" were written into this outline. The outline was sent to the whole country for trial implementation. In the past decade, the "point-axis system" model was adopted for more than 20 provincial and territorial planning, nearly a hundred regional (or municipal) territorial planning, hundreds of county-level territorial planning, and socio-economic development strategy planning (research). All these regions concentrated on building their key axis (industrial belt). The thinking habit of determining key and non-key regions by cutting blocks, which had been popular in China for a long term, was changed. The resulting economic and social benefits were incalculable. Considering that only basic models and concepts were initially proposed at that time, I wanted to have a systematic and in-depth discussion on the spatial structure and the "point-axis system" since the mid-1980s. This task is coming to an end now.

This book attempts to preliminarily clarify some of the main issues concerning the modern regional development theory in combination with China's regional development practice. The book can be divided into three parts: (I) The progress and main theoretical achievements of research on regional development issues at home and abroad, including Chaps. 1–3; (II) The relationship between industrial structure, resources, environment and regional development, including Chaps. 4 and 5; (III) The fundamental aspects of spatial structure theory and its relationship with regional development, including Chaps. 6–9. Among them, the chapters in Part II and III are the emphases of this book and are also the place where the author deliberately makes innovations by using the research findings of the predecessors.

At the beginning of my writing, this book was named "Spatial Structure of Regional Development". But very soon, I felt that it was necessary to systematically summarize the regional development theory proposed and the work done by our predecessors. Therefore, the progress of the regional development theory must be discussed. Chapter 4, "Industrial Structure and Regional Development", is a subject that scholars have studied extensively. This chapter only systematizes theoretical viewpoints and makes regular explanations of the evolution of the industrial structure in the regional development process. The issues of resources, environment, and regional development are hot topics in current social development and are the author's new research direction in recent years. This book has new insights into the environmental crisis in the next quarter of a century, the strategies for getting rid of the dilemma, the regional system of "human and nature", the establishment of a resource-saving social and economic system in China, etc. From Chaps. 6–9, I focus on the discussion of the theoretical concepts of location differential rent, point-axis progressive diffusion, point-axis spatial structure system, regional accessibility, etc., and use a large number of actual conditions of regional development at home and abroad to describe these theories in detail, in the hope of deepening the explanation of the theoretical models. The last chapter covers the discussion and prediction of the spatial effects on technological innovation of a highly informational society in the future.

This book's writing conception and data work started at the end of 1988. Only Chaps. 2, 3, 8, and 9 had been completed as of October 1992. The working environment became increasingly difficult due to the challenges for more than two years after the summer of 1991. I was encouraged by my friends at this moment. My writing accelerated due to these two reasons. Finally, Chap. 7 was completed in June 1994.

At the time of publishing this book, I still feel that some theoretical issues have not been elaborated on in this book. I hope to get corrections from teachers, colleagues, and readers and intend to make further improvements in the future research.

Beijing, China
November 1994

Dadao Lu

Introduction

This book describes the progress and prominent theories of regional development research at home and abroad in the past decades, especially in the past decade, discusses the industrial structure, spatial structure, resources, and environment, as well as a series of practical issues, and reveals the general characteristics of spatial structure evolution in the process of regional development. This book provides an in-depth and systematic demonstration of the "point-axis system" theory of regional exploitation and development as well as the T-shaped structure of China's regional economic action in theory and practice and discusses the impact of location differential rent, restricted accessibility, technological innovation, etc., on regional development theoretically. This book could be used as a reference for planning, scientific research, and teaching personnel in territorial expansion, regional economy, human geography, etc.

Contents

Chapter 1
Regional Development Is a Major Socio-economic Issue in the Contemporary World

1 Connotation of the Concepts of Regional Exploitation and Regional Development

Since the 1980s, with China's reform and opening up and the development of the national economy and regional economy, regional development and regional development have become a topic that governments at all levels and academic circles attach great importance to, and the concepts of "development" and "development" have become widely popular. In Western languages, "regional exploitation" and "regional development" is the same concept. The primary connotation is: in the context of national macroeconomic growth, the economic aggregate of the region is increased, the population and the per capita income are increased, the material and social infrastructures are continuously improved, reasonable economic relations are established among regions, the gap in the level of socio-economic development among regions is gradually reduced, and the regional policies are developed for this goal. Among China's planning departments and theoretical analysts, "regional exploitation" has been widely used, and the popularity of the concept of "regional development" is expanding. China is a developing country with a vast territory, rich and diverse natural conditions, and resources. It has a short history of industrialization, and the development level is quite insufficient. This situation determines that the primary activities in the regional development process in China account for a large proportion. The application of the concept of "regional exploitation" can more accurately reflect the essential characteristics of China's current industrialization in the early and middle stages in terms of regional social and economic development.

In China, the concepts of "regional exploitation" and "regional development" have the same connotation but also have noticeable differences.

(1) "Regional exploitation" mainly refers to the development (including exploitation) and utilization (including processing) of various natural resources in the region, such as mineral resources, water resources, energy resources, biological resources, etc., the development of new industries, the manufacturing of new

D. Lu, *Regional Development and Its Spatial Structure*,
https://doi.org/10.1007/978-981-99-7683-6_1

products, the construction of new facilities, the development of new technologies, etc. It can be said that the objects and processes involved in "regional exploitation" are material, tangible and specific. The results are an increase in output and output value, technological progress, etc. In addition to these contents, "regional development" also includes the growth in the total amount of society, economy, and industry in the region, the rationalization of internal structure and foreign economic, technological, and social relations, the spatial flow of social and economic factors, the regional equalization of the socio-economic development level, the improvement in urbanization and education and cultural levels of population, etc. Therefore, in addition to material and tangible activities, it also includes non-material and abstract contents.

(2) "Regional exploitation" aims at regional economic development. However, in addition to economic development, "regional development" also refers to the improvement of people's living and production conditions, the improvement of welfare standards, and the reasonable development of the population in terms of social development; it refers to maintaining ecological conditions and ecological balance in terms of environmental renovation. Therefore, "regional exploitation" indicates intensive economic activities in the early and middle stages of the industrialization of countries and regions. The two are different in level, and the extension of the concept of "regional development" is broader. While "regional development" comprehensively reflects the social-economic activities in various developing stages, especially the social-economic activities in the advanced industrialization and post-industrialization stages, which are development activities that improve the region's socioeconomic development and are more refined.

(3) The concept of "regional exploitation" is more embodied as the transition process from one state or stage to another state or the process from nothing to everything. It emphasizes "development," "promotion," and "mutation." But the concept of "regional development" emphasizes a gradual process and a process of improvement. That is to say, the former has the characteristics of a section, while the latter reflects a continuous process and has historical characteristics. In terms of space, "regional exploitation" emphasizes the change of the original state of the new region or the "mutation" from one state to another state; "regional development" emphasizes the continuous change, improvement, and deepening of the developed region. That is to say, "exploitation" means expanding space, including developing new regions. While "development" is generally based on the development of the developed regions. Of course, such division is not absolute. The spatial structure of "regional development" is predominantly discussed in this book, which includes spatial changes.

(4) The objects and goals involved in "regional development" fundamentally refer to economic development and "human" development. This includes the improvement of material, cultural and living standards, as well as the realization of people's value orientation. All things conducive to human development and all space-related processes shall belong to the contents of "regional development." The urbanization process here is an example. Changes in industrial structures of

the countries and regions, changes in people's lifestyles, and the improvement of living standards are reflected in this process. People's values of living and behaviors vary significantly under different cultural backgrounds and historical traditions. Therefore, there are considerable differences in the understanding of "regional development" in other countries and nationalities. This is different from what we usually call "regional exploitation" in terms of connotation.

In addition to the differences in the above four aspects, the contents discussed in "regional exploitation" are limited to issues within the region, while "regional development" also involves more inter-regional matters, that is to say, putting the researched region in the relevant region groups for investigation. Moreover, the discussion from the policy perspective often refers to the "regional development" policies, "regional policies" for short. Despite the above differences in the concepts of "regional exploitation" and "regional development" in China, their consistent meanings are still the mainstay. Moreover, the concept of "regional development" may cover "regional exploitation." This characteristic has become more apparent with the progress of industrialization and modernization in various regions of China. This is the main reason this book's title is "Regional Development and its Spatial Structure" instead of "Regional Exploitation and Its Spatial Structure." However, the things and objects concerning "regional exploitation" are still the main discussion points in the book.

2 Major Socio-economic Issues in the Contemporary World

In the last century, Western countries started the process of industrialization and constantly carried out economic plunder and political control of colonies. Relying on the plundered wealth and occupied markets, these countries achieved unprecedented domestic economic growth. Moreover, they repeatedly waged wars of all sizes to compete for raw materials and markets. This situation lasted until the 1940s. That is to say, as far as the world is concerned, a general trend had not been formed for the development issues before World War II. Many countries couldn't place the problems of comprehensive large-scale exploitation and regional development in an important position. There lacked necessary international and domestic environments for stable growth.

As far as the world is concerned, there has been a peaceful situation for nearly half a century since the 1950s. Due to changes in the world's political map, countries and regions that were initially colonies have become independent one after another, the developing countries have stepped onto the world stage as critical political forces, and regional economic and political organizations consisting of developing countries have emerged in succession. Under political independence, developing countries have begun to build their resources and national economy in a planned way. Meanwhile, the past's fierce military wars and political struggles have been gradually replaced by economic, scientific, and technological competitions among developed capitalist countries. Every country is utilizing the resources they have to develop its favoured industries, and the implementation of industry policies is the facilitator for the country's socioeconomic development. In such a context, regional development issues have also become important in developed countries. The welfare standards of these countries are high due to highly developed economies and technologies, but the development levels among domestic regions are not balanced. Therefore, it is necessary to create and implement policies conducive to narrowing the development gap among regions, that is, regional policies.

The world's most outstanding problem is how to realize sustainable socio-economic development in developing countries, narrow the gap between developing and developed industrialized countries, and establish a new order of international economic relations, the so-called "North–South Problem." Industrialized countries recovered and proliferated during the "oil crisis" from the 1950s to the early 1970s. The average annual growth rate of GNP in high-income countries and regions reached 3.7% from 1965 to 1980, including 6.3% in Japan, 2.7% in the United States, and 4.3% in the former Federal Republic of Germany. Although the growth rate of many developing countries is higher than this rate, the annual growth rate of per capita GNP is almost the same in developing and developed countries. Due to the huge differences in economic and technological foundations and economic structures, the gap between developed and developing countries has been widened. Of course, a few number of developing countries and regions have made full use of favorable resources, locations, and opportunities, changed their industrial structures promptly, participated in the international economic cycle process, and obtained many international capitals, which has further promoted domestic development.

Economic struggles and cooperation among regions and countries in the world have been closely related to the economic and even political interests of these countries for more than forty years. How to obtain the opportunities for development in such cooperation and struggles and give full play to their advantages has become the focus of industrial policies as well as foreign trade import and export policies of various countries and regions. The differences in the GDP growth and the per capita GDP are affected by the differences in natural, economic, and even political factors among various regions in some countries. How to narrow and even eliminate such differences has also become an issue of concern for the governments of all countries. This issue of regional development is quite prominent in the developed countries, it has also become one of the issues of domestic development in the developing countries (such as China).

All countries in the world generally encounter the issues of population, resources and environment in contemporary times. However, the issue of regional development is the reflection of the three issues in specific countries and regions, which are of the same universality, urgency and importance.

(1) The industrial policies of the countries and regions, including the selection of leading industries and pillar industries, technology policies, spatial distribution policies, price policies, trade policies, etc., are related to the advantages and disadvantages or the success and failure of the competition among regions and

countries. Scientific industrial policies are the basis for ensuring the success of the global competition. The new markets can be constantly occupied and consolidated by timely transferring the industrial structure and standing at the forefront of the development of new industries in the world. In order to meet the needs of the economic system dominated by manufactured goods in international trade, many countries and regions are competing to develop their own processing industries. Some countries have succeeded, while other countries are unable to achieve their goals.

(2) Gaps and conflicts in development levels, price policies and infrastructure policies among countries and regions are also quite common. If there are no acceptable regional policies for adjustment, it will also cause hazards to the development of the region and even the entire country as well as the political and economic stability. Therefore, narrowing the gap in the development level among regions has become a common concern for the governments of all countries. In this aspect, a series of regional policy measures, namely "regional policy tools" have been developed, with the implementation goal of promoting balanced development among regions. Although various countries have made many efforts in this aspect, there are still huge regional differences in the economic development level measured by the income per capita.

(3) The development issues are always related to the competition among countries and groups of countries, especially related to the "North–South Problem" on a global scale. Global and regional political and economic organizations, such as the United Nations Development Programme, World Bank, Asian Bank, Organization of Petroleum Exporting Countries, and various regional economic communities, have developed a large number of policies to promote the development of various countries and regions, and coordinate the interests of various types of countries, including regulation by means of the assistance loan. "Development" has become one of the two major goals of all activities of the United Nations. It can be seen how important the issues of national and regional development are in the world. However, due to the complex patterns in world trade, the gap in the level of economic development between developed countries and developing countries has been widened as a whole rather than being narrowed.

(4) The issue of "development" has become a worldwide socio-economic issue, thus becoming the subject of research by scientists in many disciplines, and a series of disciplines have been developed therefrom. In terms of economics, macroeconomics and development economics study the relationship between GNP growth, population growth, per capita income growth, profit distribution, investment, savings, and other factors by taking the country as a unit; and regional economics focuses on the role of factors in regional development, the flow and balanced development of factors among regions, regional policies, etc. In terms of geography, there is mainly economic geography, which develops from regional description to regional analysis until the research on the regional development strategies, exerting the characteristics of "subject methodology". In the 1950s, "regional science" was created when scholars of economics, geography, and system science cooperated to jointly solve major practical problems

in regional development. This is an interdisciplinary subject that focuses on the issues of regional development. In recent decades, fruitful achievements have been made in the research of "regional development", and systematic progress has been made in theoretical methods. It can be said that a situation where politicians and scientists cooperate to jointly solve contemporary development problems has been basically formed.

3 Huge Socio-economic Development in the World

For more than 40 years since the end of World War II, world politics has been in a relatively stable period, and great socio-economic development has been achieved in most countries and regions, which is manifested in the substantial growth of population, GNP and per capita GDP, as well as a significant improvement in the social and infrastructure level. Calculated based on the value of US dollars in 1980, the average annual growth rates of the world's gross economic output from 1950 to 1973 and from 1973 to 1979 were 5.0% and 3.5%, respectively [1].

The large-scale socio-economic development is specifically manifested in the following main aspects.

(1) Large-scale development and utilization of natural resources, especially land, water, coal, petroleum, natural gas, iron ore, forests and other resources. From 1950 to 1984, the world's energy production increased by 2.7 times from 2.54 billion tons to 9.36 billion tons in terms of standard coal. Among them, the amount of petroleum extraction increased by 4.3 times [2] from 790 million tons to 4.16 billion tons in terms of standard coal [2]. The scale of development and utilization of various natural resources in China from 1950 to 1990 was amazing. Among them, the amount of coal mining increased by 31 times, the amount of hydropower development increased by 67 times, and more than 200 million mu of uncultivated land was reclaimed. In the tropics of the world, the annual deforestation area reached more than 11 million hectares. Due to the socio-economic development, especially the rapid development of industrialization and urbanization, the scale of development and utilization of water resources by human beings has also been increasing rapidly. According to the data from the United Nations Water Conference, the annual consumption of fresh water by the world's industry, agriculture, cities, and reservoirs increased from 400 to 3000 km^3 from 1900 to 1975, with an average annual increase of 2.5%. It is estimated that the water consumption will double by 2000, reaching 6000 km^3.

    The socioeconomic development of human beings is a process of applying capital, technology, and labor to nature and various elements thereof. Through this process, more and more natural resources will be included in the socioeconomic cycle, to achieve national and regional development.
(2) Increase in population and improvement of population quality. The world population was three billion in 1960, and reached 5.2 billion in 1989, with an increase of 79% and an average annual growth of about 1.8%. The increase in population

was very prominent in most developing countries. In the recent 20 years, the annual growth rates of populations in India, the Philippines, Pakistan, Mexico, Brazil, and Turkey are around 2%, and the annual growth rate of the population in China is 1.5–1.7%, which is slightly lower than the world average [3]. In developed countries, the population growth rates of North America, Oceania, and the former Soviet Union are also higher than those of Western European countries. The improvement of the population quality level can be expressed in terms of educational level, death rate, the number of scientific and technical personnel and its proportion to the total population, etc. Scientists and engineers in the United States accounted for 1.38% of the total population in 1982 (excluding humanities and education); the proportion was 6.69% in Germany (including the former German Democratic Republic) in 1987 (including technicians); the proportion was 0.32% in India and 0.93 in China (referring to the number of natural science and technical personnel only) in 1985.

(3) Expansion of the industry scale and growth of the national income and GNP. This is the most important symbol of socio-economic development in the world for nearly half a century. In a quarter century from the 1950s to the mid-1970s, the world's economic growth rate was stabilized at a fairly high level of 4% (if this growth rate could be maintained, the economic scale will increase by 50 times within a century) [4]. The average annual growth rate was 2.2–4.3% in the United States, Federal Republic of Germany, Britain, and Italy, 6.3% in Japan, and 7.0–9.0% in India, Brazil, Indonesia, and Mexico from 1965 to 1990; and the average annual growth rate of GDP was 0.8–2.5% in the United States and other major western capitalist countries, and 1.0–6.0% in India and other major developing countries from 1980 to 1985. The world's economic growth had declined since 1975, and the main reason was the global economic downturn caused by the oil crisis. Although there was a momentum of improvement since the second half of the 1980s in some countries, the growth trend slowed down. Industrial production increased by three times worldwide from 1952 to 1985. Among them, the steel output, the crude oil output, the electricity output, and the automobile output increased by 1.6 times, over 4 times, 8 to 9 times, and 3.5 times respectively.

(4) Constant evolution and advanced development of the industrial structure. While the economic aggregate has achieved substantial growth, the economic industrialization process has also evolved in an accelerated way worldwide. A series of developed countries have basically completed the transition from the stage based on the traditional industry to the stage based on the new industry and high-tech industry. These countries include Japan, the United States, Germany, etc. Among them, emerging industries centered on electronic technology and the automobile industry account for an increasing share of the national economy in Japan. From 1955 to 1975, the product sales amount of electrical machinery (including electronic computers, integrated circuits, household appliances, etc.) increased by 42 times, with an average annual growth rate of 20.7%; however, the average annual growth rates of traditional industries, such as steel and textiles, were only 15.4% and 9.3% respectively. In major agglomeration areas in Japan,

such as Hanshin, Chukyo, and Keihin, the industrial structure has generally gone through a transformation in the process of advanced processing from the situation before the high growth rate to date (see Sect. 3 in Chapter "Resources, Environment and Regional Development" of this book).

The transformation and upgrade of the industrial structure of various countries in the process of industrialization follow roughly the same law worldwide. In the initial stage, most countries concentrate on the development of the light and textile industry; then, the fund is agglomerated through the light and textile industry for the development of energy, steel, and other basic industrial sectors. In the third stage, the raw material industry marked by the petrochemical industry and machine manufacturing is developed. In the fourth stage, the emphasis is put on the development of industries such as the electronics industry and automobiles. This regularity is reflected in the cargo composition in international trade in a centralized way. According to the World Bank's analysis of the exported commodity structures of different countries by income level, the trends of countries at different levels are roughly the same. From the 1960s to the 1980s, the proportion of resource products declined, while the proportion of mechanical products rose in terms of the total value of exported commodities (Table 1).

(5) Improvement of living welfare standards. With the improvement of national income levels in various countries, most governments of all countries focus on improving the welfare standards of the people. This includes the construction of advanced transportation and communication systems, medical and health systems, commercial service facilities, culture and education systems, etc. With the improvement of the welfare standards, the average life expectancy of the population has increased, the infant mortality rate has decreased, and the adult illiteracy rate has decreased. For example, the mortality rate in low-income countries decreased from 16 to 10‰ from 1965 to 1987. The per capita daily calorie supply increased from 8330.74 to 9965.12 J in low-income countries, from 10,295.34 to 11,933.9 J in middle-income countries, and from 12,886.94 to 14,107.5 J in high-income countries [4] from 1965 to 1987.

(6) Change of socio-economic spatial structure from one-dimensional space to three-dimensional space. The relationship between socio-economic spatial distribution and spatial agglomeration changes with the level of economic and technological development. In a relatively long period (roughly to the mid-stage of industrialization, that is, from high-speed growth to a period of stable and low-speed growth), the spatial movement (displacement) of the social economy tends to be agglomeration. The main acting force is economic benefits, that is, the company mainly seeks cost reduction, profit increase and expansion of the production scale, and the government seeks an increase in tax and employment. In the middle and late stages of industrialization, social factors (such as the principle of fairness in the regional competition, the principle of equal development opportunities, and the principle of ensuring corresponding development in remote regions and regions inhabited by ethnic groups) will play a significantly greater role than before. In this case, the role of economic factors

Table 1 Changes in exported commodity structures of countries with different income levels (%)

Country categorization	Commodity structure									
	Fuels, minerals and metals		Other primary products		Machinery and transportation equipment		Other manufactured products		Textiles and clothes	
	1965	1987	1965	1987	1965	1987	1965	1987	1965	1987
Low-income countries (regions)	22	29	53	22	1	4	23	45	11	7
Including: China	6	14	48	16	3	4	43	66		
Middle-income countries (regions)	35	23	53	20	0	16	13	43	3	12
Lower-middle-income countries (regions)	27	26	59	27	0	13	12	34	2	7
Upper-middle-income countries (regions)	40	22	46	15	3	25	13	40	4	15
High-income countries (regions)	11	9	20	12	30	39	39	39	7	5
Including Japan	2	1	7	1	31	65	60	32	17	3
The United States	8	6	27	16	37	47	28	31	3	2

Source World Bank, World Development Report 1989, Pages 194 to 195, China Financial and Economic Publishing House declines in the decision weight combination. The decision made on socio-economic location may deviate from the economically optimal scheme, but extend along with the two-dimensional space of economy and society. When the economy and technology reach a highly developed stage, generally considered the post-industrialization stage, human beings will act on nature with powerful technical means, thus strongly changing the natural structure and the structure of the ecological environment. Environmental protection and maintenance of ecological balance will become important decision-making factors. This factor is increasingly affecting the resource development direction, industry selection and spatial distribution of countries and regions. In this case, the decision on the socio-economic spatial location is affected by three factors, that is, economy, society and environment, at the same time, the changes in the spatial structure extend along the three-dimensional space. That is, the social-economic activities of human beings are carried out in a three-dimensional space.

The changes in the stages of the above factors affecting the spatial structure started in the 1930s from one-dimensional to two-dimensional in Western developed countries. Social issues were generally valued in the 1960s, and the spatial structure with the two-dimensional spatial characteristics appeared widely; the increase in CO_2 concentration in the atmospheric layer, the destruction of the ozone layer and industrial pollution were increasingly serious in the 1980s. Ecological environment protection became an issue to which the political leaders, scientific and technological circles, and economic circles in many countries attached great importance, and gradually deformed the structural graphics in the two-dimensional space again. Of course, the three-dimensional characteristics of spatial structure graphics are not universally visible, and do not commonly appear in developed countries.

Economic development in the three-dimensional space is restricted by economic benefits, as well as social justice and control of the ecological environmental degradation. Therefore, the growth rate is lower than that of the two-dimensional space and the one-dimensional space. However, it is a relatively coordinated and highly developed state from the perspective of the coordination between human and nature and the population-resources-environment (Fig. 1).

(7) Continuous progress in technological innovation. Socio-economic development is manifested in the increase in total production, the improvement of people's material and spiritual living standards, the improvement of infrastructure, etc. However, it is promoted by the development of science and technology in the final analysis. Therefore, technological innovation is one of the core contents of development. In each historical stage of human society, its development level and characteristics are determined by the state of science and technology that represents the productivity level at that time.

In the development of human society, every major scientific and technological discovery or invention has caused major changes and upgrades in the industrial structure and social structure at that time. The development of the world's navigation technology, shipbuilding technology and iron smelting technology in the

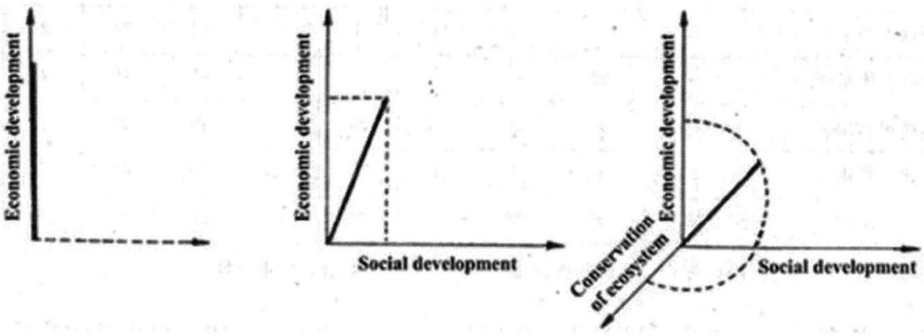

Fig. 1 Economic development in the three-dimensional space

sixteenth century promoted the prosperity of industry and commerce in capitalist countries at that time, which made the textile industry, machinery industry and steel industry gradually become the main departments of Western countries. Since the 1950s, the leading industries in developed countries have undergone several upgrades. The general rule is: steel and energy → petrochemicals and machinery → electronics and automobiles. These upgrades are also promoted by technological innovation.

Another significance of technological innovation is to promote the improvement of the economic benefits of production, which is mainly manifested in the improvement of labor productivity and quality. A huge increase in the total output value can be promoted just because of this. This is one of the main symbols of modern economic growth. It indicates that science and technology are the primary productive forces and one of the fundamental driving forces to promote socio-economic development.

(8) Expansion of difference in the level of inter-regional development. In the development of human society, the differences in the level of land and resources development and the level of socio-economic development among regions do not exist at the beginning. In other words, the differences among regions are not obvious, or even do not exist in the early stage of socio-economic development. With the expansion of the regional resource development as well as the development of economy and technology, the differences in resource endowment, location advantages and disadvantages, and people's living environment among regions start to play a role. Firstly, the differences between urban and rural areas appear. Secondly, the differences between inland and ocean, between resource-rich areas and resource-poor areas, and between transportation hub areas and remote areas with underdeveloped transportation are generally formed. Due to the accumulation of economic differences, and even the roles of different social factors, there are huge differences in the socio-economic level, industrial structure, scientific and cultural level, as well as the status in international political and economic relations among different regions and countries. Such

differences are often manifested as stability and sustainability; in certain development stages, it is difficult to reverse such differences once formed, and the gap may even be getting bigger and bigger.

In a country, the differences among regions and even the gaps in the socio-economic level may be reduced and eliminated. When the economy grows and accumulates at a speed that exceeds that of the average value for a long time in a favorable location, the negative effect of agglomeration will appear at a certain time. At the moment, the government will often implement policies that restrict further agglomeration, so as to ensure continued economic growth and social justice. The gap among regions may be gradually narrowed under such policies. Most of the gaps among regions in some developed industrialized countries are gradually narrowing. However, there is a certain degree of expansion in recent years.

References

1. L.R. Brown, *A Look to the Whole World* (China Translation & Publishing Corporation, 1985)
2. Edited by the International Information Statistics Center of the National Bureau of Statistics, *Statistical Compilation of World Industry* (China Statistics Press, 1988), p. 251
3. National Bureau of Statistics, *China Statistical Yearbook* (China Statistics Press, 1991), p. 819
4. World Bank, *World Development Report 1989* (Chinese Financial & Economic Publishing House, 1989), p. 216

Chapter 2
Progress of Foreign Research on Regional Development

"Development" is a word proposed after World War II. The background at that time was: The countries devastated by the war were recovered and restored on a large scale, and a large number of socio-economically backward countries (that is, newly called "developing countries" since the 1960s) took advantage of the relatively peaceful world to develop their economies. Economic struggles gradually replaced international military struggles. With the development of the world economy, various countries are also paying more and more attention to social development. As a result, "development," as a proper noun with today's meaning, has spread widely.

As a broad and profound socio-economic issue, the issue of "development" has attracted the attention of governments, business people, and scientists from many countries in the world, and has become an important research object in economics, regional economics, development economics, regional science and economic geography, and socio-cultural geography. A large number of empirical and theoretical research findings have come out, and great progress has been made in methodology. The guiding significance for the socio-economic progress of countries and regions is more and more obvious. The United Nations Development Program (UNDP) and the World Bank guide and support the exploitation and development of some countries in the world (mainly developing countries); the former is an official institution of the United Nations, and the latter is a large financial institution in the world. Both have powerful research institutions and play an important role in formulating development policies, preparation and consultation of development schemes, and financial support for socioeconomic development in various countries and regions.

The word, "development", as a discipline, was widely used and spread after World War II, but systematic research on the issue of "development" only appeared in the recent 30–40 years. However, the research related to "development", especially "regional development", has a very long history.

D. Lu, *Regional Development and Its Spatial Structure*,
https://doi.org/10.1007/978-981-99-7683-6_2

1 Research on International Trade and Inter-regional Work Division

With the development of capitalism and the commodity economy, international trade has become an important factor and condition for promoting the economic development of various countries. Adam Smith said in *The Wealth of Nations* (1776) that: "If an item is purchased at a lower cost than it is produced at home, you will never want to produce it at home. This is a motto that every astute parent knows. ... If foreign countries can supply us with commodities cheaper than those manufactured by us, we'd better use part of the products produced by our favorable industries to buy from them." (The Commercial Press, 1979 edition, Page 28). After a large number of case studies on international trade, Ricardo proposed the law of comparative advantage. He believed that even if one country could produce various products cheaper, work division and trade would take place between the two countries. He also pointed out: As long as the price ratios among countries that do not trade are different, each country will have a kind of comparative advantage. Even if a country is at a cost disadvantage in various aspects of production, it can still obtain a comparative advantage by exporting low-cost commodities with relatively low-cost disadvantages in exchange for relatively large cost disadvantages. The theory of comparative advantage has been regarded as the core of the theory of international work division and trade to date, and is, of course, one of the basic theories of modern development issues.

In the famous book, *The Problem of Interregional and International Trade*, published by Ohlin from Sweden in 1933, an international trade model was proposed, which is considered to have inherited and developed the classical theory of comparative advantage. Based on the difference in "factor intensity" among commodities and the difference in production factor endowments of countries and regions, Ohlin analyzed the reason for the generation of comparative advantage. The difference in factor intensity refers to the different degrees of dependence on the capital, labor, and other factors required to produce a certain product. When countries produce those commodities that intensively use their abundant production factors, there will be a comparative advantage inevitably. Therefore, each country and region ultimately export commodities that can use their abundant production factors in exchange for imported commodities that require more intensive use of their scarce production factors. Ohlin also believes that the theory of international trade is only a part of the general layout theory. The formation of industrial location results from the work division based on the comparative advantage determined by the differences in production factor endowments regardless of countries or regions within a country.

2 Research on Location Theory

1. Purpose of research on location theory

The location theory sprouted in the eighteenth century when capitalist commerce and transportation developed greatly. Four representative location theories had been formed from the beginning of the nineteenth century to the 1940s: agricultural location theory, industrial location theory, central place theory (also known as urban location theory), market area and market network theory. The goals they seek are different, but their hypotheses, research methods, expression forms, etc. are similar. That is to say, it is assumed that the researched region is an "isolated country" isolated from the outside world, and there is no difference in natural conditions within the region, which is the so-called "homogeneous region". Based on mastering certain investigations and recording data, strict geometric deduction and logical reasoning methods are used to induce and deduct the theoretical model. The other most important common point is that only a single location decision is sought. The factors considered are relatively few, and no complicated mathematical operations exist. The model shows a static and partial equilibrium state. It can be classified as classical location theory due to the above characteristics.

The fundamental purpose of location theory is to reveal the spatial laws of human social-economic activities. In the preface of the book, *Spatial Distribution of Economy*, W. Losch, a famous scholar of location theory, described the location and location laws as follows: "For us, the time of existence of things is certain, but we may actively choose its location… Everything alive needs to choose the correct location. Of course, the correct location shall also be selected for a well-run enterprise, the construction of a city, and an agglomeration area that can be developed. Not only that, once things are in the correct location, laws will also be generated" [1].

2. Agricultural location theory

In 1826, J. H. Thunen, an agricultural economist and agricultural geographer from Germany, published the book, *Isolated Countries in Agriculture and National Economy* (Volume 1) (Der isolierte Staat in Beziehung auf Landwirtschaft und Nationa-ISkonomie), in which he proposed the famous agricultural location theory. The central idea of his theory is to clarify that: The type of agricultural land use and the degree of intensification of agricultural land management not only depend on the land's natural characteristics. More importantly, they depend on the economic status, particularly depending on the distance between it and the place where agricultural products are consumed (market, specifically the city). Thunen demonstrated the existence and causes of an "isolated country" and internal structural differences from the perspectives of transportation costs, differential rents, product prices, etc. He concluded that: The types of use and the degree of intensive use (use form) of land around the agricultural product market (city) vary from one distance to another. A series of concentric circles are formed around the consumption center, which is called the "Thunen circle" (Pie Thiinenischen Kreisen). Thunen divided it into six circles, and the types of land use from the center to the outside are as follows: planting of horticultural crops and raising of dairy cows, forestry, very intensive cultivation of crops (two-year rotation), non-intensive forage grass planting and grazing, extensive three-year crop rotation, grazing, and extensive planting.

Thunen's theory points out and demonstrates the formation and mode of spatial differences in agricultural production (regional differentiation), and provides the established principles for regional agricultural development and regional structure.

3. **Industrial location theory**

European scholars continued to study various types of location models after Thunen's agricultural location theory. In the mid-to-late nineteenth century, Wilhelm Roscher made outstanding achievements. He fully affirmed Thunen's theory, and proposed a series of new supplements [2]. Thereafter, W. Launhald from Germany proposed the famous weight triangle and the "price funnel" model. In his weight triangle, two points represent the origin of raw materials, and the third point represents the market where the products are sold. The optimum location is the position where the straight lines from three vertices are intersected and the sum of ton-kilometers is the minimum in the triangle. This is the initial demonstration and early model of the industrial location theory (location triangle) [3].

A. Weber's *Industrial Location* (Part 1: Pure Location Theory) was published in 1909, which marked the advent of the industrial location theory. The core of Weber's theory is to find out the point with the lowest production cost of industrial products as the ideal location for the distribution of industrial enterprises through analysis and calculation of the interaction of transportation, labor, and agglomeration factors.

Weber first analyzed the impact of transportation costs on the choice of industrial location. He believed that the point with the minimum sum of ton-kilometers shall be sought first to achieve the lowest cost of industrial production, because the freight is directly proportional to the number of ton-kilometers. In addition to the obvious factor, haul distance, another factor that causes a difference in the freight is the raw material characteristics. In order to analyze the impact of raw material characteristics on the location, he divided the raw materials into local raw materials, ubiquitous raw materials, pure heavy raw materials, and weightless raw materials, demonstrated the most suitable corporate location under different types of raw materials with triangles and polygons, and then considered changes ("deformation") in the location pattern caused by two factors, that is, labor force and agglomeration. Weber was indeed very insightful to take the cost isotims as a tool for location analysis [4].

4. **Central place theory (urban location theory)**

The central place theory was proposed by W. Christaller, a German geographer. In the summer of 1932, he published the book, *Center Place in Southern Germany* [5]. In the book, he demonstrated urban residential areas and their regional systems from the main functions of the central residential area, urban supply, administrative management, transportation, etc., profoundly revealing the regional basis and the hierarchy-scale spatial relationship between the development of cities and central residential areas, and vividly summarized the relationship between city level and scale within a region (or country) as a regular hexagon model.

To understand the relationships between hierarchy, functions, and towns, W. Christaller utilized the concepts of "central place", "centrality" and "central goods and services", and explored that the central place assumes the function of central service for the surrounding area. The central place must be the closest to the region's location in theory. He used a geometric method to deduce that these locations should be located in the center of the regular hexagonal service area under normal circumstances. W. Christaller analyzed the three factors for the formation of city levels, that is, administrative management, the supply of goods and services, and transportation, and proposed how these three factors work, that is, how the formation of cities and systems is governed by the above three factors (principles).

The three factors commonly affect the formation of the city level and system. W. Christaller believed at the time that the principles of the market economy may be the foremost in open and accessible areas; the role of administrative management is more important in intermontane basins or other areas that are relatively isolated from the outside world; in young countries and newly developed regions, traffic lines play a "pioneer" role for immigrants, and the principle of transportation holds a leading position. He deduced the respective level-quantity series of central places under the respective effects of three factors, and drew a conclusion through comprehensive analysis: A region or a country should form the following city levels: one Level A city, two Level B cities, six to twelve Level C cities, 42 to 54 Level D cities, and 118 Level E cities (2/3 of 162-192) [6].

W. Christaller proposed the concept of the K value. The K value represents the relationship between the central place and the area covered by its service, supply, and jurisdiction under the effects of the above three factors and the quantitative relationship among them. Under the principle of administrative management, $K = 7$, that is, a center has jurisdiction over six sub-centers and regions; under the principle of market supply, $K = 3$, that is, a center supplies two sub-centers and regions; under the principle of transportation, $K = 4$, that is, a center supplies three sub-centers and regions. It can be seen from the analysis of the internal structure of the regional economic system that: The central place system formed according to administrative principles is closed, and the market economy is inactive; on the contrary, the central place system is organized based on the principles of market supply and transportation is an open structure. Generally speaking, the economy is more developed. Therefore, changing the K value from a high value to a low value is a sign of the development and evolution of the spatial socio-economic structure system.

5. **Market location theory**

A. Losch published the book, *Spatial Distribution of Economy*, in 1940, and proposed a theoretical model of market areas and market networks. The essence of his theory is still the industrial location theory, which is characterized by the combination of production location and market scope. He pointed out that production and consumption are carried out in the market area, and manufacturers aim to seek maximum profit. However, the location with the lowest cost and the smallest ton-kilometer may not always ensure the maximum profit. Therefore, to choose the location correctly is to seek the largest market and market area. Losch used the overall location equation of enterprise allocation to solve the optimal allocation point of each producer, and deduced the demand curve and sales volume from product prices, freights, etc. However, the sales volume can be converted into the area of the market area. When there is only one manufacturer that produces a certain product in the space or there are a few manufacturers that produce a certain product at the beginning, a circular market area will be formed. However, the circular market area cannot last. Because there are always empty corners that are not occupied between the circles. Competitors constantly appear, occupying empty corners, squeezing the circles together, and forming a honeycomb structure, a regular hexagonal market area.

Losch also deduced the market areas of different scales with formulas. He believed that the arrangement of market areas and market networks is not arbitrary, but depends on the economic principle. According to this principle, there must be a big city, and a series of market areas and competition points surrounding this big city. The hierarchical sequence of economic spatial distribution formed by such a market network system is called "economic landscape" (Die Wirtschaftslandschaft) by Losch.

6. **Development of location research since the 1950s**

Since the 1950s, developed countries and some developing countries have undergone great development in the social economy, which has changed the original socio-economic structure and living environment to a large extent. Such tremendous change has posed issues much more complicated than those in the past and to be solved urgently by location theorists and economists: decision-making of productivity location, the rational development of regional economy, the coordination between regional economic development and society and ecology, improvement in the spatial structure of the social economy, how to create a better living environment and working conditions for people, etc. In order to answer these issues scientifically, scholars have changed the way of observing and analyzing issues in the past, absorbed the nutrition of related disciplines, and developed new fields. The main progress is as shown below:

(1) Develop from the location decision-making of a single socio-economic object to the study of the overall economic structure and its models, so as to be more closely related to regional development issues in practice.
(2) Change from the abstract and pure theoretical model derivation to regional analysis close to practical problems and the model that can be applied in practice, so as to provide the actual decision-making standards. Develop from the static spatial location selection to the study of the spatial distribution and structural changes of various development objects and their processes, so as to provide a basis for prediction. In applied research and theoretical research, the application of mathematical statistics, input–output methods and electronic computers have been gradually expanded.

(3) In addition to industry, agriculture, cities and markets, the objects of location decision-making also include a wide range of tertiary industry facilities. Location decision-making shall not only meet the requirements of economic factors, but also consider conditions such as social justice, living environment, travel and amusement. The theoretical results of overall location decision-making are the spatial structure theory.
(4) The research of location theory has formed two main aspects that are interrelated but distinct: regional science and regional analysis of human geography.
(5) The classical location theory was mainly proposed by German scholars. Due to the world's economic development and displacement of agglomeration areas after WW II, the greatest contribution to the development of location theory is made by American scholars represented by W. Isard, B. J. L. Berry, J. Burton, D. M. Smith, C. A. Smith, et al., followed by Swedish and Norwegian scholars represented by T. Palander and W. L. Garrison, scholars from the former Federal Republic of Germany represented by E. Otremba and Boffintel, and British scholars represented by P. Haggett and F. E. I. Hamilton. Among these scholars, Isard's contribution to the comprehensive development of location theory is the most prominent. He and the research collective led by him have laid the theoretical foundation for regional science.

3 Research on Regional Policies

Regional policies refer to policies and measures formulated and implemented by national and regional government departments to adjust regional differences and macro-operating mechanisms. There are many contents, mainly regional economic policies, structural policies, landscape and nature protection policies, etc. Regional policies have become a subdiscipline of economics and geography. Its essence is the doctrine about the official influence on economic decision-making and development process regarding regional distribution. It was produced in the 1920s, almost simultaneously with research on regional development issues. It had expanded to many developed Western countries by the 1960s, and the issues had become more prominent. The background is the emergence and aggravation of the unbalanced economic development in some countries and regions such as Britain, the United States, Germany and Japan. The recession of leading industries in some previously developed regions has caused a large area of economic recession, a substantial rise in the unemployment rate, and a severe imbalance in the supply and demand of the labor market, thus leading to regional economic disorder and other social and political conflicts. Such areas are widely called "problematic areas" or "problem areas" in the western.

The research on regional policies focuses on: The kind of regional policies to be implemented in various types of areas, and the possibilities and boundaries of these policies to work. Regional policies exist and develop in opposition to the stand of classical economic liberalism. The latter believe that the market itself has the best self-control ability, and may optimize the regional distribution (spatial arrangement) and regional structure of the economy through "perfect competition".

In Western countries, the goal of regional policies is to promote rapid, stable or balanced growth of the regional economy. These three goals can also be understood as three types of modern regional economic growth. Among them, the first two circumstances are with regard to the economic growth rate, and the latter is with regard to the inter-regional relation. Under normal circumstances, if rapid growth is taken as the goal, it is difficult to achieve balanced growth or impossible to achieve the goal of stable growth. However, if stable growth is taken as the main goal, it is possible to achieve the goal of balanced growth. Due to different goals, the implementation of regional policies in the past few decades can be divided into two major stages.

1. Two development and evolution stages of regional policies

1) Stage of regional policies with the goal of reducing differences in development among regions

In those areas with a long history of industrialization, a certain "structural crisis" appeared earlier in the 1920s. Through a period of development after World War II, new industries have continued to appear. The upgrading and transformation of industrial structures have generally deepened the crises in those old industrialized areas with large "geographical inertia". In some industrial sectors, output declined, jobs were reduced, the national income was decreased, supply exceeded demand, and unemployment increased, with population outflows. The results are declines in taxation, investment capacity and welfare standards, resulting in a vicious circle: Such areas are generally depressed or recessed areas. Another type of area is undeveloped and underdeveloped areas with poor economic bases. The problem of insufficient labor shall be solved to accelerate economic growth in such areas. In both cases, the regional economic growth will be facilitated by taking policy measures to balance the supply and demand of labor employment among regions. The specific goal of regional policies is to increase capital investment in "problem areas" or the areas with "structural crises" to create new jobs and eliminate the gap in unemployment rates among regions. Due to the inherent connection between jobs and income, such a specific goal is to achieve income balance among regions, which can be said to achieve fairness and equal opportunities.

The regional policies in response to such issues require affecting the regional economic structure, such as restricting some economic activities, including the trend of agglomeration of a few economically developed agglomeration areas of industry and service industries, promoting the development of areas with economic stagnation or characterized by primary production in terms of the economic structure, reducing the gap between the economic core area and the fringe area, and helping those "areas in the state of disaster" with the so-called "structural crises" get rid of this situation (such as border areas due to political reasons, some industrial and mining areas due to exhausted resources, and areas with major economic changes).

Specific regional policies are also called tools or means of regional policies, which can be divided into two categories, that is, coercive means and induced stimulus means.

(1) Coercive means are divided into commands and prohibitions. When selecting the land for future industrial enterprises and other facilities in the planning, a certain range of land is scientifically delineated or reserved, which shall not be occupied; in addition, regulations on the number of building stories in urban areas and other specific terrains, regulations on the discharge of wastewater and waste gas at specific locations, the ratio of the construction area of a specific community to the total land area not exceeding the regulations, etc. In practice, there are fewer prohibitions than positive commands. License insurance. Sometimes, flexible licenses, rather than commands and prohibitions, are used. Once entrepreneurs and social groups have obtained licenses from government agencies, they can choose the locations they require in multiple areas.
(2) Induced stimulus means to include other contents. Those with regard to direct help are as follows:

① Information provision: That is, answering the consultation of entrepreneurs who are interested in certain location conditions, such as providing the conditions of unused factory buildings, water source, labor conditions, etc. in a certain place;
② Technical consultation;
③ Labor market policies: Such as the subsidies for separation (referring to that the place of residence and the work post are not together) and relocation of employees, and the subsidy for change of profession, which is a part of social funds (for enterprises and social institutions). This can reduce the initial difficulty of establishing new enterprises in structurally weak areas and changing the economic structure and enterprise direction within agglomeration areas.
④ Infrastructure policies: These are the policies published by the national management departments for promoting the investment and development of infrastructure in various areas. The infrastructure can be divided into economical infrastructure and social infrastructure. The former directly provides the precondition for production, such as transportation facilities, energy supply, water supply, drainage, etc. in a broad sense, while the latter only provides indirect preconditions for production, such as investment in health care, housing, education, and vacation facilities. Shortages or deficiencies in these aspects will have a restrictive effect on regional development.
⑤ National management policies: It is quite effective method to control social orders and commissions from the area, so as to take care of the region to be developed. Of course, such an external impact will inevitably affect the interests of other regions.
⑥ Distribution of social facilities among regions. For example, if the national social facilities such as colleges and universities, and barracks are concentrated in a place, it is equivalent to the state and private funds flowing only to those places. As a result, it will cause chain reactions, strengthen the economic strength of

the region, increase social orders and commissions, and promote the regional economy.

Indirect assistance, that is, stimulating locating and opening of an enterprise in a certain region:

① Loans provision: Economic difficulties in weakly developed regions are often caused by a lack of sufficient funds. The low-interest loans are generally provided from social funds, with a certain degree of subsidy significance.
② Financial assistance: There are many forms, including a one-time subsidy, which is mostly used for the development and preparation of the new industrial area and the foundation projects for new factories.
③ Reception of part of the enterprise by the state, forming a joint venture.
④ Tax preferences: Most of them are time-based, reduction, and exemption in whole or in part.
⑤ Price preferences: Including the provision of cheap public transportation, gas, water and electricity supply, etc. This was a very important regional policy in the past, but it has also caused controversy. Some people believe that preferential price policies with a subsidy nature distort the role of competition.

2) Stage of implementing regional policies on technology development

Western developed countries and some countries of other types have played a more extensive and positive role in the revival of "problem areas" and the development of undeveloped areas when implementing regional policies aimed at reducing inter-regional differences. However, since the early 1970s, the economic development conditions of major Western countries have gradually changed, so the regional policies with the main goal of eliminating regional differences have gradually been replaced by regional policies on technology development. Such regional policies focus on mining the knowledge and technical resources that exist in the region itself.

The main reasons are as follows:

(1) The oil crisis in the 1970s worsened the western economic environment, which slowed down economic growth, and lowered the corresponding investment capability, especially in the industrial sectors. Enterprise vitality and capital vitality were declining, and the number of enterprises willing to open new bases was constantly reduced. Therefore, the regional policies aimed at the "problem areas" with "structural crisis" would have a weak effect on economic recovery. Therefore, new regional development strategies must be developed, and new policies must be implemented.[1]
(2) The regional policies implemented in the past with the main goal of reducing inter-regional differences were too costly to implement, and the effects were not good. The benefit–cost ratio was low by investing an enormous sum of money in the funded "problem areas". For example, the cost of creating a job in the

[1] Zeng Gang, The Status and Role of New Technology Development and Application in the Regional Development Policies of the Federal Republic of Germany, 1992.

"problem areas" in Britain was as high as 35,000 pounds from 1966 to 1976. Canada also invested nearly 40 billion U.S. dollars in the "problem areas" in the name of regional policies from 1960 to 1984. However, the ratio of the area with the highest unemployment rate to the area with the lowest unemployment rate declined very slightly (4.74 in 1961, 2.41 in 1971, and 3.72 in 1981). On the other hand, in non-problem areas, including developed areas with combined effect, investment and other measures that expanded economic growth were restricted by regional policies. The decline in public investment levels also directly or indirectly affected the economic growth of these areas. From here, we can see the defects of the regional policies implemented before the 1970s: only paying attention to the short-term equilibrium goal of increasing the current employment rate, but ignoring the pursuit of the long-term equilibrium goal, that is to say, ignoring the industrial structure, innovation ability, as well as research and development potential factors of the funded areas.[2]

(3) Intensified international competition is not conducive to the development or revitalization of funded areas. In the past 20–30 years, many developing countries have adopted the opening policy, used cheap labor and markets, introduced foreign capital, and achieved rapid growth in the national economy. Such capital is adverse to the "problem areas" increasing the attraction of foreign capital (including underdeveloped areas, peripheral areas, etc.) in those developed countries. To compete with developing countries, the economic growth of developed countries must rely more on factors such as human capital and technical management knowledge. However, these are closely related to the inherent advantages of the agglomeration economy, that is, the willingness to decentralize the industry is weakening.

The main points of implementing regional policies for technological development are: Use new technologies to innovate traditional industrial sectors in the area, especially focusing on the development of growing industries and developing industries; implement the middle-class policy, focusing on the development of small and medium-sized enterprises with fast research and development of new technologies, and establish science and technology industrial parks and industrial technology centers with innovation as the main goal in suitable locations; the government shall provide assistance for research and development investment by region, and establish a consulting and information network with innovative suggestions as the contents. The selection of leading industries and spatial layouts in countries and regions shall be carried out in accordance with the principles of efficiency first and fairness second.

2. Regional policies

The research on regional policies by foreign scholars has developed into an independent science called "regional policies". The research contents mainly focus on the following aspects:

[2] Li Jun, Regional Policies and China's Regional Economic Growth, August 1991.

(1) Theoretical foundation of regional policies. It includes Changes and interactions between supply and demand within the region; the resources include capital, technology, knowledge, and the mechanism for the inter-regional flow of labor; how the inter-regional flow of resources leads to the development of spatial agglomeration and the formation of agglomeration areas, as well as the process of spreading from the agglomeration areas to the marginal zones; the mechanism of regional demand growth, especially the basic role of output; the spatial structure model of regional economic growth, including the dynamic changes of the central place system, the integration of growth pole, central place and growth pole theory, etc.
(2) Goal of regional policies. Among the goals and goal systems of regional policies, the basic macro pattern of regional economic growth is the most important, and the regional development stage and level, as well as economic, technological, and infrastructure foundations shall be considered to determine a such pattern. According to the views of Western scholars, the main goals of regional policies can be divided into three types: rapid growth, stable growth, and balanced growth [7].
(3) Method of evaluating the regional economic foundation. The purpose of the research is to provide the decision-making basis for regional differences in formulating and implementing scientific policies. That is to say, the corresponding regulation policies shall be adopted according to the regions with different problems and different levels before scientific categorization. It is generally believed that division based on a certain indicator system is still an important method. In terms of rigorous mechanism analysis, the most important approach to evaluate the regional economic foundation is that many scholars analyze and calculate the production function and regional multiplier of each region.

4 Research on Industrial Structure

Since scholars in Western countries began to study the issues of regional development in the 1920s, people's understanding of regional development was limited to the growth and structural changes of regional gross national product, national income, and per capita national income for a period of time. The former includes the expansion of production and capital accumulation, as well as high-speed economic growth; while the latter mainly refers to economic industrialization, that is, with the extension of the regional development process, the status of the industry in the national economy will gradually rise, and the relative status of agriculture will decline. Of course, scholars also noticed the changing laws of the internal structure of industries in regional development in the initial phase of regional industrialization. This is the basic concept of industrial structure in the early stage of research on regional development. There had been quite a wide range of content on the concept of the industrial structure

Table 1 Rostow's stage of industrial structure transformation

Economic development stage	Leading industry (cluster)
1. Traditional society	Manual manufacturing industry
2. Preparation for takeoff	Food, bricks and tiles
3. Takeoff	Textile, steel and electricity
4. Push to maturity	General machinery and chemical industry
5. Mass high consumption era	Automobiles, aviation and durable consumer goods

by the 1940s. The research of domestic and foreign scholars on regional industrial structure mainly focused on the following specific aspects.

1. Theory of industrial development stage

After investigating the changes in main industries in the industrialization process in Germany and other Western European countries, W. Hoffmann, a German economist, believed that with the development of the industrial process, the ratio of the net output value of the consumer goods industry and the means of production would gradually decline. That is to say, he believed that the light industry had a large proportion in the initial phase of industrialization. After the funds were accumulated through the light industry, the heavy industry would be developed, and investment products would be generated. This judgment was put forward in the 1930s, which was obviously limited by the scholar's country and economic development stage. Hoffman's theorem can only be used to explain the industrialization process of Germany and some Western European countries before the 1930s. The industrialization process of the former Soviet Union started with heavy industry, and it was the same in China. After World War II, the industrialization process of a series of developing countries showed a variety of patterns in the sequence of light and heavy industries, which also indicated that "Hoffman's Theorem" could not represent the universal feature. However, he is a pioneer in the research on the theory of the industrial development stage.

In 1960, W. W. Rostow, an American economic historian, studied the development process of Western Europe and North America, and proposed five stages of national development. Later, he proposed that the leading industry (cluster) at each stage was in transformation. He believed that the evolution of the industrial structure was closely related to the changes in some industrial sectors, and called such industry the leading industry.[3] He proposed five stages and their leading industries (clusters), which can be summarized as shown in Table 1.

The rapid growth of the Japanese economy since the 1950s proved the correctness of the above-mentioned stage theory, that is, three major structural transfer processes were experienced: change from a relatively light structure to a heavy structure with heavy and chemical industries as the main body in the 1950s; change from the heavy structure to the high processing degree in the 1970s; change from the capital-intensive structure to the technology-intensive structure in the late 1970s.

[3] Wang Xun, Evolution of Industrial Structure and Regional Development, New Theory of Geography, Volume 2, 1988, Page 45.

The theory of the stage of industrial structure transformation showed irregularities in some developing countries with high-speed growth after the war; the transition from light structure to heavy structure was very fast; from the very beginning, industry sectors with higher technical levels were selected as the leading industries, such as the iron and steel industry, the automobile industry, and the semiconductor industry. A higher starting point shortened the time for the evolution of the industrial structure.

2. Division of industrial type

In 1940, the economist C. Clark proposed the general trend of the three industrial classifications and the evolution of industrial structure in the book *Conditions for Economic Development* [8]. He divided the national economy into three types of industries, namely, the primary industry—agriculture, the secondary industry—industry, and the tertiary industry—service industry. After investigating the long-term changes in the economic structure of developed countries, he concluded that: with regional economic development, and the increase in the national income and per capita national income, the proportion of agricultural employment declined, and the proportion of employees in the secondary and tertiary industries increased. This was mainly caused by the difference in income demand elasticity of various industries. When the national income level increases, the income elasticity of demand for products in the secondary and tertiary industries would be higher than that in the primary industry. That is to say, the increase in demand for products in the secondary and tertiary industries was greater than that in the primary industry. In addition, the technological progress rates of the three industries were also significantly different. The technological progress rate of agricultural production was lower than those of the secondary and tertiary industries due to its own characteristics. Decreasing profits were prone to occur in the process of agricultural production. Unlike industrial production, the unit cost decreased with the increase in investment, technical updates, and output expansion. With the development of the national economy, production factors inevitably shifted from the primary industry to the secondary and tertiary industries. Clark's theory explained the intrinsic motivation of the industrialization process [9].

In the 1960s and 1970s, with the evolution and improvement of the industrial structure, some scholars divided the industries into resource-intensive, labor-intensive, capital-intensive as well as technology- and knowledge-intensive types based on the relative dependence of the industry on different production factors. Resource-intensive industries include agriculture, mining industry, and tourist industry based on natural scenic resources. The availability and quality of natural resources have become the fundamental conditions for the development of such industries. The immovability of the spatial distribution of resources determines that the location of these industries is also where the resources are located. The fixed assets required in the production process of labor-intensive industries are generally small, but a large amount of unskilled labor shall be put in. The wages of workers occupy a very important part of the product costs. Capital-intensive industries require a large amount of investment in fixed assets, and the organic composition of capital is high, especially large one-time investments. In China, large enterprises in such industries are mainly

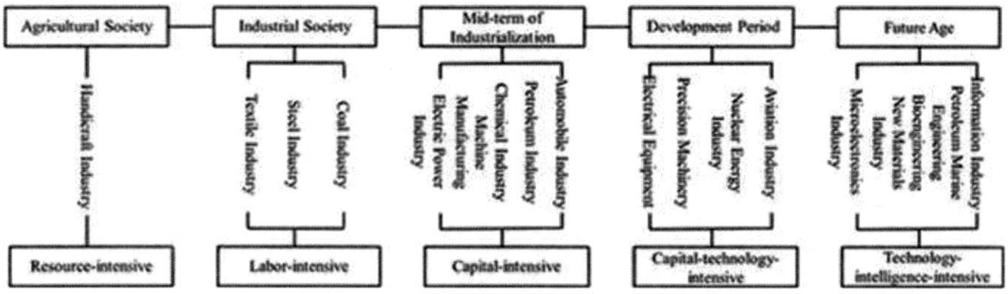

Fig. 1 Internal corresponding relationship in each stage of industrial development. The figure is from "Evolution of Industrial Structure and Regional Development" written by Wang Xun and published in "New Theory of Geography", Volume 3, 1988, Page 2

co-operated by the central government and local governments to a great extent. For technology- and intelligence-intensive industries, there are two kinds of products. One kind is the products that integrate a large number of emerging and cutting-edge technologies with the latest scientific and technological achievements of mankind. The other kind of product is knowledge software itself. The research and development costs of the products in such industries are high. More senior technical talents and a large amount of relevant knowledge and information are required during production.

In the process of regional and national economic development, there is a quite definite corresponding relationship between the above-related development stages, leading industries, and the combination features of production factors (Fig. 1).

3. Industry and product life cycles

Economists Golding and Freeman, C. put forward the theory of industry and product life cycles in 1972 and 1982 respectively. According to the market supply and demand, price and profit relationship of a certain industry (product), the life process of the industry (product) can be divided into four stages: innovation stage (birth stage), maturity stage, mass production stage, and decline stage. When the industry stayed at the innovation stage, products were only researched, trial-manufactured and developed, market demands began to appear, and the industry only appeared at a few points; at the maturity stage, a large number of manufacturers began to appear, the market demands for products increased sharply, the income demand for products was quite elastic, the prices remained at a high level, the costs gradually decreased, but market prices were high, which resulted in high profits. At the stage of standardized mass production, after high profits had been maintained for a period of time, the high-price market for products was gradually saturated, and market prices evolved into competitive prices. Demands and prices declined, and the industry entered the decline stage.

When the industry entered the maturity stage, a large number of manufacturers entered the industry, which led to the rapid development of the regions where conditions permitted. At the moment, the industry's production technologies had been standardized and open, the potential for cost reduction by technology was small, and

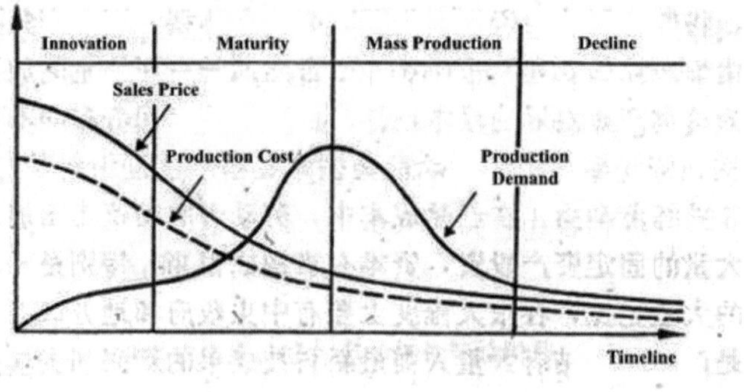

Fig. 2 Life cycle of industrial development. According to Hermut Nuhm, "Industriegeographie, Neuere Entwicklung und Perspek tiven fuir die Zukunft", Geograhpische R undschau, Sonderheft, 1987

non-technical factors affecting costs became important. Since these factors (transportation, raw materials, humanistic environment, etc.) varied from place to place, some manufacturers had increased revenue, some decreased revenue, and some quitted (production conversion or bankruptcy). Therefore, the layout of the industry experienced a period of growth and decline, and then a relatively stable pattern was formed (Fig. 2).

The principle of success for the industry to enter the maturity stage is to take the lead, the conditions for success are intelligence, bonuses, resources, high profits, and fewer competitors are obtained. The principle of success for the industry to enter the mass production stage is safety, the condition for success is the lowest potential cost, and normal profits and fierce market competition are obtained.[4]

The law of the life cycle of industry (product) production tells people that any industry has its objective process of formation-development-prosperity-recession, which has an important guiding significance for mastering market changes and determining regional leading industries.

Some scholars investigated the product life cycle and found that the effects of the main factors affecting industrial development at different stages were different (see Fig. 3) [10], and the trade volume of this product among countries also varied greatly (see Fig. 4 [11]).

In addition to the above three aspects in which the research on the regional industrial structure is manifested, many scholars have also used the input–output analysis method to reveal the internal connection characteristics of the industrial system and the interactions within the regional industry. Within the regional industry, there are unidirectional connections and multi-directional circular connections among industries. According to the intermediate demand rate and intermediate input rate among industries, industries can be divided into intermediate product-based basic industries, intermediate product-based industries, final demand-based industries, and final

[4] Wang Tuoyu, Preliminary Study on Strategic Issues of Regional Industrial Structure, July 1986.

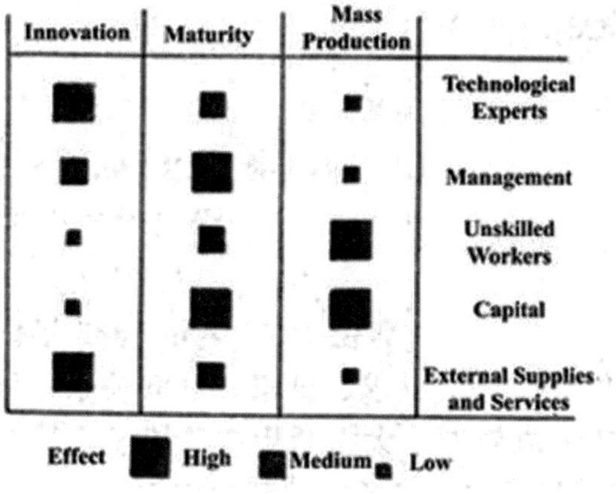

Fig. 3 Changes in the roles of main factors in the product life cycle

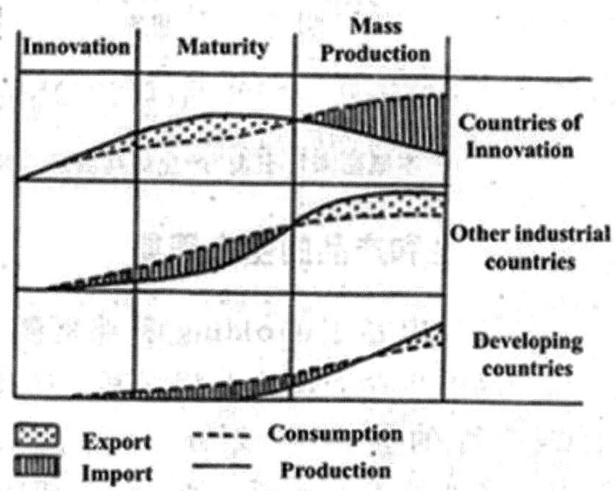

Fig. 4 Development of international trade in the product life cycle (hypothetical situation)

demand-based basic industries. The research on the characteristics of the connection relationship of the industrial structure provides a theoretical and quantitative basis for the establishment of industrial clusters, the improvement of the stability of the regional industrial system, as well as the combination of specialized and comprehensive development. W. Isard used the input industry analysis method to plan the structure of the regional industrial complex. The basic principle is based on the quantitative analysis of connection characteristics of the regional industry [12].

5 Research on the Spatial Structure

The spatial structure of social economy refers to the interaction and mutual relationship of socioeconomic objects in space, as well as the spatial agglomeration scale and agglomeration form of objects and phenomena that reflect this relationship.

The research on spatial structure began in Germany in the 1930s and 1940s. Since the 1950s, this research has been further developed in the United States, Sweden, and the Federal Republic of Germany. The spatial structure theory is generated on the basis of the location theory, and basically follows the method for location theory scholars to investigate problems, that is, the hypothesis of the regional basic conditions—geometric diagrams and simple mathematical derivation of formulas—model induction—model testing and comparison with actual conditions for effectiveness analysis. However, the goal and focus of the research on spatial structure are different from those of the location theory. In the research on spatial structure, all relevant things in a certain range shall be regarded as organisms with certain functions and shall be investigated in terms of time variation. From this perspective, the spatial structure theory can also be regarded as a dynamic overall location theory.

The research on spatial structure involves agriculture, industry, tertiary industry, and the comprehensive location of the urban residential area. Of course, its development is inseparable from the development of location theory. Its basic contents could be considered a synthesis of all classical location theories to a certain degree.

Those that made important contributions to developing the classical location theory into the spatial structure theory after World War II included E. S. Dunn, an American scholar in the 1950s [13], and E. Otremba, a scholar of the former Federal Republic of Germany [14]. Both of them put forward the concept of spatial structure respectively. Otremba believed that the projection of economic and business forms on the earth's surface would inevitably produce a certain spatial unity of agricultural economic structure. He also pointed out that such spatial unity was the research object of geography. Dunn divided all location issues into the enterprise stage, industrial stage and overall socioeconomic stage. He tried to explore the general theory of the location structure of economic activities, and developed the static partial equilibrium theory in the direction of combining reality. He further deepened the principle of the Thunen circle, introduced the concept of land rent into the derivation of the spatial structure model, and proposed that the net income of land management was the main driving force for the formation and evolution of the spatial structure. From the late 1950s to the early 1960s, Isard and Berry were the main representatives of the research on spatial structure theory. Isard advocated researching the location theory from the standpoint of "spatial economics". He quoted a lot of econometric methods to carry out a comprehensive analysis of industrial location. Of course, much of his research have gone beyond the scope of spatial structure theory, and have become regional organizations and dynamics. The "spatial system" he mentioned is extremely broad, including agriculture and land use, industrial location, retail business and service business districts, urban locations, the layout of transportation networks, etc.

Böventer [15], a scholar of the former Federal Republic of Germany, carried out a systematic theoretical analysis and model derivation of the spatial structure. He tried to integrate the location theories of Weber, Thunen, and Losch, and believed that location theories should examine and clarify not only production and goods as deep as possible, but also the geographical distribution of production factors such as place of residence, employment, and mobility. He combined the location theory with the development theory, and analyzed and demonstrated the general characteristics of the spatial structure at each stage of social economy. He analyzed in detail the most important factors that determined the spatial structure and its differences: agglomeration, freight, and the dependence of the economy on the local production factors of land, and also believed that freight was the decisive factor for the characteristics of the input–output relationship and the spatial flow of production factors. The impact of land on the spatial structure mainly depends on how it is used. Among them, productive use includes the land for agriculture, mines, industry, and infrastructure, and consumer use includes residential and recreational land. The interaction of these factors determines the characteristics of the spatial structure. In addition to the above three main factors, Böventer believed that objects of the same nature in space, such as factories that manufacture the same product, tertiary industries that provide the same kind of service, and markets that sell the same goods, always had a competitive relationship. As a result of competition, a landscape structure model of the location at the hexagonal apex was generated, that is, the theoretically optimal arrangement. He also further deduced how the production scale and consumption scale of each location depend on the production and demand mechanism of various products, especially the relationship with the scale economy of production and the advantages of inter-sectoral agglomeration. Böventer's development of the location theory and spatial structure is also manifested as follows: The scale of industrial and agricultural enterprises, town scale as well as town structure are theoretically demonstrated from the perspective of optimal land use and reasonable agglomeration.

Since the spatial structure theory is a comprehensive and integrated location theory, the objects investigated include locations and spatial relationships of industrial sectors, service sectors, urban residential areas and infrastructure, inter-regional flows of personnel, products, finances, and information, etc. Therefore, its basic problems are closely related to the classical location theory, and meanwhile, reflect its characteristics of being "comprehensive" and "integrated". Since the comprehensiveness and integrity of the object are difficult to quantify, it is of high abstractness in theoretical derivation. However, since the spatial structure theory is a product developing towards practical application on the basis of location theory, its theoretical model can better describe the actual state of object movement and distribution, and has direct reference significance for regional development and regional planning. Judging from the works of some Western scholars, the connotation of spatial structure theory covers a wide range, and scholars have different opinions. Some scholars even believe that the location theory and regional economics (also translated as "spatial economics") also belong to the category of spatial structure theory. Although the author does not agree with this viewpoint, some theories and methods in regional science and regional analysis can indeed be regarded as part of the spatial structure theory.

It is generally believed that the basic contents of spatial structure theory include the following five aspects:

(1) Spatial structure of land use centered on urban residential areas (markets). This is the development of Thunen's theoretical model and the theory of location differential rent. It uses the concept of production and consumption functions to derive the net income function of each operation mode of suburban agriculture, and the division of operation zones is obtained therefrom.
(2) Optimal enterprise scale, residential area scale, urban scale, and central place level system. Bases of theoretical derivation: One is the agricultural location theory, and the other is the agglomerate effect theory. The derivation of the optimal enterprise scale is combined with that of the reasonable scale of the urban residential area, and the city is regarded as an enterprise, which is understood as a production process. The "threshold" theory is used to apply the central place level system to the reality of regional planning.
(3) Characteristics and evolution of the spatial structure at each stage of socioeconomic development. Through the analysis of the general mechanism of action, the general trends and types of dynamics and evolution of spatial structure changes are revealed.
(4) Reasonable degree of spatial concentration of socioeconomic objects. It is manifested as how to deal with the problems of over-sparseness and over-density in practice, which is of practical significance for regional development and remediation and regional planning.
(5) Spatial interaction. Including the flow of goods, people, and finances among regions, the attraction scope of central cities at all levels, the diffusion process of innovation, information and technological knowledge, etc. These aspects are important reflections of spatial structure characteristics [6].

6 Research on Macro-regional Development and Economic Growth Mechanism

The above-mentioned research involving regional development is basically limited to the research on a single factor, single element, and certain aspects. Since the end of the 1940s, a large number of issues of regional development have appeared in many countries in the world, and social, environmental, and ecological factors affecting development have been increasingly obvious, making the issues of regional development more complicated. In this case, scientists are objectively required to analyze and answer the issues of regional development from a comprehensive perspective.

1. Macroeconomic theory

The theory believes that the actual increase in regional income mainly depends on the increase in production potential and actual demands in the region:

$$Y_1 = f(O_1, D_1)$$

where, O_1 represents an increase in output potential, and D_1 represents an increase in actual demands.

Among them, O_1 also depends on the production factors in the region, that is, labor L_1, capital K_1, and land Q_1, namely:

$$O_1 = f(L_1, K_1, Q_1, \ldots)$$

The actual demands depend on:

$$D_1 = f(C_1, I_1, \ldots)$$

where, C_1 represents the social and private demands for consumer goods, and I_1 is the demand for products used for investment caused by the increase in investment.

The theory also believes that the social production and economic growth of a region also depend on a series of other factors: degree of technological progress, spatial structure, department structure, infrastructure system, political system, social system, etc. The technological improvements can form a series of favorable factors for producers, and play a role in mobilizing the increase in demands. In addition to the influence of the existing spatial structure (such as the degree of agglomeration in the region) and the existing department structure (such as the degree of industrialization), the process of regional economic growth is also affected by the changes and reorganization of these two aspects. Infrastructure investment can increase the cost of the producer, increase labor productivity, and improve the welfare standards of consumers. At the same time, the exchange of goods and services among regions can be strengthened to enhance specialization, promote more effective utilization of production factors in the region, and greatly increase the demands for output (export).

The macroeconomic theory (model) describes the role of general factors in regional economic growth. Through a large number of theoretical and practical research, scholars have developed a large number of other models to more specifically describe the role of economic growth factors.

2. Research on structure and planning of regional development

W. Isard, an American regional scientist and founder of regional science, has made important contributions to integrated research on regional development. He had studied the issues of regional development from the following perspectives since the 1940s: First, the impact of the supply–demand relationship within the region on the regional population, taxation, investment, and economic growth; second, determining the industrial location through the analysis of comparative costs among regions; third, the agglomeration and dispersion of the overall social economy in the region, and its impact on development. He also applied the input–output analysis method to analyze and optimize the department structure within the region and the complex, and thus developed several analysis and planning methods that can be practically used in regional planning. The most famous works of W. Isard include *Introduction to Regional Science* (1957), *Analysis of Industrial Complexes and Regional Development* (1959), *Regional Analysis Methods* (1960), *Regional Input–Output Research* (1951), etc. W. Isard also applied his theories and methods to the planning practice of comprehensive regional development and achieved good results.

3. Research on regional production function and its equilibrium conditions

While W. Isard researched regional development and its structure, Western scholars of regional economics and economic geography conducted more extensive and in-depth research on the processes and factors of regional economic growth. The analysis of regional macroeconomic growth has been developed since the 1950s. The goal of this research is to reveal the conditions for the long-term stable growth of the regional economy, that is to say, to seek long-term stable growth of the regional economy, and exercise macro-control over funds, labor, supply–demand relationship, prices, taxation, and social insurance, etc. Since the 1970s and 1980s, H. W. Richardson (American regional economist), R. Thoss (German economist), et al. continued to conduct a lot of in-depth research on regional production functions and their equilibrium conditions and proposed many mathematical models that had been modified and considered more factors and balance conditions. In 1984, Professor R. Thoss, who was the chairman of the Institute for Regional Development and Regional Planning of the former Federal Republic of Germany at that time, said to the author: "His model can ensure the healthy development of the region and enable people to live better".

4. Research on the mechanism of regional economic growth

In the past 30–40 years, the Western scholar who made outstanding contributions to the research on the mechanism of regional economic growth, that is, various forces that promoted long-term growth of the regional economy, was Robert Merton Solow from the Massachusetts Institute of Technology in the United States. He proposed the production function equation in the 1950s, that is, the so-called "neo-classical economic growth theory". He conducted a lot of marginal analysis on the effects of various factors affecting regional economic growth and finally believed that the two most important factors determining regional economic growth are labor and capital investment. In the specific analysis, some other scholars would also take into account the production function factors (including technical factors) of each region. According to the analysis of Robert Merton Solow, it was believed that the following laws exist in the relationship between output, capital, and labor: If the investment in capital and labor doubles, the output will double; if capital increases without increasing labor, or labor increases without increasing capital, output increase will decrease progressively. This theory had been dominant in the West until the 1970s. Scholars who adhere to this viewpoint are called mainstream economists.

Neo-classical theory (Neoklassische Theories) is one of the theories about regional economic growth that are popular in Western countries after World War II. It is characterized by taking supply as the main body, with full employment as the presumption, product market under perfect competition, free flow of production factors, same unit factor price in all regions, and zero freight among regions. Under such assumed conditions, the regional differences in factor payments can be balanced by the flow of factors, that is, the so-called market mechanism (Marktmechanismus) can eventually lead to the balance of regional differences in per capita income. There are two neo-classical regional growth models: the single-sector model and the dual-sector model. The production function of a single sector under the above assumption is

$$Y = f(K, L, T)$$

Among them, y represents output in period t, K is total capital, L represents labor, and T means technological progress; indicating that output is a function of total capital, labor force and time. According to this formula, the following growth balance can be derived:

$$Y_i = a_i K_i + b_i L_i + t_i$$

Among them: Y_i is the growth rate of output, K_i is the growth rate of capital, b_i is the proportion of labor used for income (%), L_i is the growth rate of labor input, a_i is the proportion of capital used for income, and t_i is the growth rate of technological progress.

In the model of neo-classical theory, the functional relationship of per capita output growth is demonstrated. It is believed that only when the growth of capital exceeds the growth of labor supply can the output per worker increase. That is to say, only when the capital-labor ratio increases can the output per worker increase. There is a positive correlation between the capital per worker and the output per worker. When the capital equipment acquired by each worker increases, the output per worker will increase. However, such growth cannot continue indefinitely without technological progress. The cause is that capital, like other input factors, will also experience diminishing marginal returns. The result is that the output per worker will also increase at a declining rate. When the capital-labor ratio reaches an equilibrium level, the output per worker will also be in an equilibrium state. At the moment, capital, labor, and output will increase at the same rate [6].

The neo-classical theory also analyzes the equilibrium relationship and equilibrium conditions of the regional economic growth under the circumstances that the above factors change individually.

The neo-classical theory also involves the changes and equilibrium conditions of the spatial (regional) development pattern brought by the inter-regional flow of factors. If Region B has more labor resources than Region A, and the labor supply is greater than demands, the wage will be low under normal circumstances. As a result, labor will flow from Region B to Region A. As a result of the flow, the gap in the wage and income level between Region A and Region B becomes smaller or even disappears, and the production function mechanism of the two regions becomes the same, which eventually leads to a balanced growth among regions. This is a circumstance of the so-called spatial equilibrium theory.

In the past 10 years, many scholars investigated and compared the huge differences in economic growth in various countries and regions over the past few decades, such as the slow growth of productivity in the United States, the rapid growth of Japan, and the "Four Asian Tigers", the long-term stagnation of the European economy, etc. What's more, the capital increase is faster than the labor increase in the industrial countries, and the return on investment has not diminished; according to the neo-classical growth theory, investment in poor countries shall be able to better promote economic growth compared with the same investment in a country with sufficient fund. However, facts have not proved this point. All of these showed that the neo-classical theory is not perfect in predicting long-term trends. Therefore, many scholars believed that regional economic growth is like a black box. They conducted a lot of analysis and comparison, and tried to see what is in the black box. A new theory of regional economic growth was formed until recently.

In 1983, Paul M. Romer, a professor at the University of California, Berkeley, published an authoritative paper, taking knowledge as another production factor. He believed that there are four production factors: capital, non-technical labor, human capital (measured by the length of education), and new ideas (measured by the number of patent rights). The theory recognizes that knowledge can increase the return on investment, which explains why long-term returns grow and why the growth rates of various countries are inconsistent. It is believed that knowledge is one of the production factors. Like capital, it can only be obtained by giving up current consumption. The state must invest in knowledge as it does in machinery. Moreover, it is further believed that there is the possibility of a virtuous circle, that is, investment promotes knowledge, and vice versa. This means that the continuous growth of investment can increase the growth rate of a country for a long time, but this idea is denied by traditional theories.

Like Romer, Morris Scott of Oxford University also believed that technological progress is very important for understanding growth. He said: "Inventions are motivated and facilitated by…expected profitability". Technological progress or "knowledge" must be invested separately.

The results of the survey conducted by Robert J. Barro of Harvard University and others using Romer's method were astonishing: What prevents poor countries from catching up with rich countries is the lack of human capital (i.e., education), not the lack of tangible capital.

As the research work increases, the value of the new growth theory will become clearer. It proves that some governments are wrong to focus their attention entirely on the business cycle. If the new growth theory can lead the governments of these countries to think more seriously about education investment, research and development, trade reform, intellectual property rights, etc., it will indeed be a breakthrough.[5]

[5] See *Reference News*, December 23, 1992; for the original text, see the British *Economist*, January 4, 1992.

7 New Theory in Research on National Competitive Advantages[6]

Classical economists Adam Smith and David Ricardo studied the work division among countries and regions from the perspective of comparative advantage, which can be regarded as the initial research on the theory of national competitive advantage. In the past 30–40 years, economic and technological competition among countries reached unusual frequency and intensity. Michael E. Porter®, a famous economist at Harvard Business School in the United States, proposed the theory of "national competitive advantage" after analyzing and comparing the generation and development of more than 100 industries or industrial groups in many countries in the world. Porter's definition of national competitiveness is "the country's ability to innovate and upgrade its industries, that is, the country's ability to obtain a high level of productivity and continuously increase productivity". He believed that the competitiveness of a country does not necessarily lie in the whole national economy, but mainly depends on whether the country has some unique industries or industrial groups. That is to say, national advantages usually reside in certain unique industrial sectors, such as luxury automobiles and chemicals in Germany, the electronic industry in Japan, banking and pharmaceuticals in Switzerland, shoe making and textiles in Italy, commercial aircraft and cartoons in the United States, etc. His definition of these unique industries is "competitive advantage over the world's strongest competitors", and the measurement standard is "continuous export of a large number of commodities, skills, and equipment to many countries". Porter proposed four basic factors of national competitive advantage, which constitute a "diamond structure" (Porter's theory of national competitive advantage is also called "diamond theory") (see Fig. 5).

(1) Factor conditions. It is necessary to create skills of laborers and strengthen the scientific basis for technological innovation, which is the most important production factor, including the establishment of specialized scientific research institutions, investment of specialized venture capital, etc., stimulate a large number of innovations, and quickly transform innovations into applications.
(2) Demand conditions. The characteristics of the domestic market can enable manufacturers to perceive changes in international demand and respond quickly. Experienced and fastidious buyers in China reflect the demands of advanced consumers, forcing the company to innovate and upgrade.
(3) Relevant supporting industries. It refers to that the relevant supporting industries of a certain industry in the country are very developed and have international supporting capabilities, including the production of processing machinery, raw materials, and parts. For example, the Swiss's success in the pharmaceutical industry stems from its previous achievements in the dye industry, and the

[6] From Wang Jici's "A Unique Theory of National Competitive Advantage", 1991. Data source: Michael E. Porter: "National Competitive Advantages", Harvard Business Review, Vol. 68, No. 2, March to April 1990, Pages 73 to 93.

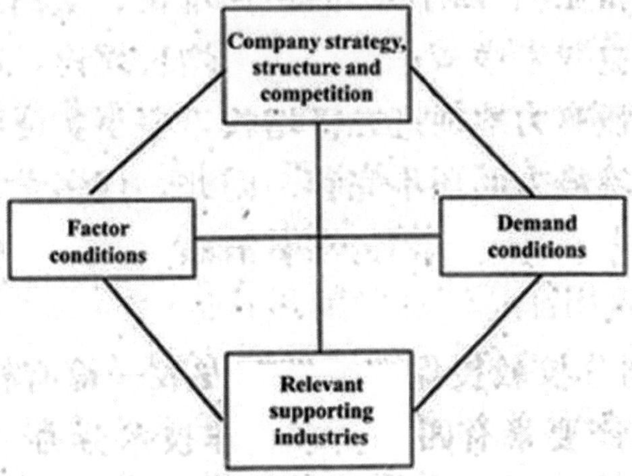

Fig. 5 Schematic diagram of the diamond theory

Japanese national competitive advantage in the electronic keyboard industry comes from technological progress in musical instruments and consumer electronics.

(4) Company strategy, structure, and competition. It mainly refers to the national management system of enterprises and companies and the organization mode of enterprises that are suitable for the characteristics of the country. In successful competitive industries in Italy, most are small and medium-sized private enterprises, which have obtained competitive advantages in footwear, furniture, wool fabric, etc. Technology-intensive enterprises in optics, chemistry and complex machinery are very successful in Germany, which is inseparable from the strict hierarchical system and extremely strict discipline requirements for all personnel. The country shall consciously train domestic competitors. The idea that a country only needs to retain one or two "national champions" and that national competitive advantages can be obtained as long as the government guarantees that it obtains a sufficient supply of raw materials, capital and technology is incorrect. In order to win in global competition, a company shall have strong competitors at home. Japan has 112 machine tool companies, 34 semiconductor companies, and 15 camera companies, which are good examples. The geographical concentration of local competitors can increase the intensity of competition and stimulate the company's innovation and upgrade.

The above four determinants of national competitive advantages promote each other, and may also restrict each other. Active domestic competitions and the geographical concentration of competitors are important driving forces for the four determinants to form a "diamond frame" system. Active domestic competitions will promote the development of other factors and the rapid formation of relevant supporting industries. An important effect of the "diamond frame" system is that the environment it creates promotes the formation of competitive industries in groups. These industries are often concentrated in groups and arranged in certain geographic regions for mutual promotion. Therefore, the country is rarely the birthplace of a single competitive industry. For example, the strength of Japan's consumer electronics has led to the success of the semiconductor industry and the application of memory chips and integrated circuits. Once the industry group is formed, various industries support each other through forward, backward and side links. In the group, the competition inside an industry will spread to the inside of another industry; research, development, the introduction of new strategies and new technologies, and other efforts of one industry will promote another industry and industrial upgrading.

After the "diamond structure" theory was proposed, Porter also elaborated on the role of the government, the efforts of the company, and the role of the leader in creating an environment that enables the company to obtain national competitive advantages. He pointed out that the government should increase productivity through factors affecting the "diamond structure"—increase investment in education, training, and technological infrastructure, encourage research and development activities, promote export-oriented economic activities, and enhance innovation capabilities.

Although the theory of national competitive advantage is aimed at the regional unit of the country, its basic idea of seeking development in competition is also suitable for the "region"—part of the country.

8 Two Methodologies for Research on Regional Exploitation and Regional Development

Since the 1950s, the issues of regional exploitation and development have gradually become widespread and profound socio-economic issues in many countries, attracting the attention of scholars in the fields of economics, geography, sociology and demography. A large number of research findings have been obtained in this field, promoting the development of sciences such as economic geography, regional economics, regional science, urban science and regional ecological economics.

Regional exploitation and development is a highly comprehensive field involving many elements such as nature, economy, society, and environment. In the long-term investigation and research practice, scholars have different disciplinary foundations and different perspectives for investigating problems, which has led to different theoretical systems. To sum up, there are mainly two dividing lines with large differences. Such a dividing line is manifested in disciplines to a greater degree. That is to say, some economists (regional economists) and geographers (economic geographers) are obviously different. The scientific and technical personnel engaged in planning practice are more inclined to be geographers in terms of space. Before the 1950s, the research on issues of regional exploitation and development was mainly affected by location theory, and both economists and geographers adopted the same investigation method for the research on location theory. It was only after the 1950s that there were obvious differences in the presupposition and focus of the research on regional issues.

(1) The issues of the spatial pattern (structure) and spatial process of social economy are the center of attention of most geographers and some economists. This part of the scholars believed that regional exploitation and development "is very natural as a spatial process".[7] They believed that the concern of these scholars in regional development is based on the fact that the development process produces the spatial format, and the spatial format can modify the regional development process through complex feedback. For example, the spatial differentiation of resources and labor will affect the growth and development stability of the regional economy. In theory, the growth pole theory, export base theory, center-edge theory, modernization theory, and industrial linkages based on economic linkages among enterprises are the theoretical contributions of these scientists.[8]

(2) In terms of regional issues, most economists are always used to assuming that it is homogeneous in the region while abstracting away the regional differences in the region. This will facilitate the application of macroeconomic theory to regional analysis. Such analysis needs to predict the short-term and long-term changes in regional economic activities from the interaction of certain parameter variables (such as the regional marginal propensity to consume, marginal propensity to import, etc.). However, these parameters can only be used as the same constant for all areas in the region to make the correct prediction. It is convenient for dissecting the interaction mechanism of socio-economic factors that affect regional development (such as labor and capital supply and demands, market, agglomeration economy, etc.) by studying the region in this way, thus revealing the direction and extent of the impact on regional economy when one variable changes or a group of variables change. Such research is of great significance for finding problems in regional development (diagnosis) and developing effective regional policies (treatment). However, some economists have been reluctant to admit that there are any regularities in the spatial organization of the economy, do not pay attention to the analysis of the distance and spatial distribution of population and economic activities, and believe that location is not an important economic variable. For example, some scholars confidently assume that international trade can be analyzed in a world with zero transportation costs. Such analysis obviously ignores the impact of distance and spatial differences within the region [16]. However, these two factors have an extremely obvious impact on much important decision-making for regional economic development.

(3) Through empirical research on national and regional development, many geographers and some economists believed that space-distance-accessibility are

[7] J. Friedmann's review of Mabogangjie's *Development Process: A Spatial Perspective*, transcript of lectures in China in August 1958.

[8] Contents of the lecture of Dr. Liang Zhijian from Hong Kong in 1986.

preconditions for regional development. Therefore, they proposed that the transportation and communication network must first be developed to achieve large-scale exploitation and rapid development of the region. This important viewpoint is often called "spatial determinism" or "spatial fetishism" by other types of scholars.

The dividing line of the above methodologies is caused due to differences in the focus and disciplinary basis for investigating the issues of regional development, which essentially reflects the complexity of the issues of regional development and the diversity of solutions. They all reflect the objective necessity of solving problems.

References

1. W. Lösch, *Die räumliche Ordnung der Wirtschaft*, 3. Auflage (Stuttgart, 1962)
2. W. Roscher, *Geschichte der Nationalokonomik* (Munchen, 1974)
3. W. Launhald, *Mathematische Begrundung der Volkswirtschaftslehre* (Leipzig, 1985)
4. A. Weber, *Uber den Standort der Industrie, 1. Teil, Reine Theorie des Standorts* (Tubingen, 1909)
5. W. Christaller, Die Zentralen Orte in Siiddeutschland, Jena. Die Hierarchic der Stadte, in *The IGU Symposium in Urban Geography*, Lund (1960)
6. D. Lu, *Location Theory and Regional Research Methods* (Science Press, 1988)
7. D. Furst, P. Klemmer, *Zimmermann KRegionale Wirtschaftspolitik* (Werner Verlag, Dusseldorf, 1976), pp. 92–114
8. C. Clark, *The Conditions of Economic Progress* (London, 1940)
9. K. Yang, *Research on China's Regional Development* (China Ocean Press, 1989)
10. C. Freeman, *The Economics of Industrial Innovation* (London, 1982)
11. R. Vernon, International investment and international trade in the product cycle. Quart. J. Econ. **80**, 190–207 (1966)
12. W. Isard, *Methods of Regional Analysis; an Introduction to Regional Science* (Cambridge, 1960), p. 510
13. E.D.G.S. Dunn, The market potential concept and the analysis of location, in *PPRSA*, vol. E (1956)
14. E. Otremba, *Raumordnung, Raumforschung und Geographic* (Institut ftir Raumforschung, Informationen, 1953)
15. E.V. Böventer, *Standortentscheidung und Raumstruktur* (Hannover, 1979)
16. H. Richardson, *Introduction to Regional Economics*, translated by J. Li (Haichao Publishing House, 1990)

Chapter 3
Progress of Research on Regional Exploitation and Regional Development in China

1 Issues of Regional Exploitation and Regional Development at Two Main Stages

Stage 1: From 1949 to the end of 1978

A planned economic management system with centralization of authority was implemented in China at this stage. Under this management system, the issues of regional exploitation and development in China did not become widespread, and profound socio-economic issues that attracted great attention like today due to the following reasons.

(1) The entity of planning management of regional exploitation and development was basically the central government. Before 1978, "under a highly centralized planned economic system and a unified administrative resource allocation mode, the investment entity was single (basically the central government), the construction projects were arranged, the layout blueprints drawn, the products purchased and sold, the raw materials allocated and supplied in a unified way, and the financial revenue and expenditure were all controlled by the central government" [1]. Under such circumstances, it was impossible for regions to plan and manage their resource development and socio-economic development as a whole. Regional national economic plans were mostly extensions of the central plan or provided regional support for national key construction. The characteristics of regions as independent stakeholders were suppressed, and regions lacked self-growth and self-organization capabilities.
(2) The regional economic development was at a low level with weak strength. By the mid-1970s, the economic strengths of various provinces, cities, and districts had developed greatly compared with those at the time when the nation was

D. Lu, *Regional Development and Its Spatial Structure*,
https://doi.org/10.1007/978-981-99-7683-6_3

founded. However, due to the poor base and small base number before, most provinces, cities, and districts were still at the preparatory stage for industrialization as a whole. In 1978, the gross national product was 358.8 billion yuan (current price), which was only 36.4% of that in 1990 when calculated at the comparable price; in 1978, the national income was 301 billion yuan (current price), which was only 38.0% of that in 1990 when calculated at the comparable price. If calculated with that in the mid-1960s, this proportion was much smaller. Liaoning is one of the few provinces with the best economic base in China. However, the gross national products were only 8.14 billion yuan and 10.26 billion yuan in 1957 and 1965 respectively, which is only equivalent to the economic scale of more developed provincially-administered municipalities today. Nearly half of the provinces and regions had a gross national product of less than five billion yuan in the mid-1960s. However, quite a few county-level cities and counties reached such a scale in the early 1990s. Under the weak regional (medium range) economic conditions at that time, it was unlikely to formulate and organize the implementation of the policies for comprehensive regional exploitation and development on its own.

At this large stage, China's national economy achieved fairly rapid growth on a weak foundation. However, such economic growth was, to a great extent, closely related to the extensive development of land resources in various regions of our country, the change in the state of backward agricultural countries, and the establishment of a complete national economic system and a distinctive regional industrial system. Coupled with the specific international environment from the 1950s to the 1970s, the issues of strategic allocation of China's productivity layout in the macro-region were raised. Therefore, the central, provincial and municipal, and district government departments, as well as some scholars did a lot of work on the issues of regional exploitation in the following aspects at this stage:

① Joint plant selection and regional planning centered on large projects;
② Macro strategy of national productivity distribution (mainly surrounding how to deal with the relationship between the coastal and inland areas and between first-front, second-front and third-front regions);
③ Agricultural division and planning of people's communes.

Stage 2: From 1978 to date

"At the end of 1978, the Chinese government established opening up as the basic national policy based on profoundly summarizing domestic economic construction experience and analyzing the international economic situation".[1] Under the guidance of this basic national policy, the following four major changes took place over the last ten years, which led to the emergence of a large number of issues of regional exploitation and development in China and promoted the development of research work.

[1] Office of Special Economic Zones under the State Council, Introduction to China's Coastal Open Zone, March 1991.

(1) The central planning system with centralization of authority gradually evolved into a management system combining planning and market regulation. Under this system, local governments and enterprises had considerable authority and capabilities to organize the preparation of comprehensive medium and long-term national economic plans and comprehensive land development and remediation plans within the region, and independently carry out the development and utilization of important resources and the demonstration and construction of major engineering projects in practice. Due to the fierce competition among regions and the emergence of unbalanced regional development, the autonomy of such planning management became increasingly strong. As a result, the issue of regional development strategy became one of the major concerns of the heads of local governments and competent departments at all levels.
(2) For the region as the entity of planning management of regional exploitation and development, its economic strength was greatly enhanced, manifested in the rapid expansion of the scale of fund accumulation and investment capacity. With the implementation of many preferential policies, many major projects could be jointly invested with the central government or even independently invested by local governments for construction. Especially for a large number of upstream, downstream, or lateral auxiliary projects related to the main project, the locality may plan to build its own distinctive regional system. Due to the changes in the foreign trade system, local governments mastered a considerable part of foreign exchange, making it possible to plan the introduction of foreign technology, which promoted the externalization of the regional economy and technical updates. More than 80% of capital construction investment in China came from state investment before 1978. This ratio has dropped sharply since 1980 (Fig. 1). Among the sources of capital construction funds in 1989, investment in the national budget and domestic loans accounted for only 39.7%, self-raised (local and enterprise) funds and other funds accounted for 46.0%, and foreign capital accounted for 14.3% [2].
(3) Due to the differences in natural conditions and the accumulation of long-term unbalanced development in history, the gap in economic strength between the east and the west, and between the Han nationality and ethnic minority areas was getting wider and wider. The results were as follows: The less developed regions strongly required that the country implement more preferential investment policies, and requested to change the unreasonable price relationship between energy, raw materials, and processed products, strengthen the construction of infrastructures such as transportation, communication, culture, and education, and implement other policies and measures to support the socio-economic development of ethnic minority areas and old revolutionary base areas, areas with large ethnic minority populations, border areas, and poor areas. For

the more developed regions in the east, it was required to realize the transformation and advancement of the industrial structure as soon as possible, develop export-oriented industries for the international market, expand the economic and technological radiation to the central and western regions, etc. Under the strong influence of these two trends, China's most typical issue of regional exploitation and development emerged: How to reduce and eliminate the imbalance among macro-regions. This issue was very prominent across the country and within the provinces.

(4) Environmental pollution and ecological imbalance were aggravated. In the 1980s, the pollution caused by "three industrial wastes" in various regions in China increased sharply, and water loss and soil erosion, salinization, seawater intrusion, and desertification accumulated for a long term became increasingly prominent. These issues cannot be ignored for the socioeconomic development of various regions. In response to these issues, specific measures must be taken in terms of rational development and utilization of resources, industrial structure, layout, and comprehensive treatment. The generalization and seriousness of environmental and ecological problems have complicated the research and planning of regional exploitation and development, indicating that scientific guidance and preparation of scientific exploitation and development planning are more necessary. At this development stage (so far), the main work carried out by government departments and the main research fields of scholars are concentrated in the following aspects:

① Comprehensive land development and management planning (national, provincial, municipal, regional, city, and county level) and macro layout of productivity. Among them, the macro layout of productivity is mainly carried out around the relationship between the three major zones and the development of the national first-level development axis.

② Regional economic development strategy. The regions involved are basically concentrated at the regional and city level.

③ Agricultural division and rural development. The "National Comprehensive Agricultural Division" and the agricultural division of various regions in the mid-1980s played an important role in the comprehensive exploitation, regulation and development of rural areas in China.

2 Progress of Main Research Fields

For more than 40 years since the founding of the People's Republic of China, some scholars in the disciplines of geography, economics and urban planning, and system science had actively participated in the research on regional exploitation and development in China. A large number of researches were conducted in response to practical issues, and many specific analyses and suggestions were adopted and implemented

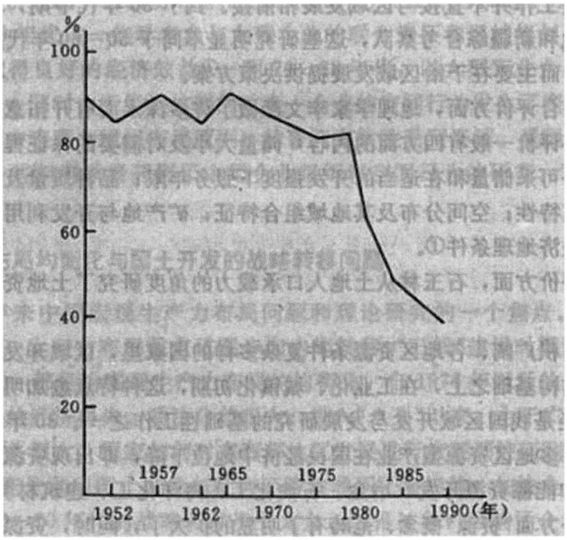

Fig. 1 The decline in the proportion of national investment in construction investment

by the government; based on digesting foreign theories and models, a few scholars put forward effective theoretical opinions on exploitation and development in China. Especially since the 1980s, Chinese scholars had occupied an important position in the world in terms of the research breadth and scale as well as the number of research findings. However, more of our research had insufficient depth, with non-standard results and fewer theoretical contributions; more in line with the trend, but less with independent opinions. Nevertheless, the achievements obtained by Chinese scholars in the following aspects were still commendable in the field of regional research.

1. Evaluation of regional resource conditions and comprehensive development

In the 1950s, China's large-scale industrialization and economic development just started, and the resource situation and the development and construction conditions of various regions were still unclear. Regional resource development and economic development objectively require a comprehensive understanding of "resources". This requirement is more prominent in those inland and remote areas. In response to such a situation, the Chinese Academy of Sciences organized a comprehensive expedition team consisting of scientists in multiple disciplines from Gansu, Qinghai, Ningxia, Inner Mongolia, Sichuan, Yunnan, Xinjiang, etc. The comprehensive expedition team for the Heilongjiang River basin, which was composed of Chinese and Soviet scientists, in the late 1950s and early 1960s also belonged to this category. The focus of this comprehensive expedition work was to find out the resources and put forward evaluation and suggestions on the conditions (resource characteristics and external environment), directions, and measures for resource development and utilization. From the 1950s to the 1960s, resource evaluation was mainly limited to agricultural resources such as land, climate, pasture, and water, as well as mineral and energy resources such as iron, nonferrous metal, coal, petroleum, and natural gas. Based on the comprehensive evaluation of resources, a series of assumptions on regional resource development and utilization were proposed. In the 1970s, the Chinese Academy of Sciences organized a large-scale wasteland expedition team, but this work was not directly linked to regional development. In the mid-1980s, the Chinese Academy of Sciences organized a comprehensive expedition team to the southwest, southern mountains, and Xinjiang. These researches mainly aimed to provide decision-making solutions for regional development and were obviously different from those in the 1950s and 1960s, which focused on understanding the natural characteristics of the region.

In terms of the comprehensive evaluation of mineral resources, Li Wenyan, a geographer, did a lot of in-depth and pioneering work. He proposed that the economic evaluation of mineral resources generally has four aspects: the size of the reserves and the assurance of demands, especially the recoverable reserves under the current technical conditions and the service life under the appropriate development intensity; variety of quality and suitability for users, including the characteristics of processing technology; spatial distribution and its geographical combination characteristics; orefield and the construction conditions for development and utilization, especially the natural and economic geographical conditions.[2]

In terms of agricultural resource evaluation, Shi Yulin researched the relationship between land resources and regional socio-economic development from the perspective of the population supporting capacity of the land.

In a country like China, which has a vast area and complex and diverse regional resource conditions, regional exploitation and development must be carried out based on the resource structure of the region. In the initial phase of industrialization and urbanization, this characteristic became more and more obvious. Therefore, the comprehensive evaluation of resources used to be part of the groundwork of the research on regional exploitation and development in China. Since the 1980s, the status of resource-based industries in many regions had declined in the national economy due to the transformation of industrial structure, that is to say, the resource-based industries (most of them are basic raw material industries, such as energy resource development, metallurgy, petrochemicals, marine chemicals, building materials, etc.) declined. In this case, the concept and scope of "resources" were significantly expanded. Meanwhile, the resource base sometimes cannot become a prerequisite for regional development.

---

[2] Li Wenyan, Pay Attention to Geographical Conditions for Resource Evaluation, Research on Industrial Town Layout and Regional Planning, First Series, 1981.

2. Productivity distribution and production area complex

China's theory of productivity distribution was first introduced from the system of the former Soviet Union in the 1950s. The research and teaching institutions at that time[3] were as follows: Productivity Distribution Research Group of the Economic Research Institute of the Social Sciences Division of the Chinese Academy of Sciences, Production Distribution Teaching and Research Section of the Planning and Statistics Department of the Renmin University of China, Economic Geography Section of the Geography Research Institute of the Chinese Academy of Sciences, Economic Geography Research Section of the Nanjing Geography Research Institute of the Chinese Academy of Sciences, Geography Departments of Peking University, Nanjing University, and Sun Yat-sen University, etc. In the 1950s and 1960s, the research on China's productivity distribution theory was greatly affected by the academic circles of the former Soviet Union. In the 1980s, the Western production distribution theory was generally introduced to the majority of scholars, which promoted the deepening and perfection of the theory. The research contents were mainly as follows:

(1) The principle of socialism production (productivity) distribution. Based on the expositions in Lenin's works, most people believed that the law of socialist production (productivity) distribution is different from that of capitalist production (productivity) distribution. The former shall serve the maximum benefits of the vast majority of the people, be as close as possible to the place of raw materials, the place of fuel, and the product consumption markets, abide by the principles of national defense security and economic development in ethnic minority areas and remote areas and focus on achieving a balanced distribution of productivity among regions. At the same time, affected by the academic circles of the former Soviet Union and the domestic political atmosphere, the "bourgeois" productivity distribution theories (mainly the location theory) were criticized as "bourgeois pseudoscience". Through the implementation of the opening-up policy, people have now fully understood how naive and ridiculous this criticism with political labels and academic prejudice is. In the preface of the book *Location Theory and Regional Research Methods* (1988, Science Press), the author of this book mentioned that "the efforts of critics are by no means comparable to the rigorous scholarship and hard work of the scholars who created the location theory".
(2) The theory and application of production area complex. The concept and theory of production area complex originated in the former Soviet Union in the 1930s. Economic geographer Baransky and economist Kolosovsky first clarified the meaning of production area complex. In the 1950s and 1960s, some scholars participated in the joint plant selection and regional planning of large enterprises, and applied the theory specifically, which achieved good economic benefits in

[3] Strictly speaking, the "Isolated Countries" written by Thunen from Germany was translated into Chinese and introduced into China in 1937.

practice.[4] In the 1970s and 1980s, attention was still paid to regional planning centered on large industrial enterprises. At the same time, the spatial behaviors of enterprises changed due to changes in the economic system, so the general validity of the theory of production area complex (manifested as the principle of group layout in practice) was questioned. Affected by the geographical research on international enterprise organizations, some young scholars carried out research on Chinese enterprise organizations and spatial behaviors and achieved fruitful results.[5]

3. The equalization of industrial production distribution and strategic shift of territorial development

This had been a focus of the issues and theoretical research on China's macro-productivity distribution for decades. The focus of productivity distribution in the "First Five-year Plan" period (1953–1957) had gradually shifted to the inland, and this shift reached an unprecedented scale during the construction period of the "third-front" movement. One of the goals is to achieve the equalization of productivity distribution. In a few years since 1976, academic circles extensively summarized the economic gains and losses of this shift. There were very different opinions. Since entering the "Eighth Five-year Plan" period, the eastern region maintained a great momentum of development due to the extensive introduction of the market economy system. The issue of China's approach to conducting macro-control over the development of the eastern and western regions became prominent once again.

(1) Acknowledge and make full use of the law of imbalance between territorial development and socio-economic development, and accelerate national economic development. Some scholars believed that the strategic shift in the 1960s and 1970s was too fast, which affected the overall growth rate of the national economy. The lessons were profound. Those holding this viewpoint were represented by Chen Dongsheng in the economic circles.[6] He pointed out: "In the area to the west of the Lanzhou-Chengdu-Kunming line, generally speaking, the natural conditions are harsh, ..., social and cultural conditions are also poor. Natural resources are abundant, but it is very difficult to develop. In this century, large-scale development is beyond the reach of national power" ("Re-discussion on China's Economic Distribution Strategy"). There was another

[4] For example, the planning of an industrial zone focusing on petrochemicals in Xigu area of Lanzhou, the planning of an industrial zone focusing on the machinery industry in Luoyang, the planning of the Beiyan chemical zone in Taiyuan, etc.

[5] The earliest typical investigation was made by Cao Yong. Later, there was Li Xiaojian's Ph.D. dissertation "Industrial Change and Company Activities—Australian Geographical Research" (May 1990), and Fei Hongping's Ph.D. dissertation "Research on Enterprise Spatial Behaviors, Industry Cluster and Spatial Organization—A Case Study of the Industrial Belt along the Qingdao-Ji'nan Railway" (April 1992).

[6] Written by Chen Dongsheng, Theory and Practice of Economic Distribution, Liaoning University Press, 1989.

Edited by Chen Dongsheng: Research on China's Industrial Distribution, Economic Science Press, 1988, Beijing.

opinion in economic circles: China's path of balanced development with "fairness" as the policy goal in the first thirty years is also the development model chosen by all socialist countries. As a result, China's economy paid a heavy price. One of them was the relative decline of coastal industries, which widened the economic gap between China and countries in the Asia–Pacific region. Scholars holding this viewpoint believed that the core of key issues of productivity distribution and territorial development strategy is how to deal with the contradiction between "fairness" and "benefits". That is to say, if the main goal is "fairness", more funds must be invested in the vast central and western regions with poor economic benefits, and the strategic emphasis shall be laid on the central and western regions. On the contrary, the eastern regions shall be taken as the strategic emphasis. These scholars were in favor of the latter strategy.

The author started to research the issue of China's macro productivity distribution after the national academic conference on economic geography in Changsha in December 1978. One of the questions discussed in this seminar was, "is there any common law for productivity distribution under different social systems?" The answer of the scholars was affirmative. This prompted me to summarize the regular issues in China based on scientific analysis and by drawing lessons from the experience and theories of foreign regional development and economic distribution. At the national symposium on economic geography and territorial management convened in Urumqi at the end of September 1984, the author read the paper "The Scientific Basis of the General Map of China's Industrial Productivity Distribution in 2000". The paper quoted the principle of "the inverted 'U' relationship between economic growth and unbalanced development". "In this century and at the beginning of the twenty-first century, the primary goal of China's economic and social development is to strive for a higher growth rate of the national economy". "Balanced development is a long historical process; due to the differences in natural conditions, geographical location as well as economic and historical basis among regions, balanced development is just relative; the greater the difference in regional stability factors, the greater the relativity of such 'balance', that is, the greater the degree of difference". "If the economic growth rate is taken as the main goal, it will have to lead to unbalanced economic development. On the contrary, if the latter is taken as the main goal, it will affect the overall economic growth rate". This paper analyzed the huge losses caused by the implementation of strategic shift and "third-front" construction from the 1960s to the early 1970s, as well as the correctness of shifting the strategic emphasis of economic development and territorial development from west to east from the 1970s to the early 1980s. In response to the widely popular call for a strategic shift to the west at that time, it was proposed that the general map of national productivity distribution in the "Seventh Five-year Plan" and "Eighth Five-year Plan" periods shall not undergo material changes, and that the general map of China's productivity distribution shall not undergo material changes from the end of this century to the beginning of the next century according to the principle of "point-axis system"; the strategic concept was formed that the eastern coastal zone and the zone along the Yangtze River shall become the axis of national first-level key development, forming a T shape. "The axis of Yangtze River connects the two most developed core regions of the inland with the axis of the coastal zone, and the two axes meet in the most economically developed Yangtze River Delta. Such spatial structure form more accurately reflects the basic framework of the spatial distribution of China's territorial resources, economic strength, and development potential, and will also help to make better use of the huge momentum stimulated by foreign economic and technical exchanges in the future. The further development and construction of the two axes can lay a foundation for quadrupling the national economy, and promote the secondary and tertiary development axes connected, so as to achieve global smooth development" [3]. After that, the author further discussed the macro-regional strategy of territorial development and socio-economic development in the next 10–20 years several times, used the change in the transportation distance between the axis of capital construction investment and the coastal zone to more accurately measure the secondary shift of China's macro-regional strategy, and further demonstrated the T-shaped key development strategy based on the principle of "point-axis" progressive diffusion [4].

The gradient theory, which is basically consistent with the above viewpoint, is very influential in the government departments and academic circles. It is believed in the theory that: "The development of economy and technology is unbalanced no matter in the world or within a country, and an economic and technical gradient shall be formed objectively. Where there is a gradient, there is a spatial shift". "First, let the high-gradient region with conditions introduce and master advanced technologies, and then gradually shift to the regions at the secondary and tertiary gradients" [5].

(2) From a comprehensive socio-economic perspective, China shall take the road of balanced development. Since the "Sixth Five-year Plan", the inclination towards the eastern coastal zone had led to adverse consequences.

In the mid-1980s, the Chinese leaders at the time stated on many occasions that they would make major adjustments to China's socio-economic development space strategy, shifting the strategic focus of development to the western region, and even more clearly proposed to shift to Northwest China. At the same time, some scholars in academic circles put forward the "anti-gradient theory", believing that China has vast central and western regions rich in resources. Among them, there are large areas where ethnic minorities live in compact communities, and their economic and technical levels are relatively low. From the perspective of strengthening national unity, consolidating border areas, and developing and utilizing rich resources, the priority of territorial development shall be given to the western region. The viewpoint of the young scholar, Guo Fansheng, was representative. In the middle and late 1980s, he repeatedly emphasized at many academic conferences and in some journals that China should greatly increase investment in the central and western regions, and should not incline towards the coastal areas. He believes: "There are limits to the preferential policies of any country. The imbalances in industrial production distribution and income distribution are restricted by the balance of reproduction of the national economy. Unbalanced development cannot cause too much trade deficit, industrial structure imbalance, or hinder technological progress..." "Researches in the past few years have shown that the simple preferential policies of opening up of coastal areas have put China's opening up into a dilemma". When evaluating the strategic shift of China's productivity distribution to the "third-front" regions in the 1960s and 1970s ("westward movement") and the inclination toward the coastal areas in the 1980s ("eastward movement"), some geographers believed that the achievements of the strategic decision of "westward movement" should be affirmative. However, in the strategic shift of "eastward movement" in the 1980s, a high rate of economic growth was achieved, but there was obvious low efficiency. "The problems were even more serious if analyzed from the national macro social benefits". It mainly refers to that policies and funds were substantially inclined to coastal areas, which widened the economic gap between the coastal and inland areas, "causing an excessive disparity in people's living standards between the inland and coastal areas", and resulting in a consequence "unfavorable to social stability and unity" [6].

(3) The disequilibrium theorists and the equilibriums had the same viewpoint of strengthening the socio-economic development of the less developed or underdeveloped regions and advocated adopting various policies and measures by the central government to support the development of ethnic minority areas and distressed areas. The difference is that the regulatory funds of the central government in the 1980s and 1990s were mainly invested in developed regions or less developed regions. At the end of the 1970s and the 1980s, China's opening started in the Eastern region of China, especially in its southern regions. This was mainly determined by the geopolitical and geoeconomic factors of these regions (the Pearl River Delta and the coastal areas of Fujian). As a result of the implementation, the eastern coastal areas and even the whole country achieved a high economic growth rate. For the last more than ten years, China's macro strategy of regional development was roughly in line with the path described by disequilibriumists:

   a. First, open up some cities and regions in the eastern coastal areas. In order to adapt to the policies for global trade promotions and the economic development of coastal areas, the central government has given preferential investment to the development of infrastructure, energy, and raw material industries in these areas.

b. Encourage and promote the east–west interaction to complement each other as the eastern coastal areas had experienced rapid development and fund accumulation. Since 1992, the State Development Planning Commission has divided the seven regions to be planned, which was in full compliance with this thought.[7] This is a successful road for regional development.
c. Since 1992, the Chinese government has expanded the opening up of land borders and inland areas, so that "simple opening up of coastal areas" was replaced by all-around opening up across the country. This process of gradually expanding the scope of opening up from the coastal areas to the border areas in the west also illustrated the development from unbalanced to relatively balanced from one perspective.

4. Regional policies

On the Ten Major Relationships issued by Chairman Mao Zedong in 1956 is a declaration of China's regional policies.[8] For a long time thereafter, this article became one of the guiding principles for the academics, especially geographers, to discuss the rational distribution of productivity. However, the actual work done by geographers within the scope of regional policies was not much. Since the reform and opening up, Chinese economists and government departments have done a lot of fruitful research, and published a large number of works and papers.[9] The main reason is that economists have always paid more attention to the research on the internal operating mechanism of the regional economy and the economic relations among regions. Once the characteristics of the operating mechanism and inter-regional relations are understood, it is easy to lead to the category of regional policies.

The research on regional policies in China's academic circles contains a wide range of contents, which can be summarized into two aspects: the first is the regional policies on how to exert regional advantages, adjust the industrial structure, promote the rational development and utilization of resources and protect the environment; the second is regional policies aiming to gradually narrow the gaps in the development level and structure among regions and promote the gradual balance of productivity distribution. The second kind of regional policy is mainly described below in this book.

[7] Wang Zhan, et al.: Development of the East and the West Should Follow the Path of Unbalanced Development, World Economic Herald, April 10, 1989.

[8] *On the Ten Major Relationships* issued by Chairman Mao Zedong in 1956 was a report at the Supreme State Conference.

[9] Edited by Chen Dongsheng, A New Starting Point for Research on Regional Economy, Economic Management Press, 1991, Beijing.

Edited by Zhou Shulian, et al., Research on China's Regional Industrial Policies, China Economic Publishing House, 1990, Beijing.

Zhou Qiye, Liu Zaixing, et al., Regional Economics, China Renmin University Press, December 1989, Beijing.

Yang Kaizhong, Research on China's Regional Development, China Ocean Press, 1989, Beijing.

(1) "Benefits" and "fairness". In the research on regional policies since 1978, "benefits" and "equity" have been the dominant themes of regional policies. The research focused on: First, national investment policies and investment distribution. The discussion mainly focused on evaluating the action of investing 35–40% of the national funds in the "third-front" areas from the 1960s to the 1970s, while more than 50% of the funds were concentrated in the eastern coastal areas in the 1980s. Surrounding the investment policies, many scholars researched the effects of different investment distributions in the east, middle, west, coastal and inland areas. According to the differences in regional investment policies suggested by scholars, they can be divided into inclined investment policies and balanced investment policies. Since the reform and opening up, the diversification of investment entities has become increasingly obvious due to the continuous growth of the market economy. Under such circumstances, capital and other production factors will inevitably flow to areas with high marginal rates of return on capital and labor, which are the areas with high economic and technical levels in China. Second, price policies and tax policies. Many scholars analyzed that the friction of economic benefits among regions and the irrational distribution of comparative advantages are mainly caused by unreasonable price policies, that is, low prices of energy and raw materials, and high prices of manufactured goods. As a result, the eastern coastal areas will obtain more funds and more development opportunities. Since investment policies and price policies are conducive to the development of the eastern coastal areas and other more developed regions, causing losses to the benefits of the regions that produce energy and raw materials and rely on agriculture, scholars proposed that regional policies should be improved in accordance with the principle of compensation. In fact, the financial subsidies and subsidized loans implemented by the state to ethnic minority areas are the embodiment of such regional policies.

(2) Regional policies and industrial policies. For more than 40 years since the founding of the People's Republic of China, including more than 10 years after the reform and opening up, the imbalance in China's industrial structure has occurred many times. For this reason, the government has made policies and regulations to adjust the industrial structure several times. However, China has a vast area, and regions have large differences. Combining industrial policies with regional policies is an issue worth studying, that is to say, the regionalization of industrial policies and the industrialization of regional policies. According to the viewpoint of Chen Dongsheng, "regionalization of industrial policies means that when determining leading industries, associated industries, related industries, as well as key supported and restricted industries in the industrial policies, it is necessary to have a general outline of the layout of various industries across the country, and to break down the development goals of all industries according to local conditions". "Industrialization of regional policies means that when determining the emphasis of regional development and various specific regional development countermeasures, these policies shall stipulate the industrial structure of various regions, especially leading industries" [7].

(3) "Poverty alleviation". For the research on the "poverty alleviation" policies, the target areas are mainly old revolutionary base areas, areas with large ethnic minority populations, border areas, and poor areas. The poverty-stricken areas are widely spread in China. Except for the lack of information on Tibet, there were 664 poverty counties across the country (1988), mainly concentrated in the border areas of ethnic minority areas and the mountainous areas of eastern and central provinces. The "poverty alleviation" policy is a significant part of China's regional policies. For a long period of time since the founding of the People's Republic of China, the policies of fiscal subsidies and material relief were mainly implemented for poverty-stricken areas. Through a large number of surveys and research, the policy of "providing work as a form of relief" was implemented over the past 10 years, that is, to change the previous way of providing relief and subsidies to support the construction of roads and enterprises in the poverty-stricken areas, and enhance the self-development capacity of the poverty-stricken areas. How to comprehensively develop the planting industry, breeding industry, processing industry, commerce, transportation industry, energy industry, building materials industry, and service industry according to the characteristics of specific regions; population policies, and education policies; environmental and ecological protection policies, etc.[10] (Fig. 2).

(4) Opening-up policy. China decided to implement opening up at the end of the 1970s, but the development of the opening-up policy and the range of implementing regions are selective and gradually expanded. Therefore, it is an important regional policy in China for the past fifteen years and for the next few decades.

China's opening-up policy is not the result of research by scientists, and the implementation of regional policies in open areas has not become the research object of many scholars. However, scientists have indeed made many contributions to the formation and development of China's opening-up policy. The main aspects are as follows: ① Introducing and elaborating the theory of international economic circulation and the theory of product life cycle in combination with the practice of China and neighboring regions and countries, and providing a theoretical basis for regional policies of various types of open areas through the research in this regard; ② Research on international export processing zones, tariff-free zones, economic and technological development zones, and their policies; ③ Transformation of industrial policies in developed countries, assessment of China's industrial development status, and conditions and direction of high-tech development; ④ Demonstrating the macro pattern and spatial shift of China's opening up from the perspective of geopolitics and geoeconomics, as well as the theory of spatial structure. According to this theory, the regions along the coastline, border lines, and important traffic arteries were opened up to the outside world first within the country and all regions, and the developed

[10] Poverty Area Development Research Group, Geography Research Institution, Chinese Academy of Sciences, March 1988.

industrial axes (belts) were established, thus driving the development of other regions. Space strategy is an important aspect of regional policies.

There are various types of open areas in China, mainly depending on geopolitical and geoeconomic factors. The policies that were basically the same were implemented in the earliest special economic zone and the economic and technological development zone (preferential policies for foreign investment). The other category is coastal open cities and the entire coastal open areas. The same opening-up policy was implemented for the border and inland open cities expanded in 1992 (Fig. 3) and the coastal open cities.[11] The tariff-free zone has the smallest geographical scope, but the most favorable policies for foreign investors and operators.

(5) Simulation of regional policies. Although Chinese scholars have completed a lot of work concerning the research on regional policies and published a lot of works, there is an obvious shortcoming, that is, the lack of simulation research and analysis of the effects of regional policies. One district or a group of districts forms a geographic unit, and its actual socioeconomic development process and state and the ecological environment evolution are always under development due to the impact of changes in its external conditions and internal factors. However, this process is relatively long in general. More importantly, once any change is not conducive to human needs, it will be difficult to save. To scientifically predict this process and results, it is necessary to rely on mathematical models and electronic computers. By giving the values (parameters) of several controllable variables that reflect regional policies, the results of changes in the entire system (regional system) can be quickly seen. In this way, the actual regional socio-economic development process (which takes many years) becomes a computer simulation and scheme experiment process (only a few hours or even a few minutes). The results of policy simulation show whether a certain policy meets the development goals and how to make corrections to meet the development goals. That is to say, regional policy simulation is a scientific method to control the regional development process according to people's wishes.

[11] The preferential regional policies implemented by the Chinese government in the economic and technological development zones mainly include: ① The equipment and raw materials used for the production of export commodities are exempt from import tariffs. ② The income tax for wholly-owned enterprises and solely foreign-owned enterprises is only 15%, and: enterprises are exempt from tax for the first two years from the profitable year, and the tax will be reduced by half for the next three years; for advanced technology enterprises, the tax reduction period will be extended for another three years, and the tax rate is 10%; enterprises that stop production of more than 70% of export commodities will be taxed at 10% each year, and the profits from foreign investment are exempt from remittance tax. ③ The export products manufactured by foreign-invested enterprises are exempt from industrial and commercial consolidated tax.

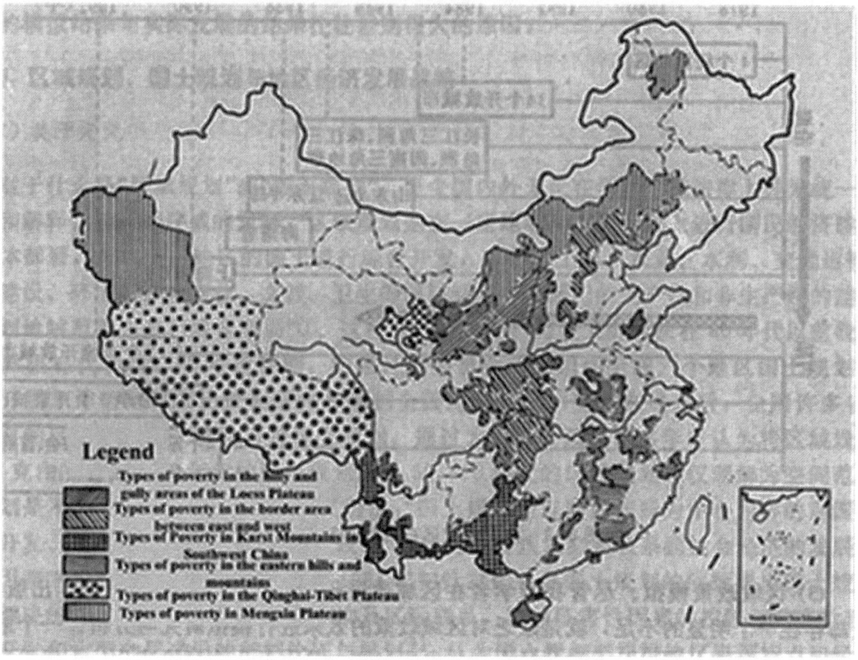

Fig. 2 Distribution and types of poverty-stricken areas in China

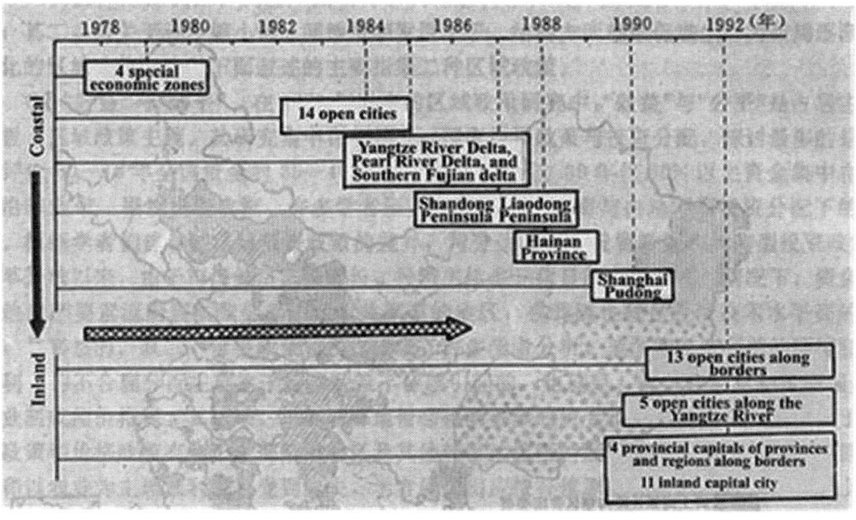

Fig. 3 The development of open areas in time and space in China

In the early and mid-1980s, the Development Research Center of the State Council and the Academy of Social Sciences organized the "Research on Shanxi Energy Base Planning" and the large research on "China in 2000". The industrial structure, resource utilization distribution, balance, income distribution, population growth, price changes, etc. of the region (including the regional unit of the "whole country") were simulated. However, there were a series of uncontrollable factors in determining the model structure and parameters no matter for a region (province) or the whole country under the planned economic system at that time. For example, it was very difficult to accurately predict prices, investment scale, resource allocation, etc. in a few years. In the first 30 years since the founding of the People's Republic of China, the economy experienced several major fluctuations, mostly caused by human factors. Therefore, it was difficult to use the results of past statistical analysis as the basis (or parameters) for predicting the future. This is why the simulation results of regional development (policy) are often very different from the actual development results.

5. Regional planning, territorial planning and regional economic development strategy

1) About the definition

As for what "regional planning" or "territorial planning" is, there is no uniform definition or interpretation in academic or legal terms both at home and abroad. According to Hu Xuwei's opinion, "Regional planning refers to the overall deployment of national economic construction on the territory within a certain geographical scope, that is, to carry out comprehensive development of the territory in a certain area and comprehensive layout of productive and non-productive construction of various sectors of the national economy including industry, agriculture, water conservancy, transportation, urban and rural construction, environmental protection, commerce, culture, education and health within the planned geographical scope".[12] This viewpoint of taking "comprehensive layout" as the definition was very representative in China's academic circles before the 1980s. In the early 1980s, the State Development Planning Commission successively organized six regions across the country as territorial planning pilots, carried out the research on territorial planning in the Beijing-Tianjin-Tangshan region, and compiled the outline of national territorial planning. After that, many provinces, cities, autonomous regions and counties across the country carried out territorial planning. Through a lot of practice, many scholars believed that it is not comprehensive to only understand regional planning (in Lyu Kebai's opinion, territorial planning in the 1980s is regional planning in the 1950s and 1970s) as spatial planning. In 1985, Hu Xuwei put forward that: "Territorial planning refers to construction planning for the development, utilization, management and protection of the land with the construction layout as the central content", and believed that such planning must be related to socio-economic development

[12] Hu Xuwei, The Nature and Tasks of Regional Planning, Research on Industrial and Urban Layout and Regional Planning, Fourth Series, December 1982.

and time factors for analysis and research.[13] At the beginning of 1984, in response to the question: what is territorial planning, the author put forward the tasks to be solved by the territorial planning, including regional economic direction and structure, inter-regional connection, control and comprehensive management of regional hazardous factors, early demonstration and planning of new district development and large-scale project construction [8]. It was believed that the territorial planning should put forward the development direction of the regional social economy and the idea of large-scale construction projects according to the characteristics of regional resources and the economic status. It put forward reasonable solutions to the energy, transportation, water supply, environmental governance and protection on the basis of prediction. Its main function is "coordination" instead of "comprehensive balance", which was the main function of medium and long-term plans of the national economy under the central planning system in the past. The object of "coordination" is the conflict of goals of various regions and departments in the utilization of resources and space. The sequence, progress and time limit of various constructions, and the development speed of the national economy are not clearly stipulated in territorial planning.

Regional planning and territorial planning need to involve the development direction, structure and inter-regional links of the regional economy, etc., and there are many similarities in contents and goals with the regional socio-economic development planning. Therefore, it is an important field of practice for scholars researching regional exploitation and development issues.

2) Practice of regional planning and territorial planning

In May 1956, the State Council passed a resolution concerning several issues on strengthening the construction of new industrial areas and new industrial cities, and proposed to actively carry out regional planning. The initial regional planning was mainly completed by the building professionals and other engineering technicians. In order to overcome discipline limitations, economic geographers began to participate in this regional and comprehensive technical and economic work, and played a major role in the process. In the 1950s, regional planning of cities with key project construction such as Maoming, Lanzhou, Baotou and Daye was mainly carried out. From the 1960s to the 1970s, economic geographers continued to carry out surveys and research in terms of the comprehensive layout of industrial bases within the scope of regional planning, such as research on the industrial layout of Southwest and North China, and research on the comprehensive industrial layout of some key development and construction areas in Shandong, Anhui, Liaoning, Hebei and other provinces. Although these researches were not regional planning, the work targets and survey methods were basically the same, with new contents added.[14]

---

[13] Hu Xuwei, Several New Understandings of the Concept and Essence of Territorial Planning, Productivity Distribution and Territorial Planning, First Series, 1985.

[14] Hu Xuwei, Research on Economic Geography of Industrial Layout and Regional Planning in the Past 30 Years, Research on Industrial and Urban Layout and Regional Planning, First Series, July 1987.

At the end of the 1970s, Zhao Ziyang, secretary of the Sichuan Provincial Party Committee at that time, led a delegation to visit Western Europe. After returning to China, he advocated the development of territorial management and territorial planning in China. In the early 1980s, the State Infrastructure Commission organized several pieces of training for personnel engaged in territorial planning. After the establishment of the Land and Resources Department of the State Development Planning Commission, it immediately started the territorial planning of the six pilot areas across the country, such as territorial planning (research) of the Beijing-Tianjin-Tangshan region from 1980 to 1985, a compilation of the national territorial planning outline since 1985, and territorial planning of most provinces, cities, and districts carried out almost simultaneously. In the planning, economic geographers and regional economists completed a lot of work and played a major role. For example, about 500 professional scholars participated in the territorial planning of the Beijing-Tianjin-Tangshan region, and the Institute of Geography of the Chinese Academy of Sciences undertook the comprehensive tasks. The planning of the industrial structure in the plan was indeed predictable, which has been confirmed by the development practice in recent years. Regarding the future spatial development strategy of the Beijing-Tianjin-Tangshan region, the difference in the spatial distribution of the carrying capacity of the resource-socio-economic system was mastered. Three directions were proposed on the basis of scientific demonstration, that is, shifting to the coastal region, focusing on the development of eastern Hebei, and the development of small towns in the outer suburbs (of large cities). In the past 10 years, this principle has become a programmatic regional policy actually applied in relevant regions. The large-scale raw material industry, electricity, transportation, and export-oriented economy in Tianjin have been placed on the coastal region, so as to shift the focus of productivity distribution; in the past 10 years, the central and local governments have spent a lot of money in Tangshan and Qinhuangdao in eastern Hebei, so as to accelerate resource development and the development and construction of raw materials, energy and transportation. With the gradual emergence of their advantageous conditions, they become one of the "hot spots" for foreign investment.

From 1985 to 1988, some geographers in China participated in the compilation of the national territorial planning outline. They did fruitful work in China's resource development and utilization, industrial structure, overall productivity distribution, and comprehensive development of key areas. "Point-axis" development and the viewpoint of taking the coast and the line along the Yangtze River as the first-level key development axis in China were written into this planning outline for the first time. The *Outline* pointed out that: "China's eastern coastal areas and the east–west Yangtze River have formed a closely combined T shape, which are the two most important axes for key development and layout in 2000 or a longer period". It was clearly stated in the national territorial planning that the "point-axis" spatial strategy has caused a huge impact on the economic development of the country and various regions.[15]

[15] It was stated in the "draft" of the national territorial planning outline 1986 that the coastal and riverside areas are taken as the first-level key development axes of the country, and the two constitute

3) Regional economic development strategy

Since the 1980s, due to the changes in China's planning system and the enhancement of regional economic strength, it is necessary for governments at all levels to clarify the future economic development strategy of the region, and it is also possible (authority and economic strength) to formulate their own development strategies, and put the strategies into effect. In order to adapt to the needs of such changes, regional economists and economic geographers have been extensively engaged in research on regional economic development strategies and corresponding planning.

The research subjects of the regional economic development strategy are the direction of regional economic development, industrial structure, and overall industrial layout. The state of the regional industrial structure formed under the central planning system did not meet the objective requirements of economic development in a long term in the past. The following problems widely appeared: insufficient supply of energy and raw materials, shortage of infrastructures such as transportation, unobvious leading industries, low level of industrial structure, backward development of emerging industries, and similar industrial structures among regions. Competition among regions appeared in the 1980s, which resulted in differences in the degree of prosperity among regions. The experience of regional development tells us that: when the investment scale and management technology are basically the same, the key to determining whether faster development can be obtained in competition is to determine the reasonable industrial structure and leading industries. That is to say, the competition among regions is essentially a comparison of the rationality of regional industrial structure and the structure level. The overall industrial distribution is to determine the overall framework of territorial development, especially the industrial axis (belt) subject to key development and central cities (or growth poles) that can drive surrounding development and have great development potential. The close relationship between the socio-economic spatial organization and regional development has been gradually recognized.

The key development axis of the region is also proposed in territorial planning for various provinces, autonomous regions, and counties. Since the 1990s, the development strategies of some provinces and regions have placed the spatial pattern in an important position, and almost all are based on "point-axis" development. For example, the governor of Anhui Province proposed that Anhui's spatial strategy focuses on one line (along the coast of the Yangtze River) and two points (Hefei and Huangshan).

Among numerous numbers of researches on the regional economic development strategy, those with large impacts include the research on the construction of Shanxi Energy Base in the mid to late 1980s (organized by the Development Research Center of the State Council and the Chinese Academy of Social Sciences), the research on the economic development strategy of Hainan Island (organized by the Chinese Academy of Social Sciences), the economic development strategy of Xinjiang (organized by

---

a "T" shape. In subsequent revisions, the Yellow River-Longhai line is also taken as the first-level key development axis.

the Chinese Academy of Sciences, named as "resource development and production distribution of Xinjiang"), and the regional development strategy of southwest regions (organized by the Chinese Academy of Sciences, named as "resource development and production distribution of southwest regions"). The basic characteristics of these large-scale researches are as follows: be able to investigate the economic development goals and directions of a region within a larger range and on a larger time scale, and be able to scientifically analyze and predict the future development potential and scale. In addition, the proposed measures to implement strategic goals are relatively reliable. Therefore, such research findings and conclusions have played a guiding role over a certain period of time.

The *Research on China's Regional Economic Structure* (written by Chen Dongsheng and Chen Jiyuan, 1988, Shanxi People's Publishing House, Social Science Press of the Chinese Academy of Sciences) and the *China's Regional Industrial Structure* (edited by Gao Chunde and Zhang Wanqing, 1991, China Planning Press) are two books on the economic development strategies of China's provinces and regions and various types of zones (economic zones, economic regions, special zones and old revolutionary base areas, areas with large ethnic minority populations, border areas, and poor areas). The economic direction, structural adjustment, infrastructure layout and other major issues of various regions have been explained more accurately from the perspective of combining theory and practice.

3 Basic Evaluation and Outlook

1. Discipline strengths and defects

Since China implemented the policy of opening up to the outside world and invigorating the economy at home, scholars and planners in economics, sociology, mathematics, systems science, geography, etc. (disciplines) have carried out a lot of work in regional development strategy and regional planning, which has promoted the democratization and scientific decision-making in the regions. In terms of research areas:

(1) The national macro development is carried out around the key issues of strategies of the east, the middle, and the west (or the east and the west). The focus of research and discussion is how to treat the relationship between economic benefits and balanced development; in the bigger scope, there are mainly provinces, cities, autonomous regions, large and middle-sized cities, and regions, and the research mainly focuses on the industrial direction and industrial distribution. The research scales are the main bodies of previous regional research, and the development of theory and method is also considerable; in the microscope, there are mainly counties, banners, and small and medium-sized cities. The number of research is small, but the practical effects and theoretical progress are not very obvious.

(2) When economists, geographers, ecologists, and system scientists carried out research on the issues of regional exploitation and development, they always felt powerless against some issues, and the synthesis issues were often incompletely resolved. For a long time in the past, Chinese economists carried out less research on regional issues and paid attention to production relations. In recent years, although attention was paid to the strategic issues of the regional economy, the focus was still on the operation mechanism of the regional economy, and less attention was paid to the natural foundation of the region and the spatial location of socio-economic factors. Some geographers tried to apply the principles of economics but the policies that were constructed by these geographers did turn out well. As for the regional work in terms of systems science, since the attention was paid to the mathematical aspect of the model, and the spatial and economic concepts were poor, many parameters and model structures could not reflect the realities, and the research and planning results were far from realities. Since regional research has become popular nowadays, even some scholars who have nothing to do with it in terms of professional quality have participated in the research, and the scientificalness and depth of the results cannot be guaranteed. Facing such realities, scholars in the above aspects shall work together.

2. Research on the mechanism of factor action

There are many factors that affect or even determine regional exploitation and regional development, such as funds, industrial structure, resources, infrastructure, technology, management level, geographic location, environmental conditions, etc. To predict and plan the future development of the region and determine the major work of territorial development, it is necessary to reveal how the above factors are related and interact with each other, and to obtain the parameters of the interaction among the factors. During the regional development in China, countless examples have proved that: there is a close relationship between investment direction and economic growth and economic structure, between economic growth and energy consumption and environmental load, between output and investment capacity, economic growth, etc. The research on regional exploitation and development requires the exploration of the essence behind the phenomenon. For this reason, it is necessary to develop and apply the analysis method that can reveal the essence of the phenomenon. The goals of using these methods are as follows: evaluation of the status quo of regional development, comparison of regional development levels and differences, deficiencies in regional development structure, development forecasts and effect simulations under different development policies and measures, etc. In other words, the purpose of using analytical methods is to understand the mechanism of regional development while diagnosing the issues of regional development, then the applications of applying those analytical methods could formulate policies that enable the region to achieve healthy development in a specific manner.

Within the scope of regional exploitation and regional development, the role of factors and causal relationships are not "deterministic", but of "probability" or "possibility". That is to say, they will not cause specific regional economic growth and social development (direction, speed and structure) under certain development conditions and policies. In other words, although the composition, scale, and action mode of the main factors are the same, there are still other schemes for regional exploitation and development. The situation is slightly different under the interaction of natural elements. For example, there is a fairly definite value for water surface evaporation under a certain temperature and wind power. During regional exploitation and regional development, natural factors and socio-economic factors interact with each other, and the situation is much more complicated. Of course, the "probability" or "possibility" here still contains numerous sound laws and rules, which are manifested as the changing trend, change range and development frequency of the phenomenon during regional exploitation or development.

The mechanism of factor action exists in various regions with different sizes and types. That is to say, there are issues of the mechanism in the micro-region and macro-region. The differences in investment effect coefficients, profit and tax rates, labor productivity, energy consumption and water consumption per unit of output value in different economic zones in China illustrate the different parameters of factor action. In the meso (provincial and district-level) regions, industrial structure has a clear relationship with economic growth, investment direction, investment area distribution, economic growth, etc.; in the micro regions, investment scale, energy consumption, and structural characteristics have a quite direct "input-industry" relationship with economic growth, income increase, and environmental load.

3. Research goals and analysis methods

In the 1990s, China's regional exploitation and development will show an extremely complex situation. In order to make the research work more targeted and constructive, the overall goal is to obtain the best development and spatial organization of the regional social economy, and to combine the two.

The best development of the region is manifested in the following aspects: moderate economic growth rate and population growth rate; expansion of demand and increase of effective supply, with a basic balance between the two; increase in output, and continuous expansion of investment with the improvement of the industrial production development and technological level; adaptation of the industrial structure and spatial structure of the economy to the socio-economic development level and natural foundation of the region; more advantageous position in the inter-regional competition, and improvement in tax and other fiscal revenues. Under the conditions of the market economy, the main tendency of labor and enterprises in the inter-regional flow is inflow. The regional infrastructure and employment conditions are improved, and the regional ecological environment remains in good condition.

The best organization of a region refers to the reasonable distribution and reasonable spatial structure of the socio-economic components of the region, including the reasonable concentration of productivity and population, the formation of a hierarchical system for the scales of urban residential areas, and adaptation of the state of spatial structure to the stage of the economic development stage of the region. The best organization of the region is mainly manifested in the reasonable concentration and the best scale in the space.

In the research on the common regional development goals: the best development and the best organization, the following aspects are emphasized: analysis of factors affecting regional development and spatial location, industrial structure in different stages of the regional socio-economic development, coordination of demand and supply, spatial interaction and diffusion, spatial structure of regional development and its schema, regional coordination of resources, population, economy and ecological environment.

For more than 20 years, regional scientists have developed a large number of regional analysis methods, which mainly have two different properties and categories. One is the theoretical analysis method, and the other is the empirical analysis method. The former mainly uses the logical reasoning model for theoretical explanation and theoretical derivation, including W. Christaller's hexagonal urban settlement system in the early stage, Williamson's model of inverted U-shaped correlation between regional economic growth and unbalanced development, "point-axis" spatial structure system model formed by the "point-axis" gradual diffusion, etc. The theoretical analysis model is based on empirical research, and uses logical reasoning as the main means to abstract the relationship between regional development factors. Although the trend and relationship represented by the analysis curve are not so precise and specific, it has a more extensive and profound interpretation ability than the specific analysis model and is an analysis method that should attract everyone's attention. The empirical analysis methods are basically mathematical. Among them, the methods the most used are as follows: correlation analysis, regression analysis, factor analysis, principal component analysis, trend surface analysis, input–output analysis, location quotient, etc. In recent years, the contents of regional analysis objects have become more and more complex. A series of advanced mathematical analysis methods are introduced into regional research, such as dynamic programming theory, dissipative structure theory, etc. Some methods of ecology have gradually played a role in regional analysis.

4. Research on the experimental zone for regional exploitation and development

In order to promote the development of regional research in China, while undertaking major practical tasks and providing proposals for regional development and industrial distribution, it is necessary to select the regions with small areas (county-level areas or urban areas equivalent to the county-level), distinct resource structure and economic structure, and large differences in characteristics from the surrounding areas, conduct observation and recording for a long term and use these reliable, specific and systematic first-hand data for multi-objective system analysis and input–output analysis, so as to accurately understand the various mechanisms in the process of regional exploitation and development. This is an important way to develop the basic theory of regional research in China. The process of regional socio-economic development is different from laboratory experiments in natural sciences. It has a long cycle, and the input–output relationship is often not obvious. It is like a "black box"—how to express the development of the previous stage in the latter stage. Such research of the experiment area lasts for three years, five years, or even longer. The results are mainly manifested in the interaction of regional development factors, with methodological results as the mainstay.

5. Establishment of a theoretical system of regional exploitation and development with Chinese characteristics

The relatively mature theoretical methods of regional research in the world today are basically established by scholars from developed industrialized countries. Since China is different from their countries in terms of the stage of socio-economic development, socio-economic system, characteristics of resources and space, and the extent of development and utilization, the theoretical system of research on regional exploitation and development to be established in China shall have distinctive Chinese characteristics. This requires serious digestion of foreign practice and theories, as well as targeted modification and development with domestic practice, so as to make them suitable for the needs of China's regional research and regional planning. China has a large land area and weak strength. For a long period of time, the primary goal is obviously to seek rapid economic growth in terms of development no matter at the macro (whole country) or meso (province, municipality and district) scale; in this process, the regional imbalance is gradually reduced; China implemented a system based on the planned economy in the past few years. In the supply–demand relationship, the dominant side often increased effective supply. The significance of the increase in demand to the regional economy is relatively small. In terms of investment channels, it is also different from some industrialized countries. The rural population accounts for a large proportion, leading to the formation of its own characteristics in the spatial distribution of the population and employment structure. As for the theoretical foundation and "tools" of regional policies, it is more necessary to explore the models suitable for China's national conditions and regional conditions.

According to the following analysis, the history, status quo and future of the research on regional exploitation and development in China can be summarized in Table 1.

Table 1 Trends of regional exploitation and development research

Project	70s and before	80s	90s
Background	National unified plan, low level of regional exploitation and development	Plan system reform, opening to the outside world; improved level of development and exploitation; Regional competition	有计划的商品经济; 外向型经济; 开发与发展水平达到较高程度; 竞争与协调 Planned commodity economy; export-oriented economy; development and exploitation to a higher level; competition and coordination
Basic characteristics of regional development and exploitation	Regional comprehensive development problems are not prominent	A large number of regional development and exploitation problems appear	Regional development and exploitation has become a wide-ranging and profound social-economic problem
The main problems of regional development and exploitation	The conditions for resources and regional development are unclear Conflicts in resource utilization between departments and regions	How to implement differentiated macro-regional development strategies The level of industrial structure is low and the structure is convergent Conflict of interests between resource sites and processing sites Emergence of environmental problems arising from development and exploitation	The gap in regional development level and economic strength is further widened The degree of resource assurance decreases Depression, underdevelopment, crowded and chaotic types of areas appear The contradiction between economic growth, population development and ecological environment intensifies

(continued)

Table 1 (continued)

Project	70s and before	80s	90s
Target	Determination and layout of departments (focusing on space aspects)	Rational development and utilization of resources; leveraging regional advantages; achieving rapid economic growth (focusing on spatial layout and structural mechanisms)	Best Development and Organization (focusing on structures, mechanisms, regional policies)
Main research topics	Industry, problem development and layout The division of various economic regions, regional planning	The macro strategy of national development and development Comprehensive development and rational utilization of regional resources Industrial direction, structure and comprehensive industrial layout Population Distribution and Urbanization	Macro-regional development strategy Coordinated development within and between regions, and the application of the three strategies of rapid, stable and balanced in specific regions Rational development and utilization of resources and sustainable development of ecological environment High industrial structure and inter-regional division of labor Population Distribution and Urbanization
Research methods	Condition analysis Technical and economic comparison Indicator system and classification	General regional analysis and forecast Model optimization	General regional analysis and forecast Model experiment Analogy method Research on Regional Development and Development Experimental Area

References

1. D. Chen, Several issues of regional economy. Territorial Dev. Manag. **1**(1) (1991)
2. Investment Institute of the State Development Planning Commission (ed.), *China Investment Report 1990* (China Planning Press, 1990)
3. D. Lu, The scientific basis of the general map of China's industrial productivity distribution in 2000. Geogr. Sci. **6**(2) (1986)
4. D. Lu, China's macro strategy of regional development. J. Geogr. Sci. **42**(2) (1987)
5. Q. Zhou, Z. Liu et al., *Regional Economics* (China Renmin University Press, 1989), p. 151
6. Y. Wu, Evaluation of the strategic transfer of "westward movement" and "eastward movement" of China's macro productivity distribution. Yunnan Geogr. Environ. Res. **2**(2) (1990)
7. D. Chen (ed.), *A New Starting Point for Research on Regional Economy* (Economic Management Press, 1991), pp. 164 and 165
8. D. Lu, Types and basic functions of territorial (management) planning. Econ. Geogr. **4**(1) (1984)

Chapter 4
Industrial Structure and Regional Development

The development processes of industrialized countries and developing countries have proved that the industrial structure is always in dynamic evolution in the development of the national and regional economy. Although the characteristics and composition of industries in different countries and regions are different, the evolution of the industrial structure mostly follows a universal law. In most cases, the economy of a region or country is like an organism composed of related elements, which are interconnected, that is, the regional or national "economic system" or "industrial system". To study and master the changes of this organism, that is, the "system", the most important way is to dissect its internal structure, that is, its components and their interrelationships. That is to understand "hierarchy" and "system" through the structure. Similarly, it is a very important way to study the regional economic system through the regional industrial structure.

The main goal of the research on the regional industrial structure is to reveal the characteristics of the regional industrial structure and its relationship with regional economic growth and explore the factors that affect the characteristics of the regional industrial structure, and how to adjust the direction of the regional industrial structure to promote the healthy development of the regional economy.

Concerning the economic management system with socialist authority centralization implemented during the first 30 years since the founding of the People's Republic of China, the issues of the regional industrial structure and its optimization were not prominent and were not taken seriously in a wide range. Among the many aspects included in the industrial structure, the relationship between agriculture, light industry, and heavy industry are easier to be taken seriously. However, what kind of relationship between agriculture, light industry and heavy industry in a province, district, or city (unit) is reasonable? Is there a basically reasonable range (quantity)? Few people have studied it. At the end of the 1970s, China began implementing the reform and opening-up policy. In terms of economic management, many authorities were lowered level by level, and regional competition became a very common

D. Lu, *Regional Development and Its Spatial Structure*,
https://doi.org/10.1007/978-981-99-7683-6_4

phenomenon. As a result, people, especially regional governments, generally recognized the extreme importance of industrial structure to regional economic development. On the other hand, for more than 10 years, economic theories popular in contemporary Western countries, especially the theories of development economics and regional economics, have been successfully introduced into China, which has improved the level of understanding of individuals about the impact of industrial structure on regional economic development and greatly promoted the research on the issues of regional industrial structure in China.[1]

In the last more than 10 years, China's economists, economic geographers and experts in the government planning departments at all levels have conducted a lot of research on the country's industrial structure, region and city, and published many books. On the whole, the academic circles and planning departments have a profound and comprehensive understanding of the significance of the industrial structure in regional economic growth, as well as a clear understanding of the measures and policies to adjust the industrial structure and promote regional economic growth. However, there is still no monograph on the mechanism of interaction between industrial structure and regional development in China. This manifests the serious inadequacy of our research in this field.

1 Basic Connotation of Industrial Structure

The term "industrial structure" was introduced by foreign countries in the late 1970s and early 1980s. However, there is no accepted definition yet. Although people use this concept extensively in practice, their understanding and interpretation are quite different. Some scholars believe that the industrial structure is the economic structure, and the industrial structure is the "proportional relationship between industries". Some articles simply understand the industrial structure as the composition of agriculture, light industry, and heavy industry in the national economy and their relationship with each other. Some scholars believe that the industrial structure "basically reflects the composition and interrelationship of various enterprises and undertakings that produce products of different natures and engage in economic activities of different natures in the national economy." A few people still believe that the industrial structure not only refers to the primary, secondary and tertiary industries, agriculture, light industry, and heavy industry, production with different factor intensities, as well as technical structure, product structure, enterprise, and company organizational structure, etc., but also includes a series of relationships and proportions in the fields of circulation, exchange, distribution and consumption [1, 2].

[1] Yang Zhi, Introduction to Industrial Economics, China Renmin University Press, 1985.
Edited by Li Boxi et al., Industrial Policy and National Economy, Shanghai Science and Technology Document Press, 1990.
Edited by Zhang Fengbo et al., China's Macroeconomic Structure and Policy, Chinese Financial and Economic Publishing House, 1988.

Scientific determination of the definition of industrial structure involves how to scientifically divide industrial sectors and understand the interrelationships between various industries and sectors in the national economy, which is beyond the author's knowledge level. However, the author believes that the core contents of the industrial structure are the sector, industry composition, and their interrelationships (including interrelationships and interactions) in the national or regional economic production system. Here, "sectors and industries in the production department" shall be said to refer to the material production departments mainly, but shall also include certain non-material production departments.

The industrial structure is like a complicated complex. A variety of structural relationships can be obtained through the analysis of different aspects of the complex.

1. Structures of the primary, secondary and tertiary industries
1) Clark classification method

The proportional relationship between the primary, secondary, and tertiary industries in the national economy is the most commonly used category of industrial structure. The classification of the three industries was first proposed by C. G. Clark (1905–), a British economist and statistician, in his book *The Conditions of Economic Progress* (1940).[2] He also summarized the law of industrial structure change accompanying the process of economic growth and opened up the field of applied economic theory, industrial structure theory. As a result, the classification of the three industries has been widely popularized. In 1957, Clark revised the classification in the third edition of the above book, replacing the term "service industry" with "tertiary industry". According to Clark's classification, the contents of the primary, secondary and tertiary industries are as follows.

Primary industry: Refer to sectors that directly use natural resources for production activities such as agriculture, forestry, aquaculture, animal husbandry, hunting, and mining.

Secondary industry: Refer to the manufacturing industry, the sector that does not directly use natural resources and produces movable material wealth through a large-scale continuous production process.

Tertiary industry: Refer to the construction, service, and all industrial sectors not included in the primary and secondary industries.

There are still some unreasonable points in applying the above division of three industries. For example, the mining industry is derived from nature and shall be classified as a secondary industry. In addition, industries such as water supply, power supply, and gas can be regarded as secondary industries and could also be categorized as tertiary industries.

---

[2] See Yang Zhi's *Introduction to Industrial Economics*, 1985. In the article "Several Issues Regarding Industrial Structure Adjustment" (Economic Research Data, Issue 34, 1981), Wang Husheng wrote that: "A. G. B. Fisher, an economist from the New Zealand, was the first person that proposed the classification of primary industry, secondary industry and tertiary industry. In 1951, C. G. Clark, a British economist, materialized it in the book *The Conditions of Economic Progress*".

2) Classification method in China

In recent years, the concept of three industries has been widely used in the planning and statistical departments as well as research work in China. Among them, the sector that takes products directly from nature is called the primary industry, the sector that reprocesses primary products is called the secondary industry, and the sector that provides various services for production and consumption is called the tertiary industry. The details are as follows:

Primary industry: Agriculture (including planting, forestry, animal husbandry, sideline, and fishery).
Secondary industry: Industry (including excavation, manufacturing, tap water, electricity, steam, hot water, and gas) and building industry.
Tertiary industry: All other industries except the primary and secondary industries. The tertiary industry is also divided into four levels due to its rich contents. First level: Circulation sector, including transportation, post and telecommunications, commerce, catering, material supply and marketing, and warehousing; second level: Sectors that serve production and life, including finance, insurance, geological survey, real estate, public utilities, residential services, consulting services and comprehensive technical services, agriculture, forestry, animal husbandry, fishery, water conservancy service industry and water conservancy industry, highway, inland river (lake) waterway maintenance industry, etc.; third level: Sectors that serve to improve the level of science and culture and the quality of residents, including education, culture, radio and television, scientific research, health, sports and social welfare, etc.; fourth level: Sectors that serve the society and public, including state organs, Party and government agencies, social organizations, the army and the police [3].

The tertiary industry is further divided into four levels, reflecting the characteristics of China's industry classification.

3) Classification method in Japan and the United Nations

The Japanese government referred to Clark's classification but made amendments. The primary industry includes agriculture, forestry, aquaculture, animal husbandry, and hunting; the second industry includes mining, manufacturing, and construction; the tertiary industry includes power supply, gas, tap water, commerce, catering, finance, insurance, warehousing, transportation, post and telecommunications, and service industries. However, power supply, gas, and tap water are sometimes included in the secondary industry.

In the "International Standard Industrial Classification" commonly used by the United Nations, industries are divided into nine major industrial sectors or 11 industrial sectors, rather than primary, secondary, and tertiary industries. For example, according to the data in the "Statistical Yearbook of International Labor Economics" compiled by the National Labor Organization of the United Nations Economic and Social Council, they are divided into the following nine sectors: ① Agriculture, forestry, hunting, and fishery; ② Mining and quarrying; ③ Manufacturing; ④

Construction; ⑤ Power supply, gas, and tap water; ⑥ Business; ⑦ Transportation, storage, and post and telecommunications; ⑧ Service; ⑨ Others. When calculating the national income of each country by industry sector, in addition to the above-mentioned ①–⑧, three sectors, namely, banking, insurance, and real estate; housing (sale and rental); and public service and defense, are added in the "World National Statistical Calculation Yearbook" compiled by the United Nations [2].

The division of the three industries provides definite standards for statistics and analysis of the national economic development of the state or region as well as the evolution of the internal structure and provides a theoretical basis for diagnosing the state and problems of regional economic development, formulating and implementing regional industrial policies, and enhancing the healthy development of the regional economy.

The proportional relationship in the national economy is analyzed according to the composition of the three industries in the GNP, which has been implemented in statistical analysis in China since the mid-1980s.

2. Structure of agriculture, light industry and heavy industry

The structure of "agriculture, light industry, and heavy industry" plays an important role in China's economic statistics and analysis for decades, and is the most important indicator to measure the main proportional relationship of the national economy. The proportion of "agriculture, light industry, and heavy industry" is investigated for two purposes. One is to provide a basis for determining the destination of fixed assets investment in the national economy, and the other is to provide a quantitative concept for diagnosing defects in the national economy. In the first 30 years after the founding of the People's Republic of China, since the goal of industrialization was in fact "heavy industrialization", the proportion of heavy industry in many regions and cities in China skyrocketed as a result.

In China, the category of "agriculture" includes five industries: crop planting, forestry, animal husbandry, sideline and fishery. Before 1984, the gross output value of agriculture included the industrial output value of villages and below. The output value was calculated according to the new caliber since 1984, and the gross output value of agriculture no longer includes the industrial output value [4] of villages and below. In order to measure rural economic development and structural changes, the industry's status in the gross output value of rural society is becoming increasingly important.

The manufacturing industry refers to the industry that provides consumer goods and produces hand tools and is an industry that meets the needs of people for food, clothing and use. In the 1950s and 1960s, the concept of "manufacturing industry" refers to the sectors of textiles, the first manufacturing industry and the second manufacturing industry; in the 1970s, some light industrial products were also manufactured in some heavy industry sectors, such as metallurgy and chemicals, and some complicated situations appeared in statistics. According to the different raw materials used, the light industry can be divided into two categories: ① manufacturing industry using agricultural products as raw materials, mainly including food manufacturing, beverage manufacturing, tobacco processing, textile, sewing, leather and fur manufacturing, paper making and printing industries; ② manufacturing industry using non-agricultural products as raw materials, referring to manufacturing industry using chemical products and a small number of metallurgical products as raw materials. With the development of industrialization, the sectors that use non-agricultural products as raw materials in the light industry have gradually increased, so the development of the regional economy has gotten rid of the dependence on the natural conditions of agriculture to a greater degree. In China, the proportion of sectors using non-agricultural products as raw materials is the highest in Liaoning when measured by province and autonomous region.

The heavy industry refers to an industry that produces the means of production, and is an industry that provides materials and technical equipment for various sectors of the national economy. According to the production and product use, it can be divided into three categories: ① Excavation (cutting) industry, including petroleum mining, coal mining, and metal and non-metallic mining; ② Raw material industry, referring to the industry that provides raw materials, power, and fuel used by various sectors of the national economy; ③ Processing industry, referring to the industry that processes and manufactures raw materials, including machinery and equipment manufacturing, metal structures, cement products, chemical fertilizers, pesticides, etc. [5].

Due to the development of the national economy and the introduction of economic statistics and analysis indicators in developed countries, the analysis of the relationship between "agriculture, light industry and heavy industry" is retreating to a less important position in the research on the regional economic development in China.

3. Relationship structure of labor, capital and knowledge-intensive industries

According to the needs or intensity of production factors in various industrial sectors of the national economy, they can be divided into labor-intensive industries, capital-intensive industries and technology (knowledge)-intensive industries. By analyzing the proportions and dynamic evolution of employment, sales amount, and gross national product of the three types of industries in a country and a region, we can see the stage and level of their economic development, and can also find defects in the economic growth of the country and the region.

(1) Labor-intensive industries: Industrial sectors that have a low level of technical equipment and require a large amount of labor for production activities. The input structure of such industries is dominated by labor, and living labor occupies a relatively large proportion of the product cost. Therefore, such industries generally have the characteristics of small investment, quick effects, more labor absorbed per unit investment, simple production equipment, easy technical operation, low requirements for cultural quality of employees, demands for skilled craftsmanship, etc. Such as the textile industry, clothing industry, toy industry, furniture industry, leather industry, etc., of which most belong to the light industry sectors that produce daily consumer goods. With the improvement of the level of science and technology, labor-intensive industries will also develop into capital- and knowledge-intensive industries.
(2) Capital-intensive industries: Industrial sectors with a high level of technical equipment and more investment required for products per unit of output value

and each employee, that is, industries with a high level of average fixed assets and equipment for each employee. According to the ratio of capital to labor, these industries have large capital investment, huge production equipment, long process, and a relatively high proportion of energy and raw material consumption in production, but less labor that can be accommodated by unit investment. Most of such industries are, what we call today, traditional industries, such as ferrous and non-ferrous metallurgical industries, chemical and petrochemical industries, building material industries, heavy machinery industries, etc.

(3) Technology-intensive industries, also known as knowledge-intensive industries. It costs more for these industries to manufacture products with high-precision technology for scientific research and product development, and the latest scientific research achievements of multiple disciplines shall be applied during production. Such as the electronic computer industry, aerospace industry, integrated circuit industry and atomic energy industry. Some categorize software design, information processing, information engineering, system engineering, etc., into technology-intensive industries. In these industries, the technology and equipment are advanced and complex, the investment is large, and the proportion of scientific and technological personnel is high. However, the consumption of labor and raw materials is small, and the pollution is small at the same time. Technology-intensive industries have undergone rapid development in many countries since the 1960s, and their scale and level reflect the level of science and technology and socio-economic modernization of a country [6], and represent the economic strength of a country or a region.

4. Traditional industries and emerging industries

Traditional industries and emerging industries are two concepts with different connotations (industries) at different times and different stages of technological progress. Traditional industries generally refer to industries (sectors or industries) that were formed and developed in the previous stage of technological and economic development in the world or industries (sectors or industries) that were formed and developed earlier but may continue mass production in the current stage. At the stage of socio-economic development in China, iron and steel, thermal power, basic chemical industries, cement, textiles, general planting industries, railway and highway transportation, commerce, etc. belong to traditional industries. In contrast, the electronics, computer, aerospace, tourism, real estate, information, finance, etc. belong to emerging industries. Emerging industries are not exactly the same as high-tech industries.

2 Basic Industries, Leading Industries and Regional Development

Basic industries, leading industries, and pillar industries are all industries or industrial clusters that play an extremely important role in national and regional economic development. During the planning and scientific research, the goal of analyzing and demonstrating the industries in these three categories is often not to analyze the quantitative relationships (like the primary, secondary and tertiary industries mentioned above) and proportional relationships in the industrial structure, but to reveal the adaptation relationship between the three types of industries and other types of industries or the entire national economy, and how to identify leading industries, and strengthen basic industries and pillar industries to accelerate the healthy development of the regional economy.

1. Basic industries

In China, agriculture, raw materials, energy and transportation, and post and telecommunications industry are the basic industries of the national economy. As the name suggests, basic industries refer to industries that play a basic and irreplaceable role in the regional and national economy. The strength of basic industries and adaption to the entire socio-economic system are key factors determining the development of the whole society [7]. The basic industry is a concept of industrial cluster, which is proposed based on the fact that China is in the early and mid-stage of industrialization and the above industries are in a state of inadaptability in China.

Basic industries play an important role in the industrial linkages of China's national economy. The sign is that the reaction coefficients of these industries are relatively large (that is, the average value of the sensitivity of each sector to a certain basic industry). Among them, the chemical industry is 2.564, the metal smelting and calendering industry is 2.195, and agriculture is 2.062, the three sectors with the highest reaction coefficients. The electricity, steam, and hot water production industry is 1.072, the coal industry is 1.054, and the transportation, post, and telecommunications industry is 1.009 [8].

Since the founding of the People's Republic of China, investment in basic industries in China has accounted for about 60% of all investment, but the total still cannot meet the needs of socio-economic development (Table 1).

Basic industries are extremely important to the economic development of China and various regions for the following reasons:

(1) China has a large population, and the agricultural populations in most provinces account for about 80%. The lack of agricultural development and insufficient grain and non-staple food affect social and economic stability. In addition, the output value of the light industry accounts for about 50% of the total industrial output value in China, while more than 60% of the output value of the light industry is produced with agricultural products as raw materials. The light industry will inevitably be affected in all provinces with weak agricultural development.

Table 1 Proportion of the investment in basic industries in total investment in China (%)

Item		Total of basic industries	Agriculture	Raw material industry	Energy industry	Transportation, post and telecommunications industry
"First 5-year plan" period		47.8	7.1	13.0	12.4	15.3
"Second 5-year plan" period		64.6	11.2	22.8	17.1	13.5
1963–1965		64.4	17.6	19.0	15.1	12.7
"Third 5-year plan" period		62.3	10.7	20.3	15.9	15.4
"Fourth 5-year plan" period		64.2	9.8	18.8	17.6	18.0
"Fifth 5-year plan" period		63.0	10.5	18.7	20.9	12.9
"Sixth 5-year plan" period		56.7	3.9	18.5	22.1	12.2
"Seventh 5-year plan" period	1986	57.8	2.2	18.5	23.1	13.0
	1987	58.9	2.4	21.3	22.9	12.3
	1988	59.6	2.3	22.0	23.7	11.6

① The proportion of investment during the "Sixth 5-year Plan" period and later is the number of entities under the ownership of the whole people, and those of the previous periods are the proportions of capital construction. ② The raw material industry and the energy industry are divided according to the economic use of the product, which is different from the mining industry, raw material industry and processing industry categorized as the heavy industry

(2) Due to the rapid growth of China's economic aggregate, especially the industrial economic aggregate, the industrial processing capacity grows rapidly, and the demands for raw materials and energy are large. Some raw material production and energy industries are still "bottleneck" industries in China and many regions until now. Due to insufficient energy supply, the power is on for four days and off for three days in some regions; insufficient supply of raw materials has caused the lack of raw materials in the processing and manufacturing industries, and the phenomenon of partial idling of capabilities is widespread. The regional development policy that puts energy and raw material industries in a prominent position will last until about 2020, that is, when the middle stage of China's industrialization is completed and the later stage is entered.
(3) The reason why the transportation, post and telecommunications industry is a basic industry is that this industry affects all aspects of China, and that coal accounts for more than 70% of China's energy consumption, and the coal transport volume accounts for over 50% of the railway transport volume in China.

There are many examples of the role of basic industries in promoting or hindering regional economic growth in China.

The economic growth of Shandong has been faster than that of other northern provinces since the 1970s. In the 1980s, the average growth rate of GNP reached 10.5%, second only to three provinces, that is, Guangdong, Zhejiang and Jiangsu, in China (Guangdong is close to Hong Kong and Macau, and achieved a great amount of foreign investment. The reason for rapid economic growth in Jiangsu and Zhejiang was the large-scale development of rural industries). The main reason is that: Shandong has been focusing on the energy and raw material industries since the 1970s, that is, using independent power grids, and always paying close attention to constructing power supply points. Taking advantage of petroleum, iron ore, salt and limestone resources within the province, Shandong has established a considerable scale of petrochemical, marine chemical, fertilizer, iron and steel, cement and other industries earlier, which basically solved the issues on the supply of bulk raw materials. Shandong's agriculture has been growing steadily for about 30 years. It is not only a major province that produces cash crops, but also a major province that produces grain. In particular, cash crops such as cotton and peanuts have supported the huge textile, oil and food industries in the province. In contrast to Shandong, the average annual growth rate of GNP in Liaoning in the 1980s was only 7.8%, and the incompatibility of basic industries such as agriculture and energy was an important reason. The average growth rate of Hebei was 8.8%, which was almost lower than that of Shandong by two percentage points. The important reasons were the weak production of raw materials for agriculture (mainly disaster and insufficient farmland infrastructure), petrochemical, etc. and insufficient power supply.

2. Leading Industries

There is no uniform explanation for the concept of "leading industries", which, however, has been widely used. Someone wrote: "Choose leading industries according to domestic market demands, resource conditions, export prospects, etc., and drive the development of other industries based on this, thereby forming a sophisticated and modern industrial structure, which is of decisive significance for economic development" [8]. According to Yang Zhi's analysis [1], leading industries are dominant industries, that is, the industries or industry clusters that can drive the industrial development of the whole region. Those similar to the concept of "leading industries" are forerunner industries, growth industries and strategic industries. Among them, forerunner industries refer to industries that represent advanced technologies; growth industries refer to those industries of which the growth rate exceeds the overall average growth rate of the national economy. The connotation of strategic industries is the same as that of leading industries. The connotation of the concept of "leading industry" and its significance to national and regional economic development are as follows:

(1) Leading industries refer to industries that can guide and drive the country's and region's overall economic development. In this sense, a leading industry shall be an industry that has the conditions to develop in the region of the

country, has a broad regional market (or even abroad), and can have driving effects on other industries. It is the dominant industry to a certain degree. The driving effects of leading industries are mainly manifested in two aspects. First, this type of industry drives the development of related industries due to a high degree of association with other industries; second, it arms other industries with new technologies and new products, so as to promote the development of other industries.

(2) The leading industries of countries and regions have different development stages following the evolution trend of domestic and foreign industrial structures and market changes. The stage theory of industrial structure development is generated due to the different leading industries (clusters) at different stages of socio-economic development. After World War II, three groups of leading industries appeared in succession during the industrial development of Japan. The first group was the electric power industry, which drove the development of the power-intensive industry (until the end of the 1950s); the second group was the petroleum, petrochemical, steel, shipbuilding and other industries. These industries provided a large number of raw materials and created conditions for the great development of the processing and assembly industries that will come a few years later; the third group was the emerging machinery industries, such as automobiles, household appliances, etc. [1]. In some areas of China, the law of evolution of industrial structure is also relatively obvious. For example, the leading industries in the 1950s were the coal industry, electric power industry, etc. in Liaoning in the 1950s; iron and steel, chemical industry, etc. had great vitality in the 1960s and 1970s, becoming the leading industries that drove the economic development of the province; the above industries had no driving effect to a great extent in the 1980s. New industries such as the petrochemical and automobile industries have begun to develop rapidly.

(3) Leading industries generally have the highest technical equipment at that time, and shall be at the forefront in developing products and opening up markets.

(4) Leading industries shall generally lead growth. That is to start from the entire national economy, create a good environment, and ensure that the actual growth of leading industries is faster than that of general sectors in terms of energy supply, technical equipment and many other factors. In order to promote national and regional economic development, it is necessary to strive for and maintain favorable regions in fierce international and inter-regional competition, determine leading industries (clusters) that meet their own characteristics and market demands, create various infrastructure conditions and policy environments, and use national administrative forces to help open up markets, so as to ensure the priority growth and long-term sustainable growth of leading industries. It shall be clear that the competition for economic strength and interests among countries and regions lies in industrial competition to a great extent. The advancement of the industrial structure is a major issue related to the long-term socio-economic development of the country and the region, and is even of vital importance.

3 Evolution of Industrial Structure in Regional Development

Regardless of countries and regions (not referring to micro-regions here. There is generally a definite law for the evolution of regional industrial structure composed of provinces and inter-provinces), and there is a general law for the evolution of industrial structure. This regularity was revealed by the German economist Hoffmann in the 1930s, the British economist Clark in the 1940s and the American economist Kuznets in the 1950s, and was proven by the process of industrial structure development in many newly industrialized countries and some developing countries after World War II.

1. Economic growth and evolution and upgrading of industrial structure

The essence of economic growth includes two aspects: One is the increase in the economic aggregate and the other is the progress of the industrial structure.

There are many reasons why the economic growth of countries or regions is different. The following factors are of vital importance: whether the industrial structure is advanced, whether it conforms to its own position in international and inter-regional economic and technological exchanges, and whether it conforms to its own natural economic characteristics. However, what are the factors that drive the evolution of national and regional industrial structures?

By analyzing the facts about the transformation and upgrading of the regional or national industrial structure at home and aboard, it can be seen that the substantial increase in the economic aggregate and the evolution and upgrading of the industrial structure occur simultaneously. The two are inter-determined, that is, the adjustment of the industrial structure leads to economic growth, and economic growth is also an important factor in the transformation and upgrading of the industrial structure.

(1) As the actions of the government and the company. Economic growth must be achieved to adjust the industrial structure that has been backward. The corresponding rapid growth of national income can only be achieved in the case of rapid economic growth. In the case of the rapid growth of national income, it is possible to concentrate more funds on industries with growth prospects, enabling them to achieve above-average growth, which will inevitably promote the evolution and upgrading of the industrial structure.
(2) In the case of rapid economic growth, the flow and shift of factors such as capital, raw materials and land can be smoothly realized in the industry, resulting in the optimization of factor combination, which can strengthen the development of certain advantageous industries.
(3) The rapid economic growth and the expansion of economic aggregate will inevitably lead to the development of urbanization and the improvement of population quality, and the development of urbanization will bring about the development of the secondary industry and the tertiary industry and the improvement of their status in the overall industrial structure. The improvement of population

quality is a prerequisite for the development of emerging industries and science and technology.

(4) The rapid economic growth and the expansion of economic aggregate will promote the increase of foreign trade as well as foreign economic and technological cooperation. As a result, external technologies will be introduced, and external commodities and equipment will be imported. In this way, the domestic or regional industrial structure will be improved, and the introduction and development of new industries and the transformation of original industries will be promoted.

Clark summed up the law of sophistication of the industrial structure, which is caused by economic development. He used the two indicators, that is, labor force and national income, to research the evolution of industrial structure. By grasping the changes in these two indicators, we have obtained a viewpoint that has a far-reaching impact on the subsequent industrial structure, that is, with the economic development and the increase in per capita national income, the labor force will gradually shift from the primary industry to the secondary industry and then to the tertiary industry. This is caused by the relative income differences among industries [9].

The above law can be clearly explained by the rapid economic growth and the transformation of the industrial structure in Japan after World War II [9].

During the post-war recovery period (1945–1960), which was a transitional period from post-war recovery to rapid growth, a rapid growth rate of 10% was almost maintained (10.0% from 1946 to 1951, 8.2% from 1951 to 1955, and 8.7% from 1955 to 1966). Basic industries, especially iron and steel, coal, petrochemicals, etc. developed greatly due to rapid growth. This is the basis for Japan's economic takeoff after World War II.

During the period of rapid economic growth (1960–1973), Japan's economy developed rapidly. The average annual growth rates of industrial and agricultural production and foreign trade were over 10 and 17% respectively. The rapid economic growth further promoted the heavy and chemical industrialization of the industrial structure, with iron and steel and petrochemical industries as the leading industries.

The above law can also be fully explained by the fact that the substantial economic growth in southern Jiangsu has led to the continuous evolution of the industrial structure since the founding of the People's Republic of China. The average growth rate of the urban and rural industries in southern Jiangsu was 12.2% from 1949 to 1980 and 21.2% from 1980 to 1988. In particular, the rural industry developed faster, with an average annual growth rate of more than 20% in many counties and cities in the early 1980s. The domestic and foreign markets for the products in southern Jiangsu were continuously expanded, and the export rose sharply. As a result, a large amount of foreign investment was introduced to southern Jiangsu, which accelerated the technical transformation of traditional industries, and a series of new products and new industries were developed. At the same time, the urbanization of the population in southern Jiangsu was quite rapid, thus forming an intensive industrial belt along the Yangtze River and along the Shanghai–Nanjing railway line. Under the effect of the rapid economic growth and the above change factors that followed, the industrial structure underwent extremely important changes[3]:

Before the founding of the People's Republic of China, the weak modern industries in southern Jiangsu were concentrated in textile, silk reeling, and grain processing and other industrial sectors, which accounted for more than 35% of all industries. Due to the rapid economic development in southern Jiangsu in the 1950s, the proportions of the food industry and the textile industry dropped from 31.2 and 57.8% in 1949 to 8.4 and 52.2% in 1965 respectively. The newly rising industries during this period were mainly machinery, chemical, metallurgy and other sectors. The three industries accounted for 28.2, 12.4 and 6.6% respectively as of 1980, 47.2% in total. Enterprise products participated in the international economic cycle to a greater extent in the 1980s. The electronic industry was separated from the mechanical industry and developed in a skyrocketed way. The urban industrial structure of Wuxi was upgraded and changed rapidly in the 1980s. The mechanical and electronic industries together accounted for 49.5% of all industries, of which the electronic industry accounted for 26.5%. This is an important sign of the development of the industrial structure towards sophistication.

2. "Rejuvenation" and "maturation" of industrial life cycle and industrial structure

Under the conditions of the market economy, products will undergo the following process: research and development, market entry, mass production, market saturation, profit decline and recession due to competition and economic factors. Before the completion of this process, some new products with similar functions are being researched and developed. When a series of products and industries in a country or a region are in this process—the industrial structure of the country or the region can be in a "rejuvenated" state.

The "rejuvenation" of the regional industrial structure is extremely important to the development of the regional economy. The so-called "rejuvenation" of the structure means that there are always new products of several industries or sectors in a region, which are constantly being developed and entering into markets outside the region (including foreign markets).

How to "rejuvenate" the regional industrial structure? The basic requirement is to produce new industries (products and enterprises) and new consumer markets created therefrom before the existing enterprises and industries and the markets related to these industries reach the mature stage according to the theory of industrial life cycle. In order to meet this requirement, national and local governments shall strongly support and encourage technology development. Measures shall be taken in terms of science and technology policies, fiscal subsidy policies, preferential tax policies, and regional economic structure policies; the banking sector shall be encouraged to strengthen financing, issue risk loans, and support the development of new products. Only in this way can it be possible to accelerate the generation of leading enterprises and promote the development of leading industries and leading industrial clusters.

[3] Xia Mingwen's Ph.D. dissertation, Research on the Development of Industrial Structure in Southern Jiangsu, June 1990.

Once this dynamic process continues, the regional (or national) industrial structure can be kept "rejuvenated" for a long time. In this regard, the market mechanism alone is not enough [9].

What is the maturity of the regional industrial structure? Scholars have not made any special opinions in this regard. However, the importance of a mature industrial structure is indeed mentioned in some works.

The so-called maturity of the regional industrial structure includes the following meanings:

(1) Have strong leading industries and leading industry clusters. Judging from the perspective of the industrial life cycle, the products of the leading industries can maintain a long-term market position, that is to say, the life of the golden period can be extended. The key here is to continuously improve the economic benefits of the product.
(2) Be able to continuously develop new products and new industries (including new technologies), maintain a certain time difference between them, and enter and occupy the market with a large number of new products in sequence. As a result, the leading industries in the region are constantly being compensated as a whole, and the economic growth rate of the whole region will not fluctuate, that is, the sustained economic growth can be kept.
(3) The industrial structure in the region has a reasonable level, which is mainly manifested between the leading industries, supporting industries and infrastructure (also supporting industries), and also between emerging industries and traditional industries. The so-called reasonable industrial level is to establish a relationship of support and adaptation among such industries. As a traditional industry, it shall generally meet a large number of production and living needs in the region. In some developed countries, such as Germany and Switzerland, many traditional products of their traditional industries have been exported for a long time, and the role in regional economic development is conceivable.

3. Structural aging and structural adjustment

In countries and regions that developed earlier in industrialization, a certain period of prosperity was achieved through recovery (reconstruction) after World War II. However, "structural crises" soon appeared in some regions. Such crises have been prominent in the Ruhr region in Germany, the northern central part of Britain, and the northeastern part and the Great Lakes region of the United States since the 1960s. The root cause of such structural crises is structural aging.

1) Sign of the aging of the regional (or national) industrial structure

(1) For industries that have played a leading role in the region for a long time, technology development is stagnated, and the products are upgraded slowly or not upgraded at all. Therefore, the product lacks competitiveness, and the market is getting smaller and smaller.
(2) Since the product market of the former leading industry is getting smaller and smaller, the production enterprises have to reduce the production scale. As a result, more and more people are unemployed. When the emerging industries in

the region cannot develop correspondingly, the unemployed and their families have to move to areas where it is easier to find jobs.

(3) Due to the massive migration of the population and labor, the tax in the region is dramatically reduced, and the national income is reduced. As a result, it is impossible to spend money to support the development of new industries, new technologies and new markets.

In the regions with structural aging, generally, there are some depressed industries, that is, industries with the above three signs. The following industries were depressed in Japan in the late 1970s, including: open-hearth furnace and electric furnace steel manufacturing, aluminum smelting, synthetic fiber manufacturing, shipbuilding, fertilizer, cotton and chemical fiber textiles, and corrugated paper manufacturing. During the period of rapid growth in the 1950s and 1960s, these industries were regarded as the leading industries, of which the development was vigorously supported.

Structural aging is quite prominent in China. Although large-scale industrialization began after the founding of the People's Republic of China, the products and production equipment introduced in the 1950s were not the latest products and equipment at that time. Even the advanced equipment and industries at that time will be in a state of structural aging after 30 years of production without large-scale equipment updates. Taking the old industrial base, Liaoning, as an example, most of the technical equipment was not updated due to the lack of sufficient funds for a long term and the full-load production of the products required by the country. In the province, the industrial equipment that belonged to the 1970s, the 1960s and the 1950s and before accounted for 38, 52 and 10% respectively as of 1987. The net value rate of fixed assets of the industries in the province was only 60%, while that in the whole country was 67.4% at that time [10].

There are many ways to solve the problem of the aging of the regional industrial structure, such as: supporting the industrial technological transformation of the region with structural aging through government or bank loans; encouraging and promoting the development of small and medium-sized enterprises through the formulation and implementation of middle-class policies. According to the international experience, the small and medium-sized enterprises have strong functions in adapting to market needs and adjusting product directions. Another important way is to make structural adjustments.

2) Structural adjustment

There are two specific ways to make structural adjustments to regional industries. First, determine a number of industries or products with broad market prospects as key development directions based on the status of the country or the region on a large scale and its own economic and technical levels, that is, to develop new industries, new products and even new industrial clusters at new growth points. At the same time, take measures to restrict the development of those industries (or products) that lack competitiveness and have poor economic benefits, or transfer them to other regions suitable for development to produce. As a result, the regional industrial structure will be "rejuvenated". In the early 1990s, the structural adjustment was put at an important position in the economic development strategy of Shanghai, which emphasized the need to gradually eliminate the production of light industrial products and general raw materials that had been generally developed in other provinces and cities, and focus on strengthening new industries such as automobiles, aerospace and biotechnology. Similar industrial structure strategies were also adopted for several large cities such as Tianjin and Beijing. Europe's largest steel monopoly group, Germany ThyssenKrupp Steel Company and Mannesmann Company, actively pursued the strategy of gradually withdrawing from traditional industries, and turning to electronics and mechanical equipment manufacturing, thus reducing the output value of traditional products to < 1/3 of the turnover [8]. Second, create conditions, and transform leading industries. When striving to make their economies develop in a higher-level technology-intensive and knowledge-intensive direction, some Western countries also use trade protectionism to maintain certain traditional industries in their countries and strengthen their technological transformation. When choosing leading industries, developing countries cannot just develop industries with high demand prices and extensive connections with other economic sectors. It is more important to develop industries with high-income elasticity and high productivity growth, so as to accelerate the pace of transformation of leading industries.

4 Resource Structure and Industrial Structure

1. Resources and regional development
1) Resources are one of the material foundations for national and regional development.

In the early stages of industrialization of developed countries in the world, the abundance and structure of resources had a major impact on the economic, military and political rise and decline of some countries, as well as wars and conflicts among countries and regions. Due to the extreme importance of resources to the development of countries and regions, the struggle for resources often led to wars among countries and regions. One of the results of the war was the redistribution of resources among countries. In modern times, owning sufficient necessary resources is an important manifestation of comprehensive national strength and an important guarantee for victory in the increasingly acute economic struggles among countries, regions and groups of countries.

With the progress of science and technology and the progress of the industrialization process, especially the development of the internationalization and grouping trend of world trade and economy, the status of resources in regional development is changing. Globally, the main trend is that resources are more important to development, and that the struggles for resources among countries and regions have become more acute. However, such struggles are carried out in an indirect way, that is, by exporting manufactured products with high technical content, low resource content and high added-value, and importing resources and resource-based products. Here, the competition for resources is carried out through technological advantages (rather than military advantages as in the past). However, the significance of resources for regional development has become less important locally. This is mainly due to the fact that the unit production of products consumes fewer resources than before. Capitalist industrialized countries have reduced their dependence on resources through the upgrading of the industrial structure during the great changes in the world economy. However, most developing countries have become more and more active in world trade in recent years, and the gaps in many resources can also be filled through trade. Therefore, when we utilize resources and develop the national economy, we must pay attention to the direct and indirect effects of resources on regional development; the indirect meaning has become greater in the process of continuous development of inter-regional trade.

2) Resources affect regional development

The main functioning link is the industrial structure. In other words, whether a region or a country has the resources required for economic development, which resources are available and which resources are in shortage will affect the economic sector structure of the region as well as the determination of leading industries and basic industries. Among many regions with oil and gas fields in the world, most have developed large-scale petroleum processing and petrochemical industries, and some have also developed huge production systems for synthetic materials and processed products. After the oil and gas resources are exhausted, petroleum processing is often greatly affected, and other processing systems may be maintained for a long time. The way is to import oil and natural gas from other regions. This process reflects the decisive significance (before oil and gas resources are exhausted) and indecisive significance (after oil and gas resources are exhausted) of regional resources to regional economic development.

3) Classification of mineral resources according to their importance to regional development

Vast and fertile land with suitable climatic conditions as well as abundant resources are of fundamental significance to regional development. This significance is most prominent in ancient civilizations and feudal society in human society. The core areas of the four great ancient civilizations in the world, that is, the middle and lower reaches of the Nile in ancient Egypt, the Tigris and Euphrates river basins in ancient Babylon, the Yellow River basin in China, and the Ganges river basin in India, belong to such regions. During the long period before industrialization, humans' ability to transform and use nature was weak, and the economic structure was dominated by agriculture due to the low level of productivity. Therefore, the natural conditions that affected agricultural development were of practical significance.

At the beginning of the nineteenth century, Western society began the process of industrialization. From the beginning of this process, mineral resources such as coal, iron, petroleum, natural gas, non-ferrous metals and non-metallic minerals have always been in the dominant and central position during regional development. The abundance of mineral resources and supporting conditions determine not only the prosperity of a region, but also the economic strength and military strength of the country. For example, at the end of the last century and the beginning of this century, the Birmingham region in central England, the Ruhr region in Germany, and the Lorraine region in France were dominated by coal and iron resources. They were once the industrial heart and military industry supporting areas of Britain, Germany and France. With the socio-economic development and the evolution of the industrial structure, the importance of mineral resources to the regional economy (in terms of the local region) is declining.

Among more than 100 kinds of minerals required for the development of the national economy, the most important ones are: iron, copper and aluminum in metal minerals, coal, petroleum and natural gas in fuel minerals, as well as sulfur (including pyrite) and phosphorus in chemical mines. In particular, coal, petroleum, iron and copper are fundamental, which can solve the most important issues concerning raw materials and fuels, and can be called "key minerals". The remaining four kinds can be called "important minerals" (Li Wenyan, 1986). Comparing the eight kinds of minerals with the world's major countries (population or area), we can see the status of China's mineral resources in the world from an important aspect (Table 2).

From the perspectives of the quantity, quality, degree of assurance and the external conditions for the development and utilization of mineral resources, we can see the impact on China's industrial development: ① China's mineral resources and energy resources can meet the needs of China's economic development as a whole, but there are also some minerals that partially depend on foreign countries; ② Judging from a comprehensive perspective, the regional differences in the development and utilization conditions for China's mineral resources and energy resources have regular changes from east to west. The conditions for resource development and utilization

Table 2 Ranking of China's important mineral resources in the world (proved reserves)

Mineral rank country	Key minerals				Important minerals			
	Coal	Petroleum	Iron	Copper	Aluminum	Natural gas	Sulfur	Phosphorus
China	3		5	6	7		2	3
The United States	1			2		2	4	4
Former Soviet Union	2	4	1	4		1	3	2
Canada			3				1	
Brazil			2		1			
Australia			4		2			

For the source of this table, see the "Theory and Practice of China's Industrial Distribution" written by Lu Dadao et al., 1989, p. 53

are good in the eastern coastal zone, followed by the central zone and the western zone.

4) About resource bearing capacity

Since the 1970s, some ecologists and resource scientists at home and abroad have put forward and researched the issues concerning the bearing capacity. However, scholars have quite different opinions on the connotation of bearing capacity. It is generally recognized that the bearing capacity is the number of population that can be continuously supported by a certain resource or total resources, a certain environmental element or total elements in a region without serious damage to a certain (or necessary) material living standard and regional ecological environment at a set stage in the future. The basic concept of bearing capacity is land bearing capacity.

The bearing capacity of total resources or a single resource depends on the quantity, quality and development and utilization conditions of the resource itself, and has the following three characteristics:

(1) Dynamics. That is, the bearing capacity of a certain amount (quantity and quality) of resources is not constant. With the improvement of the level of socio-economic development, the level of resource development and utilization as well as the management level, the resource bearing capacity also increases.
(2) Openness. The total amount of natural resources in a region is constant at a certain stage, but the system composed of people and nature in the region is a semi-open system that can exchange material, energy and information with the outside. The bearing capacity is improved through such exchange.
(3) Comprehensiveness. The bearing capacity of any resource or environmental element cannot be scientifically evaluated from an isolated perspective. The main reason is that all factors shall be coordinated with a series of other elements to achieve the bearing capacity, that is, to support the population or support the population to lead a necessary material life. For example, land resources must be coordinated with water resources, and the development of mineral resources must be coordinated with water and land resources.

Based on such a simple analysis, we can write the formula for the bearing capacity of (total) resources of a region as follows:

$$Q = f(x, y, z)$$

where: Q—Resource bearing capacity (the number of population continuously supported); x—Natural characteristics of the resource, that is, quantity, quality, and conditions for development and utilization; y—Level of socio-economic development, including technical level. Its indicators can be expressed in labor productivity, level of the equipment per employee, etc.; z—Level of socio-economic management of the region. A high management level can improve the scientificalness of resource utilization, and can dispatch the resource development and utilization on a larger scale.

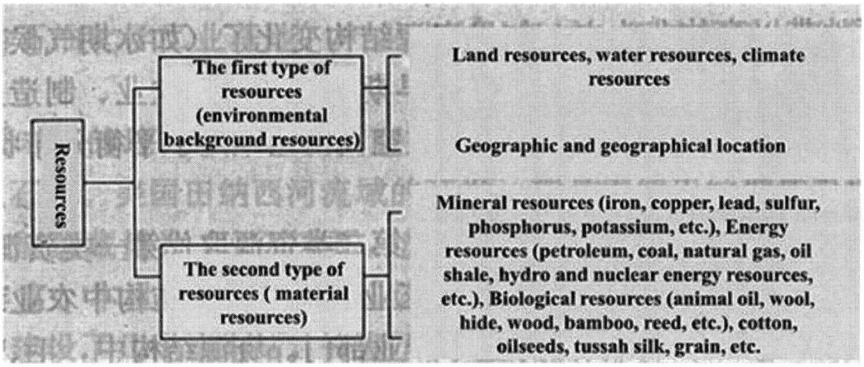

Fig. 1 Resource classification

2. Resource structure and industrial structure[4]
1) Resources and regional resource structure

Resources are the sum of various elements that can create welfare for mankind under certain technical and economic conditions. It includes not only tangible material elements such as mineral resources, land resources and fixed assets, but also intangible elements such as labor quality and technical level. The resources discussed in this book are mainly tangible resources. According to the nature of raw materials used for production activities and the external conditions of the industrial distribution, they are divided into two categories: The first category of resources is the environmental background resources in a certain region, including land resources, water resources, climate resources, marine resources that are taken as transportation channels and ecological environment functions, etc. Geographical location and geographic characteristics also belong to environmental background resources. The second category of resources includes mineral resources, energy resources and biological resources in a certain region. From the perspective of the relationship between raw materials and manufacturing production, this book also classifies the products of the primary industry as the second category of resources (Fig. 1).

Resource structure is the combined features and supporting relationships of various resources in a certain region, including the spatial relationship of distribution, quantity, quality and development conditions of various resources. Among them, it is necessary to focus on researching the relationship between this resource structure and production, such as the combination of iron, coal, metallurgical auxiliary materials, water and land, the combination of non-ferrous and energy resources, and the combination of fuel resources and chemical minerals, etc. Several resources that are closely related in development and utilization form a better supporting system in the same region (or country), which is of great significance to the development of the region, especially in the early stage of industrialization.

[4] This part is mainly extracted from Zhang Yubin's master thesis, *Regional Resource Structure and Industrial Structure*. The thesis was completed between 1988 and 1990.

Some of the world's major industrial zones are developed based on superior geographic locations, such as the Pacific Coastal Industrial Belt in Japan, the Seto Inland Sea Industrial Belt, and the North Sea Coastal Industrial Belt in Western Europe; others are formed based on the superior resource structure, such as the Ruhr region in Germany, the southern Great Lakes area in the United States, the industrial area centered on Birmingham in Britain, the Donbass in the former Soviet Union, and the central Liaoning area in China. The former has been the main direction of regional exploitation and industrial distribution for nearly half a century, that is, the large-scale development of coastal areas and the construction of large-scale production complexes and developed international central cities in the coastal areas. This trend is in the ascendant. This is one of the important contents of *The twenty-first Century—the Century of Ocean*. The latter generally has a long history of formation. Due to resource exhaustion and other non-economic factors, they are basically facing the problem of adjustments (including sector adjustment and small-scale spatial adjustment of distribution), so as to achieve revitalization. Many industrial areas in China are resource development bases, such as the central and southern Liaoning in the Bohai Sea Region, the area along Jiaoji Line, the Hanxing area and eastern Hebei. There are also some resource bases that have just been put into development, such as the Shanxi Energy Base.

2) Regional resource structure and industrial structure

Regarding the general law of the evolution of industrial structure and its internal mechanism, domestic and foreign scholars have completed a lot of research work, which have been described above in the book. From the perspective of resource structure, knowing its role in various stages of the evolution of industrial structure and its relationship with the leading industry (cluster) will help to understand the internal mechanism of regional development, and promote the rational utilization of resources and the healthy socio-economic development of the region.

(1) Evolution of resource structure and regional industrial structure

a. Stage of agricultural society before industrialization

Agriculture had an absolute advantage in the industrial structure. This stage went through a long historical period. The level of productivity development was low, and the vast majority of people were engaged in agriculture in a broad sense. The resource utilization by mankind mainly stayed at the level of the first category of resources. In the resource structure, the combined features of climate resources, water resources and land resources determined the characteristics of the industrial structure. For example, the piedmont plains, alluvial plains in the middle and lower reaches of large rivers and delta areas with a warm climate, fertile land, convenient water intake and convenient transportation are dominated by farming, and the grassland area is dominated by animal husbandry. If the resource structure changes (such as a colder climate during the ice age), mankind must migrate. The population distribution was also closely related to the resource structure. Due to the small scale of the textile industry, ironmaking industry and manufacturing industry in the early stage, the second category of resources was "widely" distributed, and such industries were basically in a state of balanced distribution.

b. Early and middle stages of industrialization

Marked by the emergence of steam engines, the social economy developed rapidly. The second category of resources replaced the status of the first category of resources. The wide use of steam engines drove the development of almost all sectors. In the early stage of industrialization, the status of agriculture declined rapidly, and industry rose rapidly in the industrial structure. Among the industrial sectors, textiles and food were the leading industrial sectors. In the resource structure, wool, cotton, oilseeds, etc. determined the development scale and structure of the textile and food industry sectors. Afterwards, railway transportation, shipbuilding and machinery industries developed, requiring a large amount of iron and steel and coal. Therefore, the two sectors, that is, the iron and steel industry and the coal industry, rose to the leading industries. The status of coal and iron in the resource structure almost determined the scale and speed of industrialization of a country. Subsequently, petroleum mining and further processing and utilization caused the decline of the coal industry. The petroleum industry and the petrochemical industry rose to the dominant position respectively.

In the early and middle stages of the industrialization process, the second category of resources in the resource structure, mainly including coal, iron, petroleum, natural gas, copper, aluminum, etc., had a decisive impact on the formation characteristics of the regional industrial structure:

The quantity and distribution of resources played a decisive role in the development scale and direction of the mining industry;

The combined features of resources affected the selection of raw material processing system and processing route;

The variety and quality of resources affected the direction of their own utilization and the technical and economic indicators of the processing industry.

For example, the industrial structure dominated by coal-steel-electricity-chemical-machinery was formed based on the large-scale coal mining in the Ruhr region in Germany. There are more than 100 billion tons of coal reserves in the Ruhr region (within 1500 m), of which the coking coal accounts for more than 90%. In the middle of the nineteenth century, the modern coal mining industry in the Ruhr region emerged, and the coking production and coal chemical industry developed on the basis of coal mining. Due to the backward iron-making technology and the high coke ratio at that time, a large number of iron and steel enterprises were attracted to develop the iron and steel industry in the Ruhr coal mine area by means of "moving the iron and steel enterprises to the coal district". The electric power industry was developed on the basis of fuels, and the mechanical industry, dominated by the coal mine machinery, was also developed. Since the 1950s, the status of petroleum in the energy production structure has risen, and the world's coal demands have fallen, which had greatly affected the coal mining and mechanical industries in the Ruhr region. The coal chemical industry also tended to shrink because its cost is higher than that of the petrochemical industry, causing the Ruhr region to suffer a serious structural crisis. Since the 1960s, the Ruhr region has started to adjust the economic structure, and developed some new sectors such as oil refining, petrochemical, automobile manufacturing, textile, food, etc. by using the superior traffic conditions, water conditions and the existing economic foundation. At the same time, a series of adjustments have also been made to the original sectors, so as to reduce costs or develop new products. At present, the mechanical industry holds a leading post in the industrial structure of the Ruhr region. The main products are mining and metallurgical equipment, boilers, steel structures, construction machinery and rolling stock, automobile manufacturing, etc.; secondly, the chemical industry. Although the raw material route of the chemical industry has been changed from coal to oil and gas, the coal chemical industry still occupies a prominent position. The iron and steel industry survived the steel crisis in the 1970s through the stages of centralization and modernization, and the production capacity accounts for 70% of the total capacity of the former Federal Republic of Germany. The electric power industry is still the largest power center in former West Germany, with a total installed capacity of 20 million kilowatts, accounting for 30% of the former Federal Republic of Germany. The coal industry has been basically stabilized, and the output of hard coal and coke accounts for more than 80% of the former Federal Republic of Germany. The output of brown coal accounts for about 70%, which is processed and consumed locally. The light industry mainly includes beer, food, textiles and clothing.

As one of the important industrial bases in China, the central Liaoning industrial area was developed on the basis of a resource structure where coal and iron are close and the metallurgical auxiliary materials are well matched. The electric power industry and non-ferrous metallurgy were established on the basis of coal mining, the iron and steel industry was developed by using the advantage that coal and iron are close, and a machinery industry dominated by heavy machinery, metallurgical equipment, steel structure, power transmission and transformation equipment and rolling stock was established. The oil refining industry was developed on the basis of oil shale. With the expansion of the scale of heavy industries such as iron and steel, coal failed to meet the needs and was transferred from Heilongjiang and Jilin, forming a layout of moving coal to the iron and steel enterprises. It was also required to transfer a large amount of steam coal from other provinces to develop the electric power industry. Daqing oil pipeline was built on the basis of the oil refining industry in the central Liaoning, forming the largest oil refining center and an important petrochemical industrial base in China. The industrial structure is dominated by machinery, petrochemistry, metallurgy, electric power and building materials.

In areas with abundant water resources, cheap power resources are often used to develop power-consuming industries, such as non-ferrous metal smelting, special steel, calcium carbide, etc. The Tennessee River Basin in the United States was developed around the abundant hydropower resources of the Tennessee River. The Tennessee River has a hydropower reserve of 4.14 million kilowatts, and the coal reserves in the basin reach 41.3 billion tons. In addition, phosphate, zinc, and bauxite are also abundant. In the early 1930s, the economy of the river basin was backward, and the industry was dominated by mining. From the 1930s to the 1940s, dozens of power stations were built in the river basin, and a large amount of hydropower was developed. Cheap hydropower attracted a large number of energy-intensive industries to enter the river basin, such as the chemical industry, electrolytic aluminum, etc. The installed hydropower capacity reached 3.76 million kilowatts. Although the hydropower resources were exhausted, the role of the agglomeration center had been formed. In the 1950s and 1960s, more than ten large thermal power plants were built on the basis of local coal, with a total installed capacity of 15 million kilowatts. The function of peak shaving was mainly adopted for hydropower. In order to prevent public hazards of the thermal power plants, two more atomic energy power stations were built in the 1970s, with an installed capacity of more than five million kilowatts. The industrial structure was also diversified, such as transportation machinery, power station equipment, metal processing, electrical appliances, rubber, papermaking, etc. Hydropower development in the Tennessee River Basin also created good conditions for the development of agriculture, flood prevention, shipping and forestry in the river basin.

Jilin, an important chemical industry base in China, also used the hydropower of Fengman Hydropower Station and the abundant limestone and coal resources surrounding to develop a chemical industry dominated by power plants. In the early 1970s, Daqing oil was used to build a petrochemical industry.

Sweden is famous for its abundant hydropower resources, high-grade iron ore and forest resources. 75% of the installed capacity of 13 million kilowatts is hydropower. The huge drop in height has made the cost of hydropower construction in Sweden low. The production of special steel is developed on this basis, which has a prominent position in the world market.

The largest petrochemical industrial center in the United States and the world is along the coast of the Gulf of Mexico in the United States. This area has a resource structure with petroleum and natural gas as the main characteristics. The crude oil reserves along the coast of the Gulf of Mexico account for 69% of those of the U.S., and the exploitation of crude oil and natural gas ranks first in the U.S., accounting for 1/4 of the total output. Hydropower resources are extremely abundant, relying on the Mississippi River. The huge resource advantages attracted a large number of investors. In the early 1950s, oil extraction, oil refining, petrochemical and electric power industries were the mainstay. In the 1960s, the status of petrochemicals rose rapidly under the influence of the large-scale and group development trend of petrochemical equipment. In the early 1970s, ethylene output was close to seven million tons, forming an industrial structure for comprehensive development of oil refining, chemical, oilfield equipment, petrochemical equipment and shipbuilding, which was dominated by petrochemicals.

By analyzing the relationship between the above several types of resource structure and industrial structure, the following law can be obtained: The industrial zones formed on the basis of resource combination generally take the development of one or two resources with comparative advantages in the resource structure as the forerunner industries. However, when most of the resources in the resource structure have been developed in succession, a comprehensive matching relationship of "resource structure *k* industrial structure" is formed. The industrial structure also shifts from single to comprehensive. The leading industries for the further development of the regional economy will generally change. On the one hand, the cost increases due to the increased difficulty in resource extraction and the decrease in reserves; on the other hand, it is also determined by the demand structure. In the later stage of development, the leading industries may have nothing to do with the resource structure in the region.

It can be seen from this that attention must be paid to the development of alternative industries when developing the region with superior resource structure, so as to avoid simple structure and timely promote the industrial structure transformation.

The two coal mining cities, Fuxin and Fushun in Liaoning, both started with coal mining. Fuxin has been focusing on coal mining for decades, and has not paid attention to the issue of comprehensive development. At present, coal mines are beginning to shrink in the later stage of mining, coal miners are unemployed (including on-the-job unemployment), and the problem of young people waiting for employment is quite serious. Although Fushun is also in the late stage of coal development, there is no structural crisis like Fuxin due to the fact that the alternative industries have been well developed, such as oil refining, petrochemical, light and textile industries.

Los Angeles in the United States and Baku in the former Soviet Union are also two opposite examples. Oil extraction and refining were taken as the leading industries in both cities, but alternative industries in Los Angeles developed rapidly. At present, the industrial structure is diversified, with comprehensive development of aerospace, aircraft manufacturing, electronic industry, food and oil refining. However, Baku has almost become a "dead city" after the decline of oil fields due to its single structure.

c. Late stage of industrialization

For resource-based industrial belts that have entered the late stage of industrialization, the resources in the region will generally be depleted, and the original layout advantages will be lost, causing the decline of resource-based industries such as the mining industry, serious environmental pollution, as well as problems with water and land use. If the sophistication of the industrial structure is not promoted in time, the regional economy will lose its vitality.

Now, we can unify the role of resources in the formation of the two industrial belts. The resource-based industrial belt is naturally developed and evolved step by step in the process of industrialization. The role of the two categories of resources is very clear. The location-based industrial belt developed at the late stage of industrialization. At the moment, the achievements of industrialization have pushed away the location restriction of the second category of resources.

The stages of industrialization in various regions in China are different, and the technical structure is also diversified. Among them, the technological achievements in the late stage of industrialization have been applied in some regions. The development of the coastal areas can enable the location-based industrial belt to better absorb foreign advanced science and technology. For resource-based industrial belts, the advantages of location-based industrial belts shall also be used as much as possible, so as to make necessary spatial adjustments.

d. Post-industrialization period

This is the stage that mankind is just approaching. Traditional industries are shrinking and being replaced by a batch of emerging industries such as microelectronics, marine engineering, aerospace, bioengineering, new materials and new energy, which take the high-tech products as the mainstay. High technologies mainly include microelectronics technology, genetic engineering, superconducting technology, optical fiber communication technology, space science, etc. At the moment, resources are "overwhelmed" in a sense. Most high-tech industries are deployed in the form of "science and technology industrial parks", such as Silicon Valley in the United States, Tsukuba in Japan, the "Isar Valley" near Munich in the former Federal Republic of Germany and Novosibirsk in the former Soviet Union. The common feature is that the environment is clean and quiet. It is composed of universities, research institutions and some companies in high-tech industries. Alvin Toffler believes that "all countries are at the same starting line on the road to the future". This is a bit idyllic. Indeed, we shall keep pace with the world level as much as possible in the pursuit of high technology, but China is not an industrialized country after all. Problems such as serious backwardness in education have plagued the existing economic structure and weakened its ability to transform. Therefore, traditional industries are still sunrise industries for us, which shall still be vigorously developed, so as to make up for the missed lessons for industrialization. Taking the advantages as a later comer, this catch-up process is not necessarily very long. The key is to grasp the opportunity.

The above mainly discusses the relationship between the industrial structure and the resource structure during the industrialization period, but does not involve agriculture and the tertiary industry in the industrial structure. Throughout the process of industrialization, the proportion of agriculture has declined very rapidly, and the main characteristic of the population is that the population in industry and the tertiary industry rises constantly. Agricultural production has undergone two transformations:

a. Traditional agriculture $\rightarrow$ industrial agriculture (also known as petroleum agriculture). Industrial agriculture is characterized by a large investment in mechanization, water conservancy, fertilizers, pesticides, agricultural films and mulching films.
b. Industrial agriculture $\rightarrow$ ecological agriculture. With genetic engineering as the theoretical foundation, ecological agriculture is characterized by biological nitrogen fixation technology, disease-resistant varieties, high-yield breeding, hybridization technology, widespread use of embryo transfer, etc., focusing on ecological effects.

The restriction of the resource structure on it is weakened. For example, the rice planting line has repeatedly moved northward and is no longer restricted by accumulated temperature, and the problem of being dependent on the weather has been solved through the water conservancy measures.

Table 3 Export structure of newly industrialized countries and regions

Item	Percentage of various export commodities (%)									
Country or region	Fuels, minerals and metals		Other primary products		Machinery and transportation equipment		Other manufactured products		Textiles and clothes	
	1965	1986	1965	1986	1965	1986	1965	1986	1965	1986
South Korea	15	3	25	6	3	33	56	58	27	25
Hong Kong	2	2	11	6	6	21	81	71	43	35
Taiwan	2	2	57	7	4	29	37	62	5	18
Singapore	21	21	44	12	11	38	24	30	6	5
Thailand	11	4	84	54	0	9	4	33	0	15

Source World Bank, *World Development Report 1988*

All tertiary industries are not directly affected by the first category of resources except for the tourist industry.

(2) Resource structure and export structure

The export structure is an important aspect of the industrial structure. Generally speaking, the export structure has a functional relationship with the regional leading industry (cluster), and the change in the regional leading industry will also cause corresponding changes in the export structure. In the early stage of regional economic development, raw materials occupy the primary position in the export structure; in the middle stage of development, products from the manufacturing industry occupy the primary position; in the later stage of development, high-tech products occupy the primary position. There is a highly positive correlation between the level of economic development (measured by GNP), people's living standards and the level of export structure. The examples of established developed countries are known to all, and emerging industrial countries also conform to this law of development (see Table 3).

The driving force for the sophistication of the industrial export structure is trade interests, that is, the upgrading of the export structure can enable the country (region) to obtain the greatest profit in world trade.[5]

At present, the income level of some countries in the world is quite high, but the level of export structure is quite low. For example, crude oil accounts for 90, 87 and 78% of the export structure of Saudi Arabia, Kuwait and the UAE respectively. There are also some developed countries also exporting a large number of resource products.[6] For example, the proportion of such export is 24% (mainly coal, sulfur, phosphate rock, grain and other agricultural products), 36% (mainly grain, oil, coal and iron ore), 63 and 16% in the United States, Canada, Norway and Sweden

[5] This issue is fully discussed in Chap. 3 of *Japan's Industrial Policy* (Chinese edition) co-authored by Ryotaro Komiya et al.

[6] Including mineral products and various primary processed products, agricultural products, etc.

respectively. Developing countries cannot help but export a large number of primary products. However, it is not true for developed countries. This indicates that resource products also have huge trade interests, which cannot be totally denied. Input–output analysis is taken as the basis for developed countries. Countries in the world have different levels of resource abundance due to geological reasons. Resource abundance includes resource reserves, quality, mining cost, etc. In terms of petroleum mining, compared with other regions, the Middle East has great advantages due to abundant reserves, low mining cost and high quality (high light oil content). Chen Chuankang once used "resource surplus value" to vividly illustrate this phenomenon. For another example, Sweden has a reserve of iron ore of 4.7 billion tons, with a grade of up to 60%. It can be smelted directly without the sintering process. This country has an annual output of 35 million tons of ore, of which 80% are exported to European countries. The iron ore in Brazil, Australia and Canada is not only of high grade, but also very suitable for large-scale mechanized open-pit mining. Therefore, there is also resource surplus value. The cost of food production in the United States, Australia and Canada is lower than the world average due to mechanized and large-scale operations. Therefore, there are benefits of export trade.

The relationship between resource structure and export structure has two forms: ① The second category of resources are exported in kind in exchange for foreign exchange income; ② The government leases the first category of resources to international capital to obtain fiscal revenue. There are many reasons for the transfer of capital from developed countries to the third world. The first is the diffusion of labor-intensive industries, aiming to obtain cheap labor to reduce costs (for example, the Pearl River Delta in China can be regarded as an industrial diffusion region where Hong Kong capital obtains cheap labor). The second is the diffusion of industries that cannot continue to develop in their own countries due to heavy pollution and large floor area.

In the long run, it shall only be a way to promote regional economic development in a relatively short period during the growth of industrial structure to obtain the fiscal revenue by using the resource surplus value. After all, resources are limited. If structure sophistication is not implemented in a timely manner, the cost will rise, and the resource surplus value will be lost in the late stage of mining, which will lead to a difficult structural crisis. Moreover, technological progress and the emergence of substitute products will also threaten the sustained economic development of resource exporting countries and regions, thus shortening the life cycle of the resource surplus value. At present, the oil reserve-production ratio in the Middle East has begun to decline, and the economic situation facing the Middle East countries in the twenty-first century will be quite severe.

Countries with low resource abundance often export resources that are not abundant to obtain technologies and equipment necessary for industrialization, because they have no other way to obtain large amounts of foreign exchange. These hard-won foreign exchange incomes shall be cherished. However, the actual situation is that most developing countries are very inefficient in using foreign exchange due to low personnel quality and backward management.

3. Declining trend of resource industry
1) Concept of resource industry

Industries focusing on resource extraction and primary processing production can be collectively referred to as resource industries. The resource industries are divided as follows:

a. Electric power industry: thermal power, hydropower and nuclear power;
b. Chemical industry: basic chemical industry (three acid, two alkalies, calcium carbide, etc.), synthetic material industry (ethylene, polyethylene, polypropylene, benzene, toluene, acetylene, etc.);
c. Metallurgical industry: mining, smelting and calendering of non-ferrous and ferrous metals;
d. Petroleum industry: petroleum extraction and refining;
e. Coal industry: coal extraction, processing and coking;
f. Building material industry: cement, bricks and tiles, building ceramics;
g. Textile industry: spinning (cotton, wool and silk) and weaving;
h. Forest industry: logging and primary processing of timber and bamboo;
i. Food industry;
j. Paper-making industry.

2) Declining trend in the resource industry

The decline of the resource industry refers to the decline in the status and role of the resource industry during the growth of the regional industrial structure. Such decline has two meanings, that is, the decline in the absolute amount of the resource industry and the decline in the proportion in the industry structure. These two declines may occur at the same time. In most cases, the relative proportion declines, and then the absolute quantity stagnates and declines.

The resource industry can be further subdivided into pure resource industries and general resource industries where the pure resources are preliminarily processed. Due to the in-depth industrialization process and technological advancement, these two industries have already had the trend of spatial separation. The development of local general resource industries may not be restricted by local pure resource industries.

Let's take steel, the main raw material for industrialization, as an example. In the 30 years after World War II, the iron and steel industry experienced a process from prosperity to decline (Table 4). It can be calculated that: From 1949 to 1960, the annual growth rate of steel output was 2.2, 19, 12.6 and 9.8% in the United States, Japan, the former Federal Republic of Germany and the former Soviet Union respectively. It can be seen that the iron and steel industry in the United States had been a mature industry at that time. Japan, Germany and the Soviet Union were in the post-war reconstruction period throughout the 1950s. The iron and steel industry developed rapidly, and was the leading industry of the national economy; from 1960 to 1970, the annual growth rate of steel output was 2.8, 15, 2.8 and 5.9% in the United States, Japan, the former Federal Republic of Germany and the former Soviet Union respectively. It can be seen that the iron and steel industry in Japan was still developing at a high speed, while that in the former Federal Republic of Germany and the former Soviet Union was close to mature; from 1970 to 1983, the annual growth rate of steel output was – 4 (declined), 0.2 (negative growth after 1980) and 2.1% in the United States, Japan and the former Soviet Union respectively.

Other general resource industries also showed a trend similar to that of the iron and steel industry.

The changes in the industrial structures of the United States, Japan and the former Federal Republic of Germany from 1963 to 1974 can also be selected to conduct analysis. Although this period of more than 10 years is very short, it represents the long-term development trend of various sectors (see Tables 5, 6 and 7).

The structure is calculated based on the sector composition of total industrial output value.

Table 5 shows that in the industrial structure of the United States, the mechanical industry obviously increased (by 2.9), while the rest of the sectors were stable with a slight decline, indicating that the industrial structure was basically in a stable state. Although the increase and decrease in various sectors also reflected the long-term development trend, there were no major fluctuations, and all sectors were mature industries.

Since the emerging industry sectors (such as the electronic industry) are not listed in the table, it is difficult to point out the newly-growing industries. However, one

Table 4 Steel output in the United States, Japan, the former Soviet Union and the former Federal Republic of Germany after World War II

Country year	The United States	Japan	Former Soviet Union	(Unit: 10,000t)
				Former Federal Republic of Germany
1949	7074	311	2329	916
1953	10,125	766	3813	1542
1960	8007	2214	6529	3410
1965	11,926	4116	9102	3682
1970	11,931	9332	11,589	4504
1975	10,582	10,231	14,133	4041
(1973)*	13,148	11,932	–	4952
1976	11,600	10,730	14,100	–
1977	11,400	10,200	14,500	–
1978	12,400	10,200	15,100	–
1979	12,400	11,100	14,900	–
1980	10,100	11,100	14,800	4381.2
1981	10,900	10,100	14,900	–
1982	6600	9900	14,700	3634.8
1983	7500	9600	15,200	3610.8

*1973 was the year with the highest steel output in the capitalist world, followed by the steel crisis. It was triggered by the economic crisis in the 1970s

Table 5 Evolution of industrial structure of the United States from 1963 to 1974 (%)

Year sector	1963	1967	1970	1972	1973	1974
Metallurgy	7.7	7.5	7.5	7.0	7.4	7.2
Electric power	2.4	2.5	2.8	3.0	3.0	2.9
Coal	2.3	2.2	2.6	2.7	2.4	2.4
Petroleum	6.3	5.9	6.2	5.6	5.5	5.3
Chemistry	9.8	10.4	10.0	9.6	9.4	9.5
Machinery	32.6	34.9	34.2	35.0	35.5	35.5
Building materials	2.9	2.6	2.6	2.8	2.7	2.6
Food	16.0	14.7	14.9	14.6	15.0	16.4
Textiles	4.0	3.9	3.8	4.0	3.9	3.7

Table 6 Evolution of industrial structure of Japan from 1963 to 1974 (%)

Year sector	1963	1967	1970	1974
Metallurgy	11.1	12.2	12.4	12.2
Electric power	3.0	2.8	2.4	3.4
Coal	0.7	0.4	0.5	0.6
Petroleum	1.8	1.8	1.7	2.2
Chemistry	11.0	11.0	10.7	9.9
Machinery	30.8	33.9	38.4	35.6
Building materials	3.5	3.5	3.5	3.8
Food	13.0	11.9	10.2	10.7
Textiles	9.8	7.8	6.5	6.0

Table 7 Evolution of the industrial structure of the former Federal Republic of Germany from 1963 to 1974 (%)

Year sector	1963	1967	1970	1974
Metallurgy	10.3	9.9	11.1	11.2
Electricity	3.9	4.2	3.7	4.3
Coal	3.4	2.5	2.3	2.7
Petroleum	2.9	3.7	3.7	4.6
Chemistry	10.6	12.4	12.4	14.4
Machinery	34.5	32.0	35.2	33.4
Building materials	3.7	3.9	3.7	3.6
Food	17.5	14.6	12.7	11.5
Textiles	5.4	4.9	4.4	3.6

thing is certain: These traditional sectors are no longer the engines of economic growth in the new historical period of the United States.

In contrast, the industrial structure of Japan and the former Federal Republic of Germany underwent major changes. For example, the mechanical industry increased by 8%, while textiles, food and chemical industries decreased by 3.8, 3.7 and 1.1% respectively in Japan. During the restoration period in the 1950s, textiles and food were the leading sectors in Japan. In the 1960s, their status declined, metallurgy increased by 1.1%, continuing the development momentum in the 1950s, and petroleum and electric power slightly increased. The mechanical industry, as a leading industry, had a prominent position.

The situation in the former Federal Republic of Germany was similar to that in Japan. However, its mechanical industry had been in a stable state, and the other sectors were also generally stable, showing a state of balanced development. All sectors advanced side by side. Food, textiles and coal declined, while metallurgy, electric power and petroleum expanded.

The post-war industrialization process in Japan and the former Federal Republic of Germany can be regarded as a concrete manifestation of the general law of industrialization. Because the original industrial foundations of the two countries were destroyed in World War II, it was necessary to retake the road of industrialization. However, the technicians of the two countries were retained, and fully played their roles, which can be regarded as a better choice for the later comers to give play to their own advantages. By superimposing the post-war roads of Japan and the former Federal Republic of Germany on the economic development roads of established industrialized countries such as the United States, Britain and France, we can get Fig. 2, Schematic Diagram of Industrialization and Resource Industry Development.

The solid line in the figure indicates that the scale of resource development in the region rose rapidly in the early stage of industrialization, and the pure resource industry was the leading industry for regional development. But soon, its development would reach its peak and then decline. The possible reason was that the resource mining cost rose, other industries developed slightly, and the fund was drawn to general resource industries; another possible reason was that resources

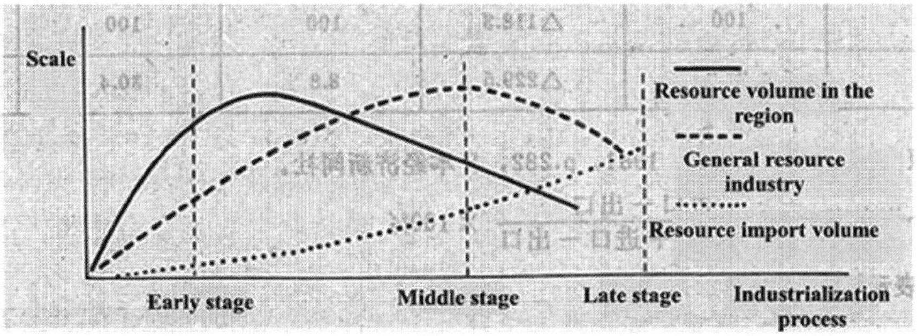

Fig. 2 Evolution trend of resource-based industries in the process of industrialization

were imported from other places under the guidance of comparative advantage, so as to meet the development of the general resource industries. With the deepening of the industrialization process, resources gradually depleted, while imports increased (such as iron ore resources in the Ruhr region). The dotted line in the figure refers to the general resource industry, whose development was very fast. The strong momentum of expansion was kept for a long period until the middle and late stages of industrialization. The point line shows that the local resources were exhausted with the development of industrialization, and the development of the general resource industries strongly depended on the input of resources from other places. The demands of developed countries for resources in the international market can be seen in Table 8. The degree of dependence has increased in recent years.

In a word, the long-term decline of the resource industry is mainly caused by the following reasons:

(1) Changes in the demand level. Demands are hierarchical. When the low-level demands are met, the high-level demands begin to become the main driving force for sustained economic growth. With the progress of industrialization, people's basic demands (such as clothing, housing, transportation and food) are solved, and high-level demands often begin to expand in the middle and late stages of industrialization. The decline in demands for resource products leads to a decline in production.
(2) The technology development makes the resource utilization develop in depth, processing procedures increase, and horizontal expansion is weakened. Accordingly, the demands for resource products decline.
(3) Through the development of the chemical industry, new materials are synthesized from a wider range of raw material sources to replace natural materials, which reduces the demand for resource products.

Degree of external dependence = (import–export)/(domestic production + import–export) × 100% Δ stands for output.

Table 8 Comparison of the degree of external dependence on major resources (1978) (%)

Country item	Japan	The United States	The former Federal Republic of Germany	Britain	France
Energy	92.4	21.9	56.9	21.2	81.4
Petroleum	99.8	37.7	95.0	45.6	97.7
Iron ore	98.4	30.4	96.4	78.7	39.6
Copper	94.1	38.0	99.9	99.8	99.8
Lead	78.6	61.3	89.4	99.5	84.6
Wood	66.4	27.0	16.8	73.8	10.2
Wool	100	27.0	90.9	59.7	70.3
Cotton	100	AH8.3	100	100	100
Wheat	93.8	Δ229.5	8.8	30.4	37.9

Source Encyclopedia of Japan's Economy, 1981, p. 282, Nikkei Shimbun

From the perspective of the whole industrialization process, the status of the resource industry constantly declines. However, the above analysis shows that the resource industry has had a strong development trend for a long period, especially in the early and middle stages of industrialization.

The process of economic development is discussed in this book based on the relationship between resource structure and industrial structure. Not considering issues within the limitation of resource factors, this book has the characteristic of comprehensive research on specific issues. China is in the early and middle stages of industrialization, and there are large differences in the resource structures of various regions. The development and distribution of the resource industry and its follow-up industries are closely related to the adjustment of the industrial structure of each region.

References

1. Y. Zhi, *Introduction to Industrial Economics* (China Renmin University Press, Beijing, 1985)
2. W. Husheng, Several issues concerning industrial structure adjustment. Econ. Res. Inform. **34**, 653 (1981)
3. National Bureau of Statistics, *China Statistical Yearbook* (China Statistics Press, Beijing, 1991), pp.74–75
4. National Bureau of Statistics, *China Statistical Yearbook* (1985), p. 238
5. National Bureau of Statistics, *China Statistical Yearbook* (1988), pp. 417–418.
6. Z. Dakang (ed.), *Dictionary of Modern Geography Science* (The Commercial Press, Shanghai, 1990), pp. 555–556
7. Z. Yinglong, Status and countermeasures of China's basic industry investment. Coast. Econ. **3**, 6212–6219 (1990)
8. Z. Fengbo (ed.), *China's Macroeconomic Structure and Policy* (Chinese Financial and Economic Publishing House, Beijing, 1988), pp. 388–389
9. L. Boxi, Q. Zhishen (eds.), *Industrial Policy and National Economy* (Shanghai Science and Technology Document Press, Shanghai, 1990)
10. L. Dadao et al., *Research on Liaoning Resource Development and Industrial Transportation Layout* (China Planning Press, Beijing, 1990), p.6

Chapter 5
Resources, Environment and Regional Development

1 Serious Challenges Posed by Environmental Degradation to Regional Development

1. **Development of the relationship between ecological environment and regional development**

The natural ecological environment is always in an unbalanced state due to its own elements. Large-scale volcanic eruptions, strong earthquakes, tropical storm surges, etc. may change the local balance of nature quickly and even bring great damage. Rainwater and runoff can fundamentally change the geomorphology of some areas in the geological age; and may also cause the loss of topsoil in some areas within the scale of human history. Soil erosion can be regarded as the most serious natural disaster on the earth today, because it is the most common and occurs almost all the time. If the topsoil that provides (produces) a large amount of food required by human beings is lost, disaster is about to come soon.

There is a long history of ecological and environmental issues faced by human society. However, a series of ecological and environmental issues brought about by the social and economic activities of human beings, such as the destruction of the ozone layer and sea level rise caused by CO_2 emissions, industrial pollution, the extinction of a large number of biological species, soil erosion, deforestation, reduction of the vegetation coverage rate, etc., only appeared after the Western developed countries began the large-scale industrialization more than 100 years ago. During a period of 100 years before World War II, the main ecological and environmental issues were basically localized industrial pollution. As early as the 1970s, Engels foresaw that the development of European industry would inevitably lead to social nuisance in his *Anti-Duhring Theory*. At the same time, he pointed out: "Only a society that arranges its own productivity in a coordinated manner according to the unified master plan can fundamentally solve the problem of public nuisance." He

D. Lu, *Regional Development and Its Spatial Structure*,
https://doi.org/10.1007/978-981-99-7683-6_5

also pointed out: "Only through the integration of cities and villages can the current pollution of air, water, and land be eliminated." [1].

From the 1950s to the oil crisis in 1973, most developed countries in the world achieved sustained economic development. However, with the rapid economic recovery and the substantial increase in social welfare standards, industrial pollution became more and more prominent, causing serious consequences. The most famous was the Minamata disease that occurred in Niigata, Japan due to mercury poisoning (1953), and the air pollution incident in London, England. As industrial pollution continued to be serious and expanding, a series of growing rivers, lakes, and oceans all began to be polluted. Environmental pollution was more and more serious, and gradually became a global problem from a local and regional problem. At this stage, many developing countries were also committed to developing their own economies, and a series of processes that endangered the ecological balance had been continuously expanded and accelerated. The United Nations Conference on the Human Environment was held in Stockholm in 1972, and the *Declaration of the United Nations Conference on the Human Environment* of historical significance was passed. Since then, more and more people all over the world had begun to attach importance to environmental protection. In the 1970s, people had clearly realized that the root causes of various pollution and environmental and ecological issues were population growth and economic development. In the 1980s, the international community had clearly considered various environmental issues and social and economic development in close connection and put forward environmental and development issues.[1]

2. **China is facing dual pressures of promoting development and protecting the environment**

1) Serious environmental crisis

In recent decades, the rapid growth of the population and economy and the application of powerful technical means have strongly changed the socio-economic structure and ecological environment structure of various regions, thus promoting economic industrialization and social urbanization. At the same time, environmental degradation is increasingly worse. "Three wastes" pollution, ozone layer destruction, greenhouse effect, acid rain, forest destruction, soil erosion, resource depletion, desertification, marine pollution, etc. are becoming more and more serious. The "environmental crisis" is threatening the survival and development of mankind. These ecological problems are quite prominent in China (Fig. 1).

The issues of the environment and sustainable development have attracted great attention from the United Nations and the governments of many countries. "Getting out of the predicament", "saving the earth" and realizing "environment protection and sustainable development" has become the common voice of politicians and scientists, which is also one of the major themes facing human society in the 1990s and after entering the twenty-first century.

[1] Song Jianghua, Some Methodological Issues in Environment and Development Research, December 1991.

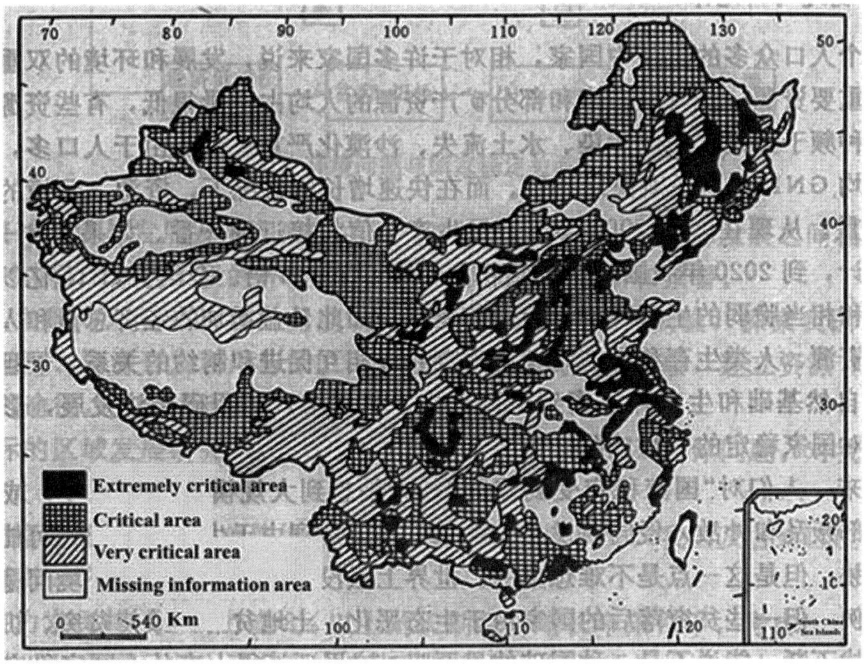

Fig. 1 Fragile situation of China's ecological environment and the distribution of critical areas. From Yang Qinye's "China's Environmental Fragile Situation and Critical Regions", 1992

Today, while facing tremendous pressure for development, human society is also caught in a serious environmental crisis. Therefore, the contradiction between sustainable development and environmental protection must be resolved, that is to say, to seek a method that "meets the current needs and pursuits of mankind, without causing harm to the needs and pursuits in the future". In this regard, scientists from various countries have conducted a lot of research in the field of environment and development, and various research institutions and disciplines have emerged as the times require. The research on the environment and sustainable development has become a "hot spot" in many disciplines and at the leading edge of development.

2) Worrying resource shortage

China has a large population and a large economic scale and is in the early and middle stages of industrialization. The industrial structure is obviously dominated by sectors and products that consume a lot of resources. In the case of the rapid growth of economic aggregate, resources are being consumed in an accelerated way, and some resources are becoming exhausted. There is a serious shortage of resources that are important to human survival and national development, such as arable land and pasture, fresh water, energy, etc. Although the per capita arable land in China is still 1.6 mu, the per capita arable land in some populous and economic provinces is below 1 mu, which is very low. The north is severely arid, the average water shortage in North China has reached 10–15 billion cubic meters, and the pasture is seriously degraded. Although there is enough coal, China's energy structure has a fatal weakness (coal accounted for 74.2% of the total primary energy consumption in 1990), which brings tremendous pressure and burden to transportation and the environment. China is rich in hydropower resources, and the natural conditions for development and utilization are superior. However, the problem of immigration is a serious obstacle. There are quite a few mineral resources that can meet domestic needs in China, while some cannot. The shortage of resources will become more and more serious in China.

3) New thinking on national stability

China is a developing country with a large population. Compared with many countries, the dual pressure of development and the environment are more prominent. Many important resources, such as arable land, water, and some mineral resources, have very low per capita availability. Some resources are running out, and some species are on the verge of extinction. Industrial pollution, soil erosion, and desertification are serious. However, rapid development must be achieved due to the large population, the low level of economic development, and the low per capita GNP. In the case of rapid growth, the consumption and destruction of resources and the environment are even more serious. The gross national product will almost double from now to 2000. If calculated at an average annual growth rate of 8–10%, China's economic aggregate will be 9–14 times larger than the current value by 2020. The national population will reach more than 1.5 billion. How can China's limited resources and quite fragile ecological environment system maintain such an increasing economic aggregate and total population?

There is a relationship of mutual promotion and restriction between the environment, resources, human survival, and economic growth. Once the natural foundation and ecological environment on which the Chinese nation relies are completely destroyed, it is bound to seriously hinder economic development and affect social development. This will be a threat to national stability.

For a long time, when it comes to "national stability being threatened", people always think of large-scale invasion by a foreign enemy, internal rebellion, or both, but have never thought that ecological and environmental issues might threaten national stability. However, this is not difficult to imagine. Although there is no instance that a country is destroyed purely due to environmental problems in the world, deteriorating ecology and barren land in some poor and backward countries cause poverty and hunger to occur year after year, thus leading to continuous civil wars. Isn't this a situation where the country is ruined? Not to speak of national stability.

Since effective development policies and environmental policies are adopted, the coordination between development and the environment has been maintained, long-term sustainable economic development has been maintained, and the ecological environment has been maintained in a good state (sustainable). Such a prosperous and beautiful country can maintain national stability in today's world, and it is unlikely to cause domestic disturbance. This is extremely beneficial to national security (Fig. 2).

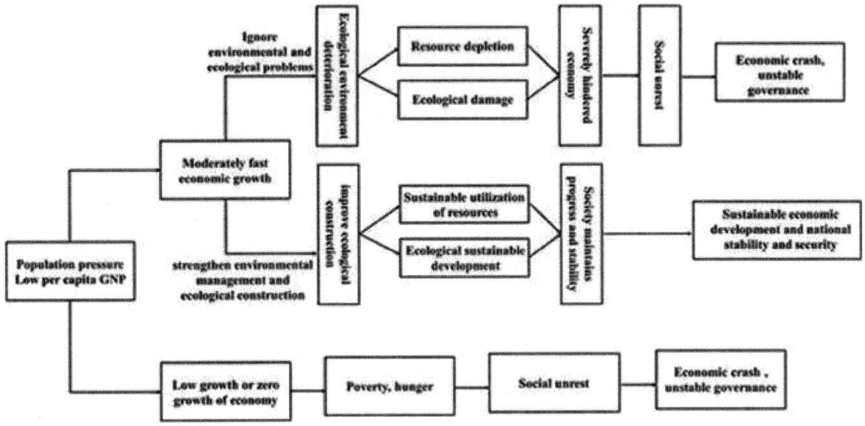

Fig. 2 Block diagram of new thinking on the relationship between ecological and environmental issues and national stability

3. Basic tasks and main issues of the research on the relationship between resources, environment and development

The research on resources, environment, and regional development mainly focuses on the relationship between them, rather than the elements they include respectively. That is to say, from the viewpoint of mutual coordination between them, research the major theoretical and practical problems in promoting sustainable development and protecting the environment and facing contemporary human society, such as the diversified consequences of human activities on the environment, the interaction between population processes, natural resources, and life-supporting systems, the stability of the human ecosystem, the population carrying capacity, economic growth, industrial structure adjustment and resource protection policies of countries and regions, etc. According to our country's national conditions, China's theoretical system concerning environmental and development issues shall be established. The basic tasks are as follows:

(1) Research the impact of human social-economic activities on the environment and resources within the framework of China and even the world, and the general law on the evolution of the ecological environment under different social and economic development strategies and models, as well as different production and consumption modes.
(2) Research the mechanism of interaction between environment and development and their elements, as well as the value standard of the mutual relationship between environment and development, and make diagnoses and early warnings of the environment and development status of China at various stages in the future.
(3) From the perspective of the coordination between development and environment, and through system simulations and program tests, explore the basic

strategies of China's long-term economic growth, structural transformation, resource development, utilization, and protection, as well as ecological environment governance, establish the approach to building a resource-saving social and economic system, and put forward the recommendations on policies.

In actual research on regional development, a series of specific issues are derived from these three tasks. For China, these issues are as follows:

(1) What is the long-term economic growth rate to be sought in china and various regions according to the requirements for per capita GNP growth as well as the resource and environmental conditions?
(2) The degree of resource assurance and the adaptability of the ecosystem in China and various regions at different time scales in the future under the condition of striving to maintain a long-term economic growth rate of more than 6% or 8%, and taking into account the upgrading of industrial structure, technological progress, as well as appropriate increase in the share of ecological environment improvement in the redistribution of national income.
(3) The dependence on the international market on the premise of maintaining the sustainable use of water, land (mainly arable land), energy, major minerals, biological resources, etc., and the sustainable development of the ecological environment in China, and taking into account technological progress, as well as the economic aggregate and total population in China and major regions at different time scales in the future.
(4) The status quo and forecast of the impact of various resources and ecological environment on social and economic development in China.
(5) The carrying capacity of regional resources under the conditions of different levels of social and economic development.
(6) Coordination of the relationship between humans and nature in different types of regions (including different types of natural areas, economic areas, and other areas).

2 "Human and Nature" Regional System and Its Sustainable Development

1. The regional system with the relationship between "human and nature"

How to find ways to maintain economic growth without severely deteriorating the ecological environment in the face of severe resource shortages, environmental pollution, and ecological destruction challenges is a big concern of the public. There have been many theories and successful experiences on how to promote economic growth in the process of human industrialization for more than 100 years, and there are also many measures on how to manage the environment. However, if the two are coordinated, there will be obvious conflicts at a certain stage of social and economic development, especially in the early and middle stages of economic industrialization and social urbanization.

To deal with the relationship between economic growth and ecological environment protection at the stage of rapid economic growth, the choice of goals cannot be unilateral. That is to say, we cannot only seek rapid economic growth without paying attention to ecological environment protection; we cannot only emphasize the ecological environment, thinking that the ecological environment must be maintained at the initial state (it shall be pointed out that this state is not necessarily balanced). However, comprehensive and diversified goals shall be established. To achieve the comprehensive and diversified goal of coordinating economic growth and the ecological environment, it is necessary to dissect how the elements of economic growth and the elements of the ecological environment co-exist in unity, and how the two interact with each other. On this basis, the unity, that is, the regional system is regulated to achieve the sustainable development of the regional social economy and ecological environment. The main elements of "economic growth" are industrial scale and structure, infrastructure, capital investment, technical level, consumption, input, and output, etc. The main elements of the ecological environment are water and land resources and their sustainable use, mineral and energy resources and their protection, biological resources and diversity, space and rational use, disaster prevention and mitigation, etc. These elements are in unity, forming a regional system. The core of "economic growth" is human economic activities. In order to carry out economic activities, social activities must be carried out. The human social and economic activities can be simplified as "human" here. The elements of the ecological environment essentially represent the "nature" side (Fig. 3).

China has the largest population in the world, and its annual growth rate is astonishing. Compared with some powers, the quantity and quality of a series of important resources in China have many defects, and the per capita resources are even less in the world. There are many areas with fragile ecology and natural areas with harsh conditions in China, which are among the worst in the world's powers. In order to ensure the survival and development of the Chinese people, it is obvious that there is more work to do in transforming and utilizing nature and coordinating the relationship between humans and nature. The conditions are quite harsh.

Like other systems, it is also required to dissect the mechanism of interaction among the main elements of the system during the research on the regional system with the interrelationship between humans and nature. From the perspective of

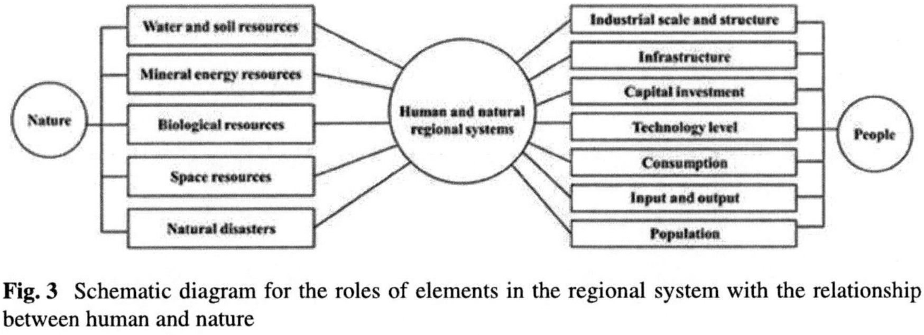

Fig. 3 Schematic diagram for the roles of elements in the regional system with the relationship between human and nature

regional development, Fig. 3 summarizes the elements of the regional system with the interrelationship between humans and nature.

2. **The regulation target of the regional system**

For the regulation of the regional system with the interrelationship between human and nature, the main focus shall be on the mode and intensity of regulating the interaction between all elements and the structure state formed. Combining the characteristics and problems of the regional system with the interrelationship between humans and nature in China, the regulation targets shall focus on the following aspects:

(1) Economic and intensive use of limited resources and space. That is, to take economic and technical means to improve the utilization of water, soil, heat, energy, minerals, and other resources and increase the assurance degree and the bearing capacity.

    Within a certain space–time range, the resource-bearing capacity is a function of technical and economic levels. For this reason, it is necessary to research and determine the scientific principles of resource-carrying capacity, and develop the resource and space utilization standards and corresponding technical and economic policies suitable for China's national conditions, so as to allocate resources among regions and replace scarce resources with abundant resources.

(2) Determine the economic development speed and economic aggregate. In the past decades of development in China, the excessive population and overhigh economic growth rate are the main factors that have caused issues in the relationship between humans and the natural system. However, if the development speed is too low, another type of socio-economic issue may also occur. Is the economic growth rate of about 6%, which was proposed by China in the early 1980s, the best rate? From a high-level perspective, there are three strategies for national and regional economic development: high-speed growth, stable growth, and balanced growth. The economic development of China shall be at the early stage of industrialization or in the transition to the middle stage. The general law of this stage is high-speed growth, with the basic raw material sectors as the mainstay. Science and reason are required to determine the growth rate suitable for China's national conditions.

(3) Adjust the industrial structure, and promote the high processing degree of the industrial structure. With the improvement of the industrial structure, the dependence of production on resources and space will decrease, and the adverse effect on the environment will also be weakened. However, the high-tech industry only has a considerable scale in very few developed countries nowadays. In China, basic industries will need to be strengthened and expanded for a long time. If we can catch up with certain developed countries and regions quickly and change the situation of exporting resources and raw materials and importing

manufactured products, it would symbolize the progress of the industrial structure in terms of high processing degree. Due to the large population in China, grain planting accounts for 3/4 of the planting area; the production of basic raw materials, such as energy, steel, synthetic materials, chemical raw materials, and building materials, occupies a major position in industrial production. There is an incompatible opposition between this industrial structure and the ecological environment.

(4) Adjust the spatial structure of the social economy. The purpose is to attract more resources and space to the socio-economic cycle. However, large-span adjustments to the spatial structure will be restricted by economic strength. In China, the nationwide "east–west problem" will exist for the long term, but the "east–west problem" in the province can be resolved quickly.

(5) Strengthen infrastructure construction, and improve space accessibility. Due to the improvement of accessibility, the difference in location differential rents of different regions can be reduced, so that the resources and space of each region can be fully utilized. The resource-bearing capacity is improved on the whole, which is equivalent to adding new resources and space. Of course, this process also lasts for a long term.

(6) Increase input and output, so as to expand the degree of openness of the regional system of humans and nature, and make full use of rich resources and vast space of foreign regions and countries to support China's socio-economic system as far as possible.

(7) Control the urbanization process. A large number of people have rushed into cities, forming numerous urban agglomeration areas, which is one of the types of regions with the most prominent ecological and environmental issues today. At present, more than 200 million people live in cities across the country. The urban population may increase by 150 million to 200 million in the next 15–20 years. What impact does this process have on the ecological environment? Is it economically affordable? How to solve the food problems?

Sustainable development of the regional socio-economic and ecological environments could be finally ensured by achieving the above regulation goals. There is an inherent correlation among the aspects as mentioned above. There are interactions of several elements in each aspect. These problems cannot be truly solved by only paying attention to any one or a few elements. For the effects of various elements on the regional system with the interrelationship between humans and nature, some can be quantified, while some cannot be quantified. However, they all have a quantitative impact on the development and evolution of the system. This requires combining qualitative and quantitative research and regulation, and the regulatory measures in the above seven aspects shall be correctly determined so that the system can develop towards a virtuous cycle. However, it is challenging to predict the development of the system quantitatively after adjusting one or even several aspects. Of course, systems science can play a unique role in this regard.

3. **Basic characteristics of the regional system with the interrelationship between humans and nature**

In order to solve the major problems during the evolution of the regional system with the interrelationship between humans and nature, the basic characteristics of the regional system shall be investigated first.

(1) It is a semi-open system from the perspective of the regional system and its external interrelationship. Since each region has its own unique natural foundation, the characteristics of socio-economic and administrative management will also have a strong impact on the humanities and nature of the region. Starting from the administrative center or economic center of a specific area, the policies, the strength of the economic field, and the infrastructure system that spread out will have a boundary, which is the dividing line where the applied force is obviously weakened. In practice, the characteristics of the semi-open system can be easily seen in the national network and local networks that reflect the direction and network of the traffic connection force. The capitals of most provinces have connecting networks and trunk lines with Beijing and large regional cities, while there is also a dense highway network and local shipping branch line network in the province. For other aspects of industrial structure, cultural education level, and development degree, there is also a clear "barrier" at the border location. The semi-open feature of the regional system shows that it is much more complicated to solve its problems compared with the closed system and the open system.

(2) It is an unstable system from the perspective of stable performance and changing conditions of the system. That is to say, the external or internal force can strengthen or accelerate the movement of the regional system, but the change of control causes the system to deviate from the marginal value of balance, which leads to the destruction of the unstable system. When people take measures that meet objective needs, the unstable system can last for a long time.

(3) It is a probabilistic system from the perspective of system development factors and trends, and the opposite is a deterministic system. The elements and association of the deterministic system and its input–output relationship with the environment are all known to people. People can build it, precisely calculate its trajectory, and predict its future development. Both machines and technical systems are deterministic systems. The solar system is also deterministic in terms of the motion and orbits of its planets and satellites. For the regional system with the interrelationship between humans and nature, when people set and control several indicators, the development results of the system do not show a determined state, or in other words, the development of the system is not linear. However, there are still sound laws for the development of the regional system, which are manifested as probability trends, amplitudes, etc. For the characteristic of the development of the regional system, probability, there are two points worthy of our attention: First, the research and regulation of regional systems are much more complicated than those of technical systems (such as a machine); second, the problems cannot be solved solely and even mainly by relying on mathematics and computers.

In the field of natural sciences, the flight track of satellites and the docking of spacecraft need to and can be accurate to several millionths of a second. In the field of social economy, such a precise forecast is impossible and is not required in most cases. For example, it is impossible to forecast the population of a certain area at a certain time point to the single digits, although there will be certain data when this time comes. However, population forecast is possible and meaningful, and there is no doubt about it.

To forecast and regulate the regional system in the context of interrelationship between humans and nature, it is required to consider more parameters and constraints, build more equations, and conduct multi-scheme simulations in general. However, the model is not necessarily as complicated as possible, and the equations are not necessarily as many as possible. In this regard, an example can be given. In the early 1950s, the former Federal Republic of Germany was required to forecast car ownership in the whole country in 1980, and the task was entrusted to several research units. After more than 20 years, the actual growth process line deviated greatly from the one they forecast but was very close to another forecast curve. A scholar proposed this curve by using an analogy. He used the actual process of car development in a state in the United States. He revised several data reflecting the socio-economic characteristics of the former Federal Republic of Germany, thus obtaining the curve.

4. **Research on the mechanism and decision-making**

1) Basic theory and major practical issues

The research on the regional system with the interrelationship between human and nature is to investigate the interrelationship between resources, environment and population, and economic growth from the perspective of the coordination of environment and development. In this field, major basic theoretical issues in three aspects can be included:

(1) Demands resources of the scale, structure, and level of socio-economic development, as well as the direction and intensity of the effects on the ecological environment within a certain geographical scope;
(2) Counteraction of regional resources and ecological environment to socio-economic development, and the stability of the human ecosystem;
(3) Targeted control of the regional system of population, economy, resources, and environment.

The major basic theoretical issues in these three aspects are directly related to the strategic goals of the mid-and long-term socio-economic development in China and a series of related major decision-making issues, such as:

(1) What is the long-term economic growth rate to be sought in China according to the requirements for per capita GNP growth as well as the resource and environmental conditions?

(2) The degree of resource assurance and the adaptability of the ecosystem in China at different time scales in the future under the condition of striving to maintain a long-term economic growth rate of about 8% or the growth rate of more than 6% for others, and taking into account the upgrading of industrial structure, technological progress, as well as appropriate increase in the share of ecological environment improvement in the redistribution of national income.
(3) The dependence on the international market on the premise of maintaining the sustainable use of water, land (mainly arable land), energy, major minerals, biological resources, etc., and the sustainable development of the ecological environment in China, and taking into account technological progress, as well as the appropriate economic aggregate and population scale in China at different time scales in the future.

Many practical issues can also be proposed according to different targets and variable indicators set. They all involve the comprehensive balance of the interaction of a large number of elements in the social economy, resources and environment. Every indicator related to socioeconomic development (such as economic growth rate, per capita GNP value, average life expectancy, population urbanization rate, industrial structure, consumption structure, technical level, import and export trade, etc.) and resources and ecological environment indicators (such as arable land, productive energy; quantity and quality of water and mineral resources, forest coverage rate, etc.).

2) Research on mechanism

In order to reveal the evolution law of the regional system, it is very necessary to research the mechanism of the interaction of system elements. The mechanism often refers to the impact of the element (including changes in direction, value, shape, etc.). Since the elements of the system are relatively complicated in general, when studying the mechanism, the system and some elements can be assumed, and only the interaction between two elements or a few elements is observed and analyzed. The result is the parameter of the system change. In the natural world, water surface evaporation is strictly related to temperature and air velocity. Mechanism relationships like this are very common. It is impossible to have such a strict correlation within the scope of the social economy and the regional system of humans and nature. However, there are still different degrees of correlation between many phenomena and elements.

The research on the mechanism of the element functions is generally to forecast and regulate the system. With regard to the regional system of humans and nature, the regulation requires research on both mechanisms and decision-making. The research on decision-making is to coordinate and balance the development, scale, and structure of the regional system and its elements on the basis of multi-faceted mechanism research, and propose the ways and measures for regulation. The goal is different from that of the research on mechanisms. One is vertical penetration, while the other is horizontal integration. The research on decision-making needs comprehensive analysis and judgment ability and requires rich knowledge and overall concepts.

The importance of the research on decision-making shall be emphasized, so as to make our achievements more constructive and operable. Several parameters are obtained from the research on mechanisms. However, decisions cannot be made on adjustments to relevant elements and the whole system by only referring to such parameters, because other relevant factors and the environment shall also be considered in the decision-making process.

3) Cultivation of the ability to solve major practical problems from many aspects

Scientists need relevant basic knowledge, basic methods, and basic theories to solve the problems concerning the regional system with the interrelationship between human and nature. In recent years, great progress has been made in the use of methods and computers, but some deviations have also appeared. People who research any subject always like to use words like "model", "system", and "optimization". People who research the "dissipative structure" of a certain thing do not understand the basic situation of the thing. People who research models do not investigate the basic structure and basic parameters of actual things summarized by the model. For some large-scale research, the "models" and "systems" are proposed based on only a few statistical manuals. Many issues of exploitation and development in China involve the relationships between nature, economy and society. The gains and losses are complicated. Some individuals simply deny the reality, and develop a scheme for "optimization" without going deep into the realities for specific research. Some people who research models and "systems" do not spend their energy researching and understanding the actual characteristics, structure, correlation, and parameters of the thing, but use a simple way to imagine the structure and parameters in reality, which are entered into the computer in a "serious" attitude. Experiments are made on the scheme, and the results are audited repeatedly.

What we want to advocate is the combination of practice and ideals, and the combination of qualitative and quantitative. Due to the above-mentioned trends that have emerged in recent years, it shall be emphasized to attach importance to solving major issues of exploitation and development in China, and performing investigation and research in practice.

3 China's Environmental Crisis in the Next Quarter of a Century and the Strategy to Get Rid of the Predicament

In the absence of major political or economic fluctuations, China's economic aggregate will double from now to 2000, starting to become one of the world's economic powers. As of 2020, China will basically achieve the development goals at the middle stage of industrialization, and the per capita GNP will reach 3000–5000 dollars, so that the degree of socioeconomic development and living standards will basically catch up with the moderately developed countries. Within the 1/4 century, before this goal is achieved, it will be difficult to reverse the trend of gradual deterioration of the environment in China. However, the country and society shall implement the correct strategies, which can alleviate environmental crisis and pressure; after entering the 2020s, China's ecological environment may be truly improved, and China will begin to enter a process of real coordination between economic development and the ecological environment.

1. **China faces the serious predicament of environmental pollution in the 1/4 century**

Within the 1/4 century from now on, it will be difficult to reverse the trend of gradual deterioration of the environment in China. This conclusion is drawn on the basis of analyzing China's national conditions and the actual process of the evolution of the interrelationship between development and the environment in the world (especially in developed countries). The specific reason is as follows.

1) China must maintain a moderately rapid economic growth for a long term

For the last more than ten years of reform and opening-up, China's economy has achieved tremendous development. Among them, the annual average growth rate of the gross national product was 9.0% in the 1980s. At the same time, China's environmental pollution and ecological damage were also very serious. From the 1950s to 1973, most developed countries in the world achieved sustained economic development. However, industrial pollution was becoming more and more prominent. At this stage, many developing countries were also committed to developing their own economies, causing a series of processes that endangered the ecological balance to be continuously spread. The United Nations Conference on the Human Environment was held in Stockholm in 1972, and the *Declaration of the United Nations Conference on the Human Environment* was passed. Individuals had gradually realized that the root causes of various pollution and ecological and environmental issues were population growth and economic development. In the 1980s, the environment and development issues were clearly raised internationally, and the United Nations convened the Rio de Janeiro Conference on Environment and Development on this theme.

In order to deal with the increasingly serious ecological and environmental issues facing human, some Western scholars believed that the economic growth of various countries should be inhibited, and put forward the proposal of "low growth" and "zero growth". However, China could not accept "low growth" (according to the view of Western economics, the growth rate of GNP is 2–3%), let alone "zero growth". Because the per capita GNP in China was only a few hundred dollars, the Chinese government was facing tremendous pressure to accelerate development, expand employment, and rapidly increase people's income and living standards. According to China's national conditions, only by accelerating development can it be more likely to gradually solve the ecological and environmental issues.

Japan and other emerging industrial countries and regions in Asia have achieved sustained and rapid economic growth in the past 40 years. By analyzing the living standards of the Chinese people, the existing economic foundation, and the scale of foreign capital introduction, it is not only necessary but also possible for China to maintain an economic growth rate of 6–8% for the long term (it can be assumed to be 8% before 2000, and 6% from 2000 to 2020). If the national economy continues to develop at this speed, the per capita GNP will be 3000–5000 dollars by around 2020. After that, China's economic growth will gradually slow down, entering the later stage of industrialization. Accordingly, the pressure of economic growth on the environment will also be reduced. At the early and middle stages of industrialization, industries with rapid economic growth, high resource consumption, and large waste emissions will inevitably occupy the dominant position.

Iron and steel, petrochemical and marine chemicals, building materials, thermal power, paper making, etc. are all industries with high resource consumption and large waste emissions, that is to say, they belong to the category of "three major sectors and six enterprises" with "high pollution". However, due to the late start of industrialization in China, domestic demands are under great pressure. These industries have to increase their scales at a faster speed during a quite long historical process.

Take the steel industry as an example. From the late 1970s to the late 1980s, "the proportion of steel output in developing countries increased from 25 to 32.8%, and this trend will continue until the beginning of the next century" [2]. According to the research results of the World Steel Association on steel consumption in various countries, when a country develops from a low level of per capita GNP to a medium level of per capita GNP, steel consumption will rise constantly. After becoming a developed country, steel consumption will stagnate and even decline slightly: When the per capita GNP is less than 300 dollars, the growth rate of steel consumption is not too high; when the per capita GNP reaches 300–1500 dollars, the steel consumption almost rises linearly; when the per capita GNP reaches 1500–2500 dollars, the steel consumption reaches the peak; when the per capita GNP is more than 2500 dollars, the steel consumption declines significantly [3] (Figs. 4 and 5). As of 2000, China's steel output will reach about 90 million tons, and the per capita GNP will be 1000 to 1300 dollars. According to the above-mentioned laws, China must continuously increase steel output, expand the existing large and medium-sized steel plants, and build several extra-large steel bases until around 2015.

The petrochemical industry can provide a large number of raw materials and consumer goods for the national economy, and is also a sector that consumes a lot of water and discharges a lot of pollutants. At present, China has nearly 20 large petrochemical bases (the emissions of water consumption, "wastewater" and "waste gas" of each base are basically equivalent to those of a medium-sized city). The ethylene production capacity of those established and under construction is less than three million tons (the ethylene production capacity in the early 1980s was 15 million tons in the United States and five to six million tons in Japan), which is far

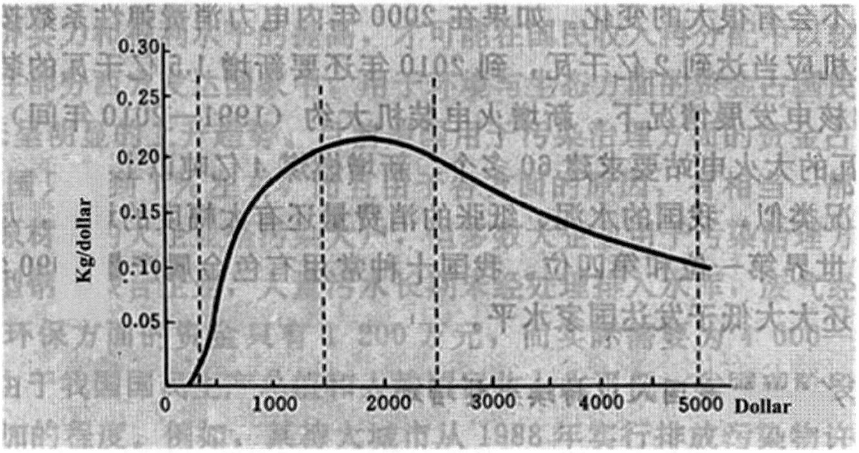

Fig. 4 Curve of steel demand intensity (The longitudinal coordinate represents the steel consumption of 1 dollar of GNP)

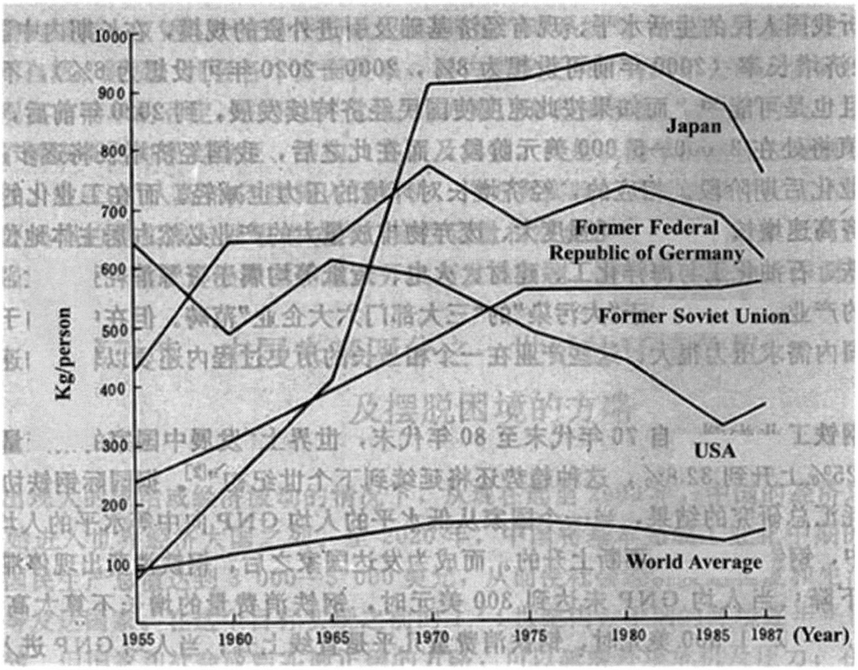

Fig. 5 Changes in per capita steel output in the world and major countries

from meeting the needs of more than 1.2 billion people; in addition to the large-scale expansion of the existing bases, more than ten new bases will be built by 2020.

2) The level of electrification in China is not high

Although China's installed capacity of electric power has already ranked fourth in the world, the electric power industry shall still be developed on a large scale in the future. Moreover, the status that coal and electricity account for 70% of the resources that are used for electricity generation is still a harsh truth to face at this point. Suppose the elasticity coefficient of electricity consumption is calculated as 1.0 in 2000. In that case, the installed capacity of electric power in the whole country shall reach 200 million kilowatts by 2000, and the installed capacity will also increase by 150 million kilowatts as of 2010. Considering the strengthening of hydropower and nuclear power development, the installed capacity of thermal power will increase by about 130–150 million kilowatts (from 1991 to 2010). It is required to build more than 60 large thermal power stations with an installed capacity of 2.4 million, and add more than 400 million tons of coal.

Similar to the above situation, there is still room for substantial growth in the consumption of cement and paper in China, although the output of these two products in China has already ranked first and fourth in the world respectively. The output of ten commonly used nonferrous metals in China reached 2.39 million tons in 1990, far below developed countries' level by per capita availability.

3) The population in China will continue to increase substantially in the next 1/4 century

The total population in China will reach its peak from 2010 to 2020; large-scale population urbanization will change the consumption structure to a great extent, and increase the consumption of energy and water sources. Wastes will also increase accordantly.

According to the current population policies in China, China's population will increase by more than 10 million every year, and will reach its peak from 2010 to 2020, exceeding 1.5 billion. The domestic energy consumption of the urban population in China is 3–5 times that of the rural population, and the consumption of fresh water is much larger. Moreover, urban water consumption and water intake are concentrated, which often has a great impact on the balance of water resources in local regions. If the population movement is not controlled, about 500 million people will live in urban areas as of 2010. Among them, 200 million people may live in the urban–rural fringe zones of large and medium-sized cities. The area of the urban–rural fringe zones will continue to expand. Coupled with crowded space, poor infrastructure, and industrial and urban pollution, prominent ecological and social problems will occur in such regions (Fig. 6). After 2020, China's total population and urbanization rate will slowly decline, and the increase in pressure on the ecological environment will also slow down.

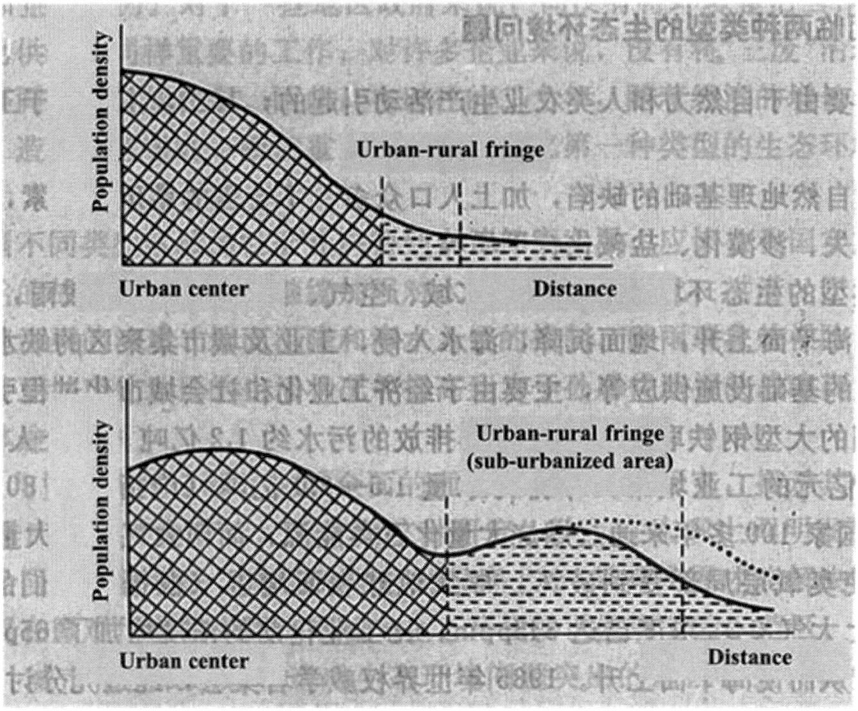

Fig. 6 Development of sub-urbanization of the urban–rural fringe zones in big cities in China

4) China's per capita national income is still very low, and it is impossible to use a large amount of funds for environmental governance and ecological construction for quite a long time.

Only with the improvement of economic strength and welfare standards can it be possible to use a larger share in the redistribution of national income for ecological environment protection. In some Western developed countries, the proportion of funds for environmental and ecological protection in the gross national product has been on the rise since the 1980s. At present, the funds used for pollution control account for 2–3% of the gross national product in the United States, but only about 1% in China. Due to various reasons, a considerable part of the funds have not played a role. Large enterprises that produce raw materials are major polluters, but most of them have a serious shortage of funds for pollution control. For example, in a large iron and steel complex, a large amount of sewage is discharged into the reservoir without treatment for a long term, and the exhaust gas often covers the whole city. However, the annual fund for environmental protection is only 12 million yuan, while 40–50 million yuan is required actually [4]. From an overall analysis, due to the low level of gross national product and per capita national income in China, the pollution control capacity at the current stage cannot keep up with the increase in pollution.

For example, a megalopolis implemented the pollutant discharge permit system since 1988. As of March 1993, only 82 pollutant (waste water) treatment projects were completed. The investment was 51.5 million yuan, which could reduce pollutants by 74,000 tons/year. However, there were 577 enterprises that applied for the permit in the city from 1988 to 1989, with a pollutant discharge volume of 348,000 tons per year, and a fund of 300 million yuan is required for treatment. As the economic growth stays at above 8%, dozens of large and medium-sized enterprises are put into production every year.

In the next 10–20 years, China's industrialization will continue to spread to rural areas. After a period of accumulation, rural industrial pollution will gradually reveal its severity and impact. From 1978 to 1990, the number of rural industries nationwide increased from 790,000 to 7,220,000, and the output value increased from 42.3 billion yuan to 605 billion yuan, with an annual increase of 25%, and most of them did not have control measures. The widespread pollution caused by rural industrialization has become quite prominent in the Yangtze River Delta and the Pearl River Delta due to the high degree of industrial intensiveness. In other vast areas of the country, the problem of industrial pollution in rural areas will become more and more serious in the next 10–20 years, then be possible to enter the stage of comprehensive control.

2. **China faces two types of ecological and environmental issues**

First, it is mainly caused by natural forces and agricultural production activities of humans; second, it is mainly caused by the process of industrialization and urbanization.

The shortcomings of China's natural geographical foundation, coupled with factors such as a large population and unreasonable cultivation, have caused quite serious soil erosion, desertification, salinization, drought, etc.

Other types of ecological and environmental issues, such as water, air, and soil pollution, acid rain, greenhouse effect caused by CO_2 emissions, sea level rise, ground subsidence, seawater intrusion, water shortage in industrial and urban agglomeration areas, space congestion, lack of basic infrastructure supply, etc., are mainly caused by the process of economic industrialization and social urbanization. A large iron and steel complex with an annual output of 3 million tons of steel discharges about 120 million tons of sewage each year; an industrial city with a population of 500,000 and an industrial output value of 6–8 billion yuan has an annual sewage discharge of 150–200 million tons (all refer to the general situation of China in the 1980s). Industrialized countries have emitted a large amount of CO_2 into the atmosphere by burning a large amount of fossil energy for more than 100 years. The accumulation of a large amount of CO_2 destroys the ozone layer locally, causing the earth's surface to warm up (according to the observations and calculations of scholars, the atmospheric CO_2 concentration on the earth reached 345 ppm in 1986, increased by 65 ppm compared with that before industrialization, with an annual growth rate of about 0.4%), which then raises the sea level. In a gathering of authoritative scholars in the world in 1985, it was believed through full discussion that if CO_2 doubles, the average global surface temperature will increase by 1.5–4.5 °C, and the precipitation will increase by about 7% [5]. The total consumption of primary energy was 980 million tons of standard coal in 1990, accounting for 9.1% of the total consumption in the world; among them, the coal consumption was 1.055 billion tons, accounting for about 18% of the total coal consumption in the world. It could be seen that China is indeed a country with large emissions of CO_2 in the world at the current stage due to the coal-based energy structure, backward utilization mode (a large amount of coal is used for direct combustion), and lack of advanced CO_2 recovery system, etc.

3. **The strategy of rectifying two types of ecological and environmental issues at the same time is implemented**

Over the past 40 years since the founding of the People's Republic of China, society has reached a consensus on the ecological and environmental impact caused by natural forces and agricultural production activities of humans. The state and local governments use huge funds, manpower, and material resources for the control, prevention, scientific research, and experiments of these ecological and environmental disasters, which slows down the deterioration of the above ecological issues to different degrees. According to the predictions by interested parties, desertification in arid and semi-arid regions in China was on the rise from the 1950s to the 1970s. However, since many effective measures have been taken, some areas have begun to reverse, and it is preliminarily estimated that about 10% of desertified areas have been controlled. According to the estimation of the average annual reversal rate of several typical areas, after measures are taken, the desertified land in China will be reduced by 26,000 km^2 as of 2000 [6]. In the early days of the founding of the People's Republic of China, the area of soil erosion was about 1.5 million square kilometers, accounting for 1/6 of the country's area. As of the mid-1980s, 460,000 km^2 of areas with soil erosion had been controlled. Of course, due to population growth and increasing reclamation of slopes and land prone to soil erosion, there is newly increased soil erosion in some areas, and the area of soil erosion is still about 1.3 million square kilometers [7]. In order to prevent serious soil erosion and desertification, the state and local governments have insisted on afforestation for the long term. Among them, significant achievements have been made in the construction of "Three North" shelterbelts.

China has paid attention to the pollution and other ecological and environmental issues caused by industrialization and urbanization since the early 1970s, which only received widespread attention from society in the mid to late 1980s. Even so, national consciousness has not been achieved. It shall be admitted that the speed at which such pollution increased during the rapid economic development in the 1980s was surprising. Public opinions, local governments, and specific companies have expressed the will to control pollution. In fact, due to financial, institutional, technical, equipment, and management reasons, universal and standardized government behaviors and enterprise behaviors have not been formed for the regulation of such ecological environment phenomena. For some regional governments, environmental improvement has not been regarded as a task as equally important as employment settlement, income increase, food supply, etc.; for many enterprises, the control of "three wastes" has not been regarded as a task as equally important as product manufacturing and product sales. Therefore, in the past two decades, the scale of environmental pollution has also expanded with economic growth, and the harm caused has been increasing. This speed is much faster than that of the first type of ecological environment deterioration process.

In order to achieve sustainable and healthy development of the social economy in different types of regions in China, the strategic policy of simultaneously rectifying two types of ecological and environmental issues shall be implemented in China, that is, continue to strengthen the comprehensive control of soil erosion, desertification, flood, drought, etc., and raise the following to the strategic height of benefiting future generations, such as control of environmental pollution, control of ground subsidence and seawater intrusion, eradication of acid rain, improvement of living conditions in sub-urbanized areas by opening up new places of residence, etc. The specific ways to implement this strategic policy are as follows:

(1) Strengthen comprehensive theoretical and practical research on the interrelationships between humans, resources, and the environment, reveals the mechanism of interaction among them, and conduct extensive social publicity on this. In order to carry out this kind of research, in addition to clarifying the law that development causes environmental pollution, environmental degradation will hinder economic growth (on a larger time scale), and only development can solve environmental problems, it is important to establish experimental demonstration areas for the coordinated development of economic growth and the ecological environment on a micro-scale. Such demonstration areas can first be established in areas where industrialization and urbanization are developing rapidly and prominent ecological and environmental issues are caused. Its task is to collect and observe natural, socio-economic continuously, and waste discharge data in the demonstration area, establish corresponding information systems and models, reveal the mechanism of interaction between the two elements of economic growth and ecological environment, and propose several models of coordinated development of economic growth and environmental protection for the government to choose. During specific implementation, such experimental demonstration areas are established in the Pearl River Delta, the Yangtze River Delta, and the Bohai Sea Region. They are different from each other in terms of natural foundation, industrial structure, and level.
(2) Increase the percentage of national income used for environmental governance. At present, China's GNP only ranks eighth in the world (1990), and the per capita GNP is only about 1% of that of the United States and Germany. However, our industrial waste emissions are not much less than those of the United States (mainly because of our large production scale of energy and raw materials). Due to the weak economic strength and the low-income level of the people, it is impossible to allocate a lot of funds for environmental protection and ecological construction. However, with the rapid economic development, the government and enterprises may use more funds to improve the environment and control industrial pollution. As of 2000 and 2010, China's GNP may reach 3.82 trillion and 6.84 trillion (measured in 1990). Suppose 2% is used for the investment

in the ecological environment. In that case, the amount will be 76 billion and RMB 137 billion, respectively, together with the funds for enterprise governance (the total amount may exceed this number), the impact of increasingly serious industrial pollution can be mitigated.

(3) According to the characteristics of the natural environment and the environmental load-carrying capacity of various regions, macro-control is implemented for the national industrial layout on a medium and long-term scale. Northwest China and North China are dry and in a shortage of water, and have a low capability for bearing industrial pollutants (wastewater) and poor dilution capacity. Most of the pollutants infiltrate the ground in dense residential areas, thus polluting groundwater. According to the combination of dilution and self-purification capacity and other natural elements, the large plants of energy (referring to thermal power) and raw material industries in China shall be constructed and expanded in the eastern coastal zone and along the Yangtze River. On the micro and meso scales, a reasonable industrial layout also has a certain effect on reducing the hazards of pollution.

For the sub-urbanized areas of large and middle-sized cities, measures such as planning and construction of infrastructure shall be taken to improve the environmental conditions and living conditions.

(4) Clarify the responsibility for pollution control, and reform the product cost accounting system. At present, there is no clear scope of responsibility for pollutant control. If industrial emission of the "three wastes" does not conform to the standard, a fine will be imposed only. However, the urban environmental protection department and the urban construction department cannot build many control equipment for various pollutants. The fundamental method shall be that the factory handles the "three wastes" by itself until the emission standards are met (not without harm), and the urban construction department and the environmental protection department build unified urban waste treatment facilities. For other types of environmental problems caused by industrialization and urbanization, the responsible person shall also be clarified, and the corresponding regulations shall be published, so as to truly achieve that "those who cause pollution shall be responsible for control". If an enterprise causes harm to society due to waste discharge. In that case, all necessary governance funds and other losses shall be included in the production cost of the product, and the government shall extract part of the funds as a special environmental protection fund for the construction of environmental protection projects.

(5) Save resources, and establish a resource-saving social and economic system. Widespread industrial pollution not only causes hazards to the environment in which humans live and work but also wastes many precious resources. For example, improve the reuse rate of water resources, which enables enterprises to discharge little or no "wastewater"; the recover SO_2 tail gas from power plants, which can reduce the harm of acid rain and increase the output of sulfuric acid. The waste emissions will be inevitably reduced by establishing an industrial structure with low resource consumption intensity and developing process technologies that save and reuse resources.

4 Establish a Resource-Saving Social and Economic System[2]

1. One of the basic national policies

The issues of resources and the environment have different connotations and are interrelated. Certain environmental and ecological problems can be gradually solved or tend to be alleviated through the economical and intensive use of resources.

Through the conservation of freshwater resources, ecological water can be increased, and the ecological cycle can be improved; seawater intrusion in coastal areas can be alleviated. The discharge of wastewater can be reduced by saving industrial water, which facilitates the recovery of harmful substances in the wastewater.

Through energy conservation, the emission of harmful substances such as SO_2 and CO_2 can be reduced, and air pollution can be reduced; the pollution during the mining, transportation and processing of energy such as coal and petroleum can be reduced.

Through the conservation of land, especially the arable land, the pressure on the population and food supply can be reduced to a certain degree, and unreasonable reclamation and disasters such as desertification and soil erosion can be reduced. The conservation of land also means that more space for human activities and living can be provided for the agglomeration areas in industrial cities, and individuals' living conditions could be improved.

Establish a resource-saving consumption structure that is different from the Western society, including food structure, vehicle, energy consumption for living, and living standards that are suitable for China's characteristics. This will fundamentally determine the resources and environmental conditions in China, and have a fundamental significance in aggravating or mitigating resource scarcity and environmental crises in China in the future. Therefore, it is also a fundamental issue of national policy, which is related to the sustainable development of various regions in China and the sustainable survival of all nationalities.

That is to say, the establishment of a resource-saving national economic system in one country and one region will help the sustainable use of resources, and is also an important foundation for promoting a virtuous material cycle in the ecological environment. Therefore, some developed countries have established a national economic system that saves such resources in a targeted way according to their own resource structure and utilization conditions.

From the early 1990s to 2000, the total industrial and agricultural output value of China will double, and China will become one of the world's economic powers. Accordingly, the energy consumption will increase by 50–60% (that is, reaching

[2] The "regions" referred to in this section shall include countries and large administrative regions (provinces, autonomous regions and cities in China). In most micro-regions with small area, it is generally impossible to form an "economic system". However, certain cities and smaller regions under special natural and economic conditions can also be taken as "regions" for establishing a resource-saving national economic system referred to in this article.

1.5 billion tons of standard coal), the land occupation will increase by about 50 million to 60 million mu, and the per capita arable land in some regions in Jiangsu, Fujian, Zhejiang, Guangdong, and other provinces will fall below 0.8–0.6 mu; the annual water consumption will increase by about 200 billion cubic meters, and the water resources in North China will be fully utilized; the annual demands for grain (only according to the current consumption level) shall increase by 60 million to 70 million tons; the turnover of passenger and cargo transportation will increase by 80% and 60% respectively, reaching 1010 billion person-kilometers and 4190 billion ton-kilometers each year. The annual sewage discharge may reach more than 50 billion tons. After passing the first 1/4 of the next century, if we continue to develop in accordance with the social and economic model based on the massive consumption of natural resources, the degree of resource assurance will be extremely reduced, and long-term food assurance will become a serious problem. The space will be crowded and chaotic, and the environment will be unbearable. In turn, it will be bound to hinder sustained economic development seriously. That is to say, we have reason to conclude that: If human beings cannot develop and utilize resources in a controlled manner, and even destroy the natural foundation on which human society depends for survival and development, the consequences will be catastrophic.

China is facing an extremely severe situation in terms of resource assurance and environmental conditions. In addition to the impact of general factors in the process of industrialization, the main reason lies in China's special national conditions. This is: Due to the large population and the low level of economic and cultural development, the pressure of development is high; due to the rapid economic growth and the low level of China's social structure and industrial structure, the consumption of resources is large, and ecological damage and pollution are serious. The natural conditions of about 1/3 of China's territories are harsh, which is unsuitable for the daily social-economic activities of humans. In some areas, water and land resources as well as water and mineral resources are seriously unbalanced. All of these have aggravated the incoordination between population, nature, and the environment in China (Fig. 7).

In the past 40 years, the huge socio-economic development in China was based on the extensive development and utilization of natural resources such as arable land, pasture, water, and minerals from the perspective of interrelationships between human and nature. Some basic resources have been greatly reduced, and some resources are running out. This tells us that this situation can no longer continue for a long term. The annual import of grain, wood, metallic mineral ore and basic raw materials is gradually expanding in China. Among them, 13.72 million tons of grain were imported in 1990, accounting for 3% of the total domestic grain output; 1.13 million tons of sugar were imported, accounting for 19.4% of the domestic output; 4.15 million cubic meters of logs were imported, accounting for 7.5% of the domestic output; 342,000 tons of pulp and 376,000 tons of textile fibers were imported, accounting for 23% of the domestic output; 14.34 million tons of iron ore were imported, accounting for more than 10% of the total domestic output; 4.19 million tons of steel material were imported, accounting for 8.1% of the domestic output. The imports of the above-mentioned basic living materials and production materials have shown a rapid increase in the past few years. For example, the increase

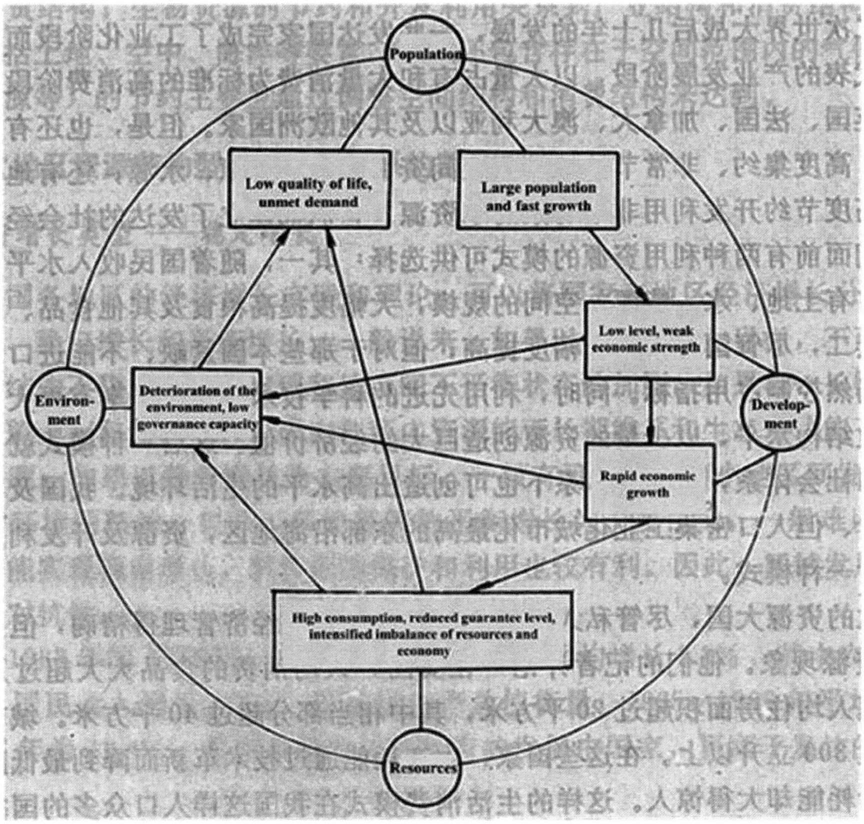

Fig. 7 Non-benign interrelationships between population, economy and environment in China

in imports in 1990 compared to 1985 is as follows: 7.7 million tons of grain and 4.28 million tons of iron ore. We could assume that when the population and economic aggregate continue to grow, and the assurance degree of domestic resources and space is lower and lower, it is required to import the above-mentioned living materials and production materials several times or even dozens of times compared with the current. How can the world market be able to afford it? How can we afford it? How can China's huge social and economic system be maintained for a long time? The answer can only be as follows: determine a model with Chinese characteristics and uses resources to achieve sustainable social and economic development according to China's national conditions.

2. **Mode selection and basic framework**

1) Two modes to choose from

Due to the extremely serious shortage of resources mentioned above in China, there are two modes for us to choose from to use resources to develop the economy.

Through decades of development after World War II, some developed countries have completed the industrialization stage, and entered the industrial development stage represented by high-tech industries and the high-consumption stage based on large-scale possession and large-scale consumption, mainly the United States, Germany, Britain, France, Canada, Australia, and other European countries. However, there are also a few industrialized countries, such as Japan, using space resources, energy resources, and water resources in a highly intensive and economical way; and Israel, which is located in the arid zone, develops and uses very limited water resources in a highly economical way. They all have established a developed social and economic system.

There are two modes for us to choose from to use resources. First, as the national income level increases, increase the scale of land, water, energy, and space per capita accordingly, and substantially improve the consumption of grain, other food, and basic raw materials; second, despite the substantial increase in national income, still control the occupation indicators of those resources that are in scarce need, cannot be imported or are too expensive to import. At the same time, use advanced science and technology to enable people to enjoy greater spatial mobility; improve the level of industrial structure, and create huge economic value with a limited amount of resources. This latter model is a resource-saving social and economic system, under which a high-level living environment can also be created. In China and various regions, especially the densely populated eastern coastal areas with few resources and the highest industrialization and urbanization, the resources and their development and utilization conditions determine that only the latter model can be accepted.

In some developed countries with large resources, although private entrepreneurs may manage their economy accurately, there is a huge waste of resources in the whole society. Their reporters commented that: In the United States, the per capita consumption of food far exceeds the physiological needs of people. The per capita housing area of many families exceeds 20 m^2, of which a considerable part exceeds 40 m^2. The daily water consumption per capita in cities is generally up to 300 L or more. In these countries, the energy consumption of production is reduced to a minimum through technological innovation, but the energy consumption of social and private life is astonishing. Such a living consumption mode is really impossible to work in a country with a large population, such as China.

2) Overall framework of resource-saving social and economic system

According to the serious incoordination between population, resources and environment in China and various regions, the specific goals of building a resource-saving social and economic system are as follows: determine the national production industry group and corresponding scale according to the resources and resource structure advantages and disadvantages of the country and various regions; build an accumulated consumption system based on long-term stable economic growth; improve the employment structure, and vigorously develop the tertiary industry; strengthen transportation, post and telecommunications, and increase spatial mobility; adjust the spatial structure of the social economy, and attract more resources and space into the socio-economic cycle; focus on the development of medium-sized cities; build the management, monitoring, engineering planning and design systems for resources and environmental protection; improve the population quality, and enhance the ability to transform, utilize and protect nature.

To implement the above specific goals, all regions are finally required to establish an industrial structure, agricultural planting structure, scale structure of urban residential areas, technical structure, foreign trade structure, consumption structure, and socio-economic spatial structure that uses land, water, energy, biology, minerals, and other natural resources in an intensive and economical way, so as to achieve the sustainable development of the national economy, gradually improve the living quality of people in China on a large time scale and various spatial scales, and coordinate the interrelationship between human and nature.

As all "systems" are expressed in a certain structural state, the structural state of the resource-saving national socio-economic system shall be described, analyzed and forecasted.

Land and its economic use are the fundamental elements for establishing a resource-saving socio-economic system. Its economical and intensive use is the foundation for establishing a resource-saving planting structure, residential area structure, consumption structure, and socio-economic spatial structure. The conservation of water and energy is mainly reflected in the aspects of industrial structure (the main content is the leading department of the national economy, industry selection, and its departmental relations, etc.), planting structure, technical structure, foreign trade structure, etc. The main direction of saving the consumption of mineral resources is to adjust the industrial structure and foreign trade structure; the conservation, development, and utilization of biological resources are related to the industrial structure and consumption structure. The conservation of space resources (including not only elements such as land, air and ocean but also climate, topography, soil, underground resources, etc. that exist in space) shall be mainly achieved by adjusting the spatial structure and consumption structure.

3. **Basic approach to establishing a regional resource-saving socio-economic system**

1) Type of economic growth—steady growth

According to the practices and theories of economic growth in various countries and regions, the economic growth of countries and regions can be divided into three types, that is, rapid growth, steady growth, and balanced growth. Generally speaking, if rapid growth is taken as the goal, it will cause the internal structure of the economic system unstable and the unbalanced state among regions aggravated. If rapid growth is sought over a long period of time, it would be a question for general countries and regions whether domestic and regional resources can be maintained for a long time and whether the ecological environment can be maintained in a benign state; if steady growth is taken as the main goal, it is possible to gradually reduce the imbalance of regional development and better ensure the environmental improvement; if balanced growth among regions is taken as the main goal, it is generally difficult to achieve rapid growth, but there is a possibility to achieve steady growth, which is also beneficial to the protection and utilization of resources. Therefore, the three main goals of regional development are confrontational.

Within the 30 years from 1953 to 1983, the annual average growth rate of the total industrial and agricultural output value in China was 8.2%, 4.0% for agriculture, 10.1% for industry, and 6.2% for national income. If measured in the gross national product, the annual average growth rate was 6.4% from 1965 to 1980 and 10.4% from 1980 to 1987, which was significantly higher than that of almost all developing countries and even higher than that of developed industrialized countries (Table 1).

According to the goal of doubling the total industrial and agricultural output value from 1980 to 2000, the average annual growth rate of GDP will be 6.6%. In fact, the annual average growth rate of GNP was 9.0% in the 1980s, and 13% in 1992 and 1993, which clearly exceeded the long-term development momentum of most countries. Therefore, it shall be considered rapid growth. What model will China's economic growth follow in the next century? If China continues to maintain a growth rate of 10% or more than 8% without major adjustments to the industrial structure, it would lead to high-intensity utilization of resources and serious ecological impact. It is specifically manifested as follows:

(1) The total industrial and agricultural output value in 2000 will be quadrupled compared with that in 1980, and the demand for primary energy was 1.7 billion tons (high scheme) and 150 million tons of standard coal [7]; according to calculations by the World Bank, the minimum energy consumption will reach 1.27 billion tons when the acceptable adjustment is made to the industrial structure [8]. In 2020, the national industrial and agricultural output value is equivalent to four times that in 1980, reaching 11.2 trillion yuan (at the constant price in 1980), which is equivalent to 15.8 times and 5.5 times that in 1980 and 1990 (actual value) respectively. To ensure such a huge economic aggregate, the primary energy shall reach about 3.5 billion tons of standard coal. If there is no major breakthrough in oil–gas exploration (no such prospect can be seen now), it would have to expand the amount of coal mining several times, even if the length and capacity of the railway network have doubled compared with the current, it would not be able to adapt to it. There are also land occupation and environmental problems caused by a large amount of coal mining. The main way out can only be energy saving.

(2) An area of 180 million to 200 million mu shall be added for the production and urban–rural construction. Based on the total population of 1.5 billion in 2020, the per capita arable land will be reduced from the current actual number of 1.7 mu to 1.2 mu. The per capita arable land in some provinces and regions in the eastern and central regions will drop to less than 0.6 mu and 0.8 mu respectively. Due to the land occupation for cash crops, vegetables, and fruits, the planting area of grain has to continue to shrink. To ensure that the per capita consumption of grain can reach 450 kg (388 kg in 1990), the total grain output shall reach 675 million tons. Compared with that in 1990, the domestic grain output will increase by 229 million tons, which is equivalent to 1/2 of the current national grain output. This amount is equivalent to 16.7 times that of China's

Table 1 Annual growth rate (%) of gross domestic product (GDP) of China and other types of countries

Country type	Gross domestic product (GDP) growth rate		Where							
			Agriculture		Industry		Manufacturing industry		Service industry, etc	
	1965–80	1980–87	1965–80	1980–87	1965–80	1980–87	1965–80	1980–87	1965–80	1980–87
China	6.4	10.4	3.0	7.4	10.0	13.2	9.5	12.6	7.0	7.6
Low-income countries	5.4	6.1	2.7	4.0	8.7	8.6	8.1	10.3	5.7	5.1
Middle-income countries	6.2	2.8	3.4	2.5	6.0	2.9	8.1	3.0	7.3	3.1
Upper-middle income countries	6.7	3.4	3.4	2.6	5.8	3.7	9.2	4.7	8.2	3.8
High-income countries	3.7	2.6	0.8	2.8	3.2	2.3	3.6	3.3	3.7	2.7

Source World Bank, *World Development Report 1989*, Pages 166 to 167

grain import in 1990. It is impossible for the world food market to provide such a huge amount of food. The only way out is to increase output and control consumption.

(3) If it grows at a rate of 8%, the total economic scale will double in less than 10 years. As of 2020, the total amount of water consumption for the national economy (increasing at a rate of 2.5%, the factors of industrial structure adjustment have been considered) will increase by 30%. If no major adjustment is made to the social-economic structure progressively, the steel consumption will increase by 70%. The iron mines with better mining conditions will be exhausted. Most of the mine resources of ten commonly used nonferrous metals will enter the peak period of mining, and some will enter the decline period.

2) Investment scale and investment direction—accumulation and consumption

Since the 1980s, China has implemented a series of reforms in various management systems and investment systems. As the new control mechanism has not been fully established, the investment scale has been repeatedly expanded. The situation of high accumulation rate in the first 30 years after the founding of the People's Republic of China has not changed significantly. This is an important reason why the growth rate of the national economy is still too high these years. When summing up the experiences and lessons of the previous 30 years of national economic development in the early 1980s, it was concluded that the excessive accumulation rate was an important reason for the proportional imbalance of the national economy. During the "First Five-year Plan" period and the adjustment period from 1963 to 1965, the accumulation rate was lower than 30%. In the 1980s, the accumulation rate was generally higher than 30%, which was slightly higher than that in the later period of the "Great Leap Forward" and the "Great Cultural Revolution" (Table 2). The investment in fixed assets of the whole society was 254.319 billion yuan in 1985 and 442.929 billion yuan in 1990. It almost doubled in five years, with an average annual growth rate of 11.7%.

Judging from the redistribution of national income, capital construction investment, and investment in main industries, the actual investment direction in these years is still concentrated in heavy industry. In 1990, heavy industry investment accounted for 48.0% of total investment in fixed assets, while light industry investment accounted for only 5.0–8.8%. The investment direction of the national and local governments has a significant impact on the industrial structure and the development, utilization, and consumption intensity of energy, raw materials, transportation volume, and land resources. A large number of funds are invested in heavy industry, which has accelerated the consumption of resources in China, and also caused a waste of human resources such as slow development of the tertiary industry and insufficient employment.

3) Industrial structure with low resource consumption intensity

The industrial structure of China has a low level and slow conversion, which are manifested as the characteristics in the early stages of industrialization in many developed

Table 2 Proportions of consumption and accumulation in national income

Period	Amount used in national income (100 million yuan)	Consumption rate (%)	Accumulation rate (%)
"First Five-year Plan"	4122	75.8	24.2
"Second Five-year Plan"	5616	69.2	30.8
1963–1965	3578	77.3	22.7
"Third Five-year Plan"	7785	73.7	26.3
"Fourth Five-year Plan"	11,054	67.0	33.0
"Fifth Five-year Plan"	15,024	66.8	33.2
"Sixth Five-year Plan"	26,182	68.7	31.3
"Seventh Five-year Plan"	58,411	65.7	34.3

Source China Statistical Yearbook, 1991

countries. Such an industrial structure is the root cause of high resource consumption intensity, unreasonable utilization, and large emissions of "three wastes" in China. It is specifically manifested as follows:

(1) Industry accounts for a large proportion of the national economy. In the past 40 years, industrial investment has accounted for more than 50% of total investment, and the tertiary industry has been seriously underdeveloped. According to the analysis of the World Bank, the composition of China's gross national product is as follows: In 1965, primary industries such as agriculture accounted for 39%, industry accounted for 38%, and tertiary industries such as service industry accounted for 23%; in 1987, agriculture accounted for 31%, the proportion of industry rose to 49%, and the proportion of tertiary industry was reduced to 20%; the composition of the United States in the same year was: 2% for agriculture, 30% for industry and 68% for the tertiary industry. For a long time, the total industrial and agricultural output value has been taken as the most important indicator to measure economic development, which has led to a continuous increase in the proportion of industry and poor development of commerce, service, finance, and other industries. Compared with the tertiary industry, the energy consumption, water consumption, land occupation as well as traffic volume and discharge capacity per unit of industrial output value are several times larger or even more.
(2) Among the industries, heavy industries and basic raw material industries account for a large proportion. Most of the heavy industries are capital-intensive and resource-intensive, and consume large amounts of energy and water. Compared

with the light industry, the number of employed people is small. The situation where heavy industry accounts for a large proportion of resource-saving socio-economic system must be changed to build a resource-saving socio-economic system. From 1982 to 1990, heavy industry accounted for more than 56% of total industrial output value. According to the analysis, the labor absorbed by the same capital per unit output value of the heavy industry is only 1/3 of that of the light industry. In 1989, the national industrial energy consumption was 662.91 million tons of standard coal, of which light industry only accounted for 21%. That is to say, the energy consumption of the light industry was only 1/4 of that of heavy industry in terms of unit national income. According to preliminary calculations, both the energy consumption and the pollutant discharge volume of heavy industry have reached about 80% of the total industrial volume. In the industry, basic raw material industries such as metallurgy, heavy chemical industry, building materials, paper making, etc. account for a large proportion. At present, China has become a major producer of these industries in the world. If we develop according to the curve of growth and decline in the output and consumption of raw materials (steel, basic chemical products, cement, etc.) per capita in developed industrial countries, the peak demand for steel output in China shall reach 500 million to 600 million tons, and other basic raw material products shall also be several times larger than the current scale. It is extremely difficult for the financial resources of China to bear, and cannot be supported by national resources (see Fig. 4).

4) Population urbanization and urban scale structure

With the development of industrialization, China's social urbanization (including population urbanization) is also accelerating. In 1949, the non-agricultural population in China's urban areas accounted for 11% of the total population. In 1990, the proportion reached 20.61%, an increase of 9%. The non-agricultural population quadrupled from 60 to 236 million. If measured by the urban and rural population, the proportion of the urban population in the total population increased from 18% in 1965 to 38%. From 1980 to 1987, with the implementation of the policy of "reform and opening up, reinvigorating the economy" and the cancellation of quantitative tickets for grain, the original household registration system almost lost its function. A large number of rural populations flowed into cities, and the average annual growth rate of the urban population reached 11%. During the eight years from 1982 to 1990, the number of cities increased from 239 to 476, and the total urban population increased from 145 to 332 million, which is more than doubled. Such an urbanization rate was extremely rare in the world (Fig. 8).

Cities are the product of industrialization and modernization development, an important manifestation of modern civilization, a concentration of national economy, technology, science, education, and art, and the core and commander of the exploitation and development of vast regions. At the same time, cities are also the places where resource consumption and environmental damages are the most concentrated. According to estimates, the amount of energy and other resources consumed per capita (taking into account industrial factors) and the amount of pollutant discharged

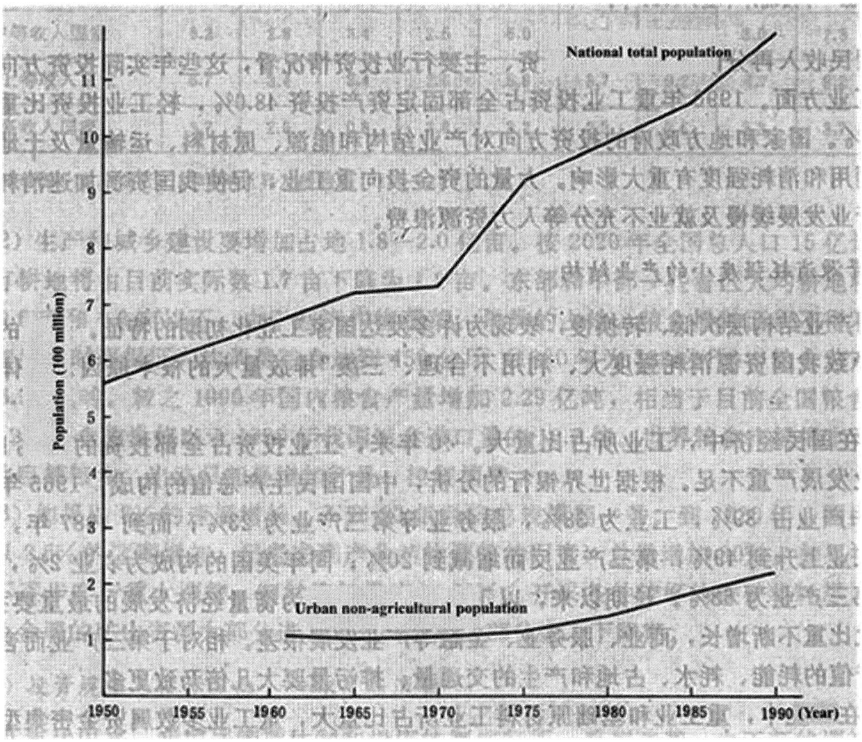

Fig. 8 Development of China's population urbanization

in cities are dozens of times those of rural areas, and food consumption is much larger. The land occupied by urban construction is mostly high-quality arable land.

Determination of China's urbanization development strategy is closely related to the establishment of a resource-saving socio-economic system in China and various regions, which is mainly manifested in the following two aspects:

(1) Basically stipulate the speed and scale of population urbanization. In the process of industrialization in developed industrial countries, the influx of large numbers of people into cities was not controlled. As a result, most people live in cities today.[3] Should China continue to develop according to this goal? The rural population accounts for the majority in China and is in scattered distribution. The urban construction funds and infrastructure provision are insufficient, and the level of mechanization in the vast rural areas is not high. Under such circumstances, on the one hand, cities must continue to develop; on the other hand, if the process of urbanization is not controlled, the pressure on urban development

[3] According to the World Bank's *World Development Report 1989*, the proportions of urban population in the total population in 1987 (%) are as follows: 92% in the United States, 92% in Britain, 86% in the former Federal Republic of Germany, 74% in Norway, and 77% in Japan.

needs will be too high to develop urban modernization. It may also affect the healthy economic development of the vast rural areas. Therefore, we must take the road of population urbanization with Chinese characteristics.

(2) The urban scale and structure not only concern economic benefits, but also affect the protection of land resources and pollution. For a long time, there have been different opinions on China's urban development policy (that is, strictly control large cities, appropriately develop medium-sized cities, and actively develop small towns). From an economic point of view, the larger the scale, the higher the economic benefits generally. It can be seen from Table 3 that the per capita GDP and profit tax per RMB 10,000 in the mega cities with a population of more than two million exceed those of small and medium-sized cities with a population of less than 200,000 by two times and over 40% respectively [9].

However, the larger the city, the higher the operating costs. For example, the investment in urban construction (roads, water supply,, and drainage, heat and electricity, housing, culture and education, sanitation, other public facilities, etc.) for adding a basic worker in large cities is 5–8 times that of medium-sized and small

Table 3 Development indicators of cities of different scales in China (467 cities)

Item		Scale				
		Population over 2,000,000	Population 1,000,000 to 2,000,000	Population 500,000 to 1,000,000	Population 200,000 to 500,000	Population less than 200,000
Number		9	22	28	117	291
Non-agricultural population (10,000)		3444	2814	1899	3644	3245
Gross domestic product (Current price)	Subtotal (100 million yuan)	2072.28	1478.03	985.15	2134.02	2391.05
	Primary industry (%)	2.5	4.8	4.4	14.0	32.2
	Secondary industry (%)	58.2	61.7	68.0	57.8	43.2
	Tertiary industry (%)	39.3	34.5	27.6	29.2	24.6
Per capita GDP (yuan)		5112.07	3855.77	3948.98	2750.85	1600.29
Industrial enterprises with independent accounting	Original value of fixed assets (100 million yuan)	2223.07	2007.95	1478.92	2863.25	2000.14
	Profit tax per RMB 100 (yuan)	15.91	14.60	12.92	10.44	11.23

cities. That is to say, if the emphasis is placed on the development of large cities and megacities, a huge amount of funds shall be used for urban construction, but the economic benefits will be better. Per capita, energy consumption (referring to living energy consumption) and water consumption in large cities are generally 30–60% higher than those in small and medium-sized cities. The larger the city, the more fully utilized the land and spatial resources. On the contrary, the smaller the scale, the larger the land occupation (Table 4) [10]. For cities with a population of less than 100,000, the land area per capita is twice that of large cities with a population of two million or more. The per capita floor area is more than 150 m^2 in counties, and 200–300 m^2 in towns and rural areas. According to the analysis of the urban land use and urban population in the Sixth Five-year Plan, the urban built-up area will increase by 700 km^2 each year in the next ten years, and the land will increase by 400 km^2 each year on average in small towns. There is no exact calculation for houses built by farmers. According to the statistics of Jiangsu, farmers built 546,100 mu of houses from 1985 to 1989, with an annual average of 91 km^2. It can be concluded from these figures that if the land occupation of residential areas in small towns and vast rural areas is not strictly controlled, it would be catastrophic.

From the perspective of resource conservation, environmental protection, and development of regional economic centers, it is required to use the systematic simulation method to find the urban scale and structure that can not only exert the economic benefits of the city, but also use land and other resources rationally. It can be seen from the above figures that medium-sized cities shall be the targets of key development. This is an important measure to establish a regional resource-saving socio-economic system, and is also a direction and model of social development.

5) Consumption structure

In a certain sense, the consumption structure is the most concentrated reflection of the social-economic structure. To establish a regional resource-saving socio-economic system, a resource-saving society and private consumption structure must be determined.

Since the Western developed countries entered the post-industrialization stage, the high-consumption society emerged at the same time. According to the popular "values", every family and even every adult must have a car and a spacious house,

Table 4 Per capita land in cities of different scales in China from 1981 to 1985

Urban scale (10,000 persons)	> 200	100–200	50–100	20–50	10–20	< 10	Number of cities in statistics
1981	53.5	71.6	69.2	83.2	87.3	155.7	225
1982	52.9	71.9	73.0	83.4	90.3	163.7	236
1983	53.4	67.0	73.5	97.4	86.6	131.3	256
1984	52.6	70.5	73.6	77.4	93.1	120.3	286
1985	52.2	71.2	74.5	79.5	87.6	117.9	317

Unit: Square meters/person

and all physical work must be replaced by mechanical power. As a result, each person has to consume more than ten tons of standard coal and even more life energy every year. Table 5 lists that there was one manned car for every two people in the United States in 1981. Do China's national conditions allow us to take the same road?

(1) Dietary structure. Individuals in China have taken plant-based products as their main food since the beginning of history, and the current consumption structure is characterized by a high level of direct grain consumption per capita. According to the calculation by the World Bank [11], the per capita consumption was 205 kg for food, 12.8 kg for pork, mutton, and beef, and 4.2 kg for vegetable oil in 1982. It can provide more than 1.13×10^7 J of calories, 68 g of protein, and 38 g of fat every day. This is not very low in the world. However, to what extent will dietary diversification be achieved in the future? Shall we follow the pattern of some developed countries (with the increase in per capita national income, the proportion of animal food will be greatly increased)? The conditions for the development of grassland animal husbandry in China are quite difficult. The substantial increase in meat output mainly depends on the vast agricultural areas, that is, the exchange grain for meat products to a considerable extent. If measured by heat, a certain amount of animal energy can be exchanged with several times of plant energy regardless of pork, mutton and beef. In addition, more energy and other inputs will be consumed during meat production and processing.
(2) Housing and clothing. With the improvement of living standards, the per capita housing area shall gradually increase. However, what is the appropriate area for housing in cities in China? "A human body is only a few liters in size. Is it necessary for a person to live in a house with an area of several hundred square meters", "is it necessary for a person to have dozens of suits"?[4] In terms of housing, we shall learn from the Japanese who have a small land area but a very developed economy. During the period of high growth from the 1950s to the 1970s, the housing area in Japan only changed slightly. It is only in the past ten years that the welfare system has been manifested in the area of household housing due to the great increase in per capita national income. Even so, the per capita housing area is still much smaller than that of Western society.
(3) Vehicles. Since the 1920s, cars have been widely used in a few countries; the output of cars in the world had almost risen sharply by the 1970s. From 1950 to 1973, petroleum production increased at a rate of 7.6% in the world, and the output of vehicles also increased at an almost equally significant rate of 5.8% every year. In 1978, the world's automobile production reached an unprecedented 31.8 million, of which about 70% were manned vehicles.

The number and popularity of cars in a country or region have an extremely important impact on the socio-economic structure. There must be a large amount of stable fuel supply, so that a large number of cars can enter the families. For example,

[4] Three Speeches Made by Comrade Yu Guangyuan at the Symposium on Socio-economic Strategic Issues, Research on Technical Economy, No.8, 1981.

Table 5 Popularity of manned vehicles in 20 countries with larger populations

Country	Population (1,000,000)	Number of manned vehicles (1000)	Number of persons per vehicle
The United States	232	121,724	2
Former Federal Republic of Germany	62	23,236	3
France	54	19,150	3
Italy	57	17,696	3
Britain	56	154,348	4
Japan	119	23,660	8
Brazil	128	8213	16
Mexico	71	3360	21
Former Soviet Union	270	8255	33
Iran	39	1028	38
Turkey	48	711	68
Philippines	52	550	95
Egypt	45	428	105
Thailand	50	435	115
Nigeria	82	500	164
Indonesia	152	637	239
Pakistan	93	264	352
India	714	930	768
Bengal	93	22	4227
China	997	55	18,137
Around the world	4585	320,513	14

Source Panoramic View of the World, Page 271, Editor-in-Chief: Lester R. Brown, China Translation and Publishing Corporation, 1985, Beijing

suppose each car travels for 50 km every day on average. In that case, the annual mileage will be 18,000 km, and the annual fuel consumption will be about 1620 L (an average fuel consumption of 9 L per 100 km). If the long-term goal is to have one car for every 30 people, there will be 50 million cars in total in China. The cars will consume 70 million tons of gasoline each year. There are also freight trucks and other power consumption. The annual gasoline consumption will be about 150 million tons. According to the current composition ratio of product oil, the national crude oil processing volume shall reach 550 million to 600 million tons, so that the need of producing 150 million tons of gasoline can be met. It seems impossible. Secondly, a large number of roads and parking lots are required. According to an incomplete statistical analysis of several large cities, a car occupies an average of 150–200 m^2 of parking lots and roads, that is to say, the area is 2–3 times larger than the average per capita floor area in the city. In addition, the popularization of cars has greatly expanded the scope of people's travel, which will have a series of impacts on the distribution of rest, tourism, employment, and residence places. Inevitably, a great amount of exhaust emissions will make the urban environment deteriorate.

According to the above analysis, the solution to the issues of urban traffic in China cannot be mainly achieved by cars; a more effective way is to vigorously develop urban public transportation (subway, highway, branch railway, etc.); in the metropolis and some large cities, the three-dimensional transportation shall be developed as much as possible, so as to fully utilize the urban land and space.

People's consumption structure and level often change in the process of economic development. The key is to start from the resources of one's own country and region, and choose a consumption level and structure with one's own characteristics. Even in developed countries like the United States, not all people support that life should be "extravagant and wasteful". According to an American material published at the 1980 International Futurology Conference: The results of an American opinion poll showed that 79% of Americans want to "educate people to live on basic needs", while only 17% of Americans want to "reach higher living standards". 55% of Americans believed that "it is a good thing to live a more serious and simple life". 76% would rather get happiness from non-material experiences than ask for more goods and services.[5]

6) Spatial structure—gradually solve the problem of unbalanced exploitation and development

The distribution of territorial resources and socio-economic development in China is very unbalanced. The degrees of development and utilization of 9.6 million square kilometers of land and about three million square kilometers of territorial waters, inland seas, exclusive economic zones and continental shelves are very different. Among them, a considerable part of the land is unsuitable for human to carry out daily social-economic activities. Although China has paid great attention to the less developed regions since 1949, however, if measured by per capita GDP and economic strength, there are still significant gaps between the east and the west, between the inland and remote areas, and between Han residential areas and ethnic minority areas. In recent years, in some areas with very developed social economy, such as the Yangtze River Delta, Beijing-Tianjin-Tangshan, central and southern Liaoning, the Pearl River Delta and some areas where the provincial capitals and cities specifically designated in the state plan are located, there are more and more large and medium-sized projects that are constructed or expanded. This situation indicates that the gap in economic strength between macro and meso areas will be further widened for a long period of time in the future.[6] That is to say, the density of the social economy will become more unbalanced from the perspective of the spatial structure.

---

[5] Three Speeches Made by Comrade Yu Guangyuan at the Symposium on Socio-economic Strategic Issues, Research on Technical Economy, No.8, 1981.

[6] Such situation does not mean that the gap between the poor and the rich of the regions is getting wider, that is, the rich get richer, and the poor get poorer. However, various regions will develop on the original basis. Those underdeveloped regions, that is, areas with low per capita national income, are constantly improved, not "getting poorer". However, even if the same growth rate is

In a general sense, "imbalance" indicates that the use of space and resources in China is not yet sufficient. The resources and space of relatively sparse areas can be attracted to the socio-economic cycle by gradually alleviating the imbalance of the spatial structure. Such exploration of resource potential is equivalent to resource saving. The approach is to strengthen the construction of regional infrastructure and increase the accessibility of underdeveloped regions.

7) Import and export trade structure—expand the use of resources and living space in the world and other regions

In the long run, the limited resources and space of China are insufficient to support and sustain China's large population and huge economy. It is necessary to constantly expand the volume of international trade, adjust the structure of import and export trade, increase the import of resource-based products, and reduce the export of resource-based products, so as to alleviate the constraints of resource shortage on China's socio-economic development. Many developed countries have mainly relied on this approach to achieve domestic development for a long time. Some developing countries and regions (especially countries and regions with restricted land resources, such as South Korea, Singapore, Taiwan Province, etc.) have made every attempt to open up such an approach in recent years.

China began industrialization in the 1950s, decades to nearly a hundred years later than some developed countries in the West. China is in a backward position in occupying the international market with manufactured products, especially those high-tech products with low resource consumption and high added value, and then importing a large number of resource-based products. According to a rough analysis, more than 60% of the foreign trade volume of the former Federal Republic of Germany takes place within the European Community. However, these are mainly trades between industrial products, which are carried out in accordance with the provisions of the Community on the trade balance. Trade with developing countries accounts for more than 30%. Although this part of the trade volume is smaller than that in the Community, it is even more important for such a country with poor resources, highly developed industries and high levels of welfare, since Germany imports 80% of iron ore, more than 70% of wood, about 70% of oil, 100% of rice, bananas, etc. from developing countries (all refer to the proportion of consumption). Japan is a small country. Its huge economy relies on importing most of the coal, oil, wood, iron ore, chemical raw materials, construction materials, etc., and 45% of its food is imported. The production of these materials requires occupying and consuming a large area of land and resources. On the contrary, the large number of high-tech products exported require fewer resources and space. That is to say, the societies of a few developed industrialized countries are built on global natural resources and living space. On a global scale, the average conditions of resources, population, and economic aggregates are still looser than those of China. China must

achieved, the gap in economic aggregate and strength with developed regions will also be widened. Of course, the government will develop various policies, so that the living standards of the people in the underdeveloped regions and those in the developed regions can roughly keep balanced.

resolutely take the road of expanding the use of foreign resources and living space. Specifically, it is necessary to gradually increase the import of primary products and the export of manufactured products.

In the past 10 years, with the improvement of China's industrial technology, the structure of import and export trade is undergoing changes that are conducive to the conservation of China's resources. From 1981 to 1990, the ratio of manufactured products to primary products rose from 1.75 to 5.13 in the composition of the total value of export products. Over the past ten years, the export of manufactured goods increased by an average of 14.7% annually; on the contrary, the export of primary products increased by an average of 4.5% annually. At the same time, among imported products, primary products increased very little (with an annual average of 1.3% only), while the manufactured products grew faster, with an annual average of 11.3%. The total import value of manufactured products reached 43.49 billion yuan in 1990, which was 4.4 times that of primary products (mainly food and raw materials).

In terms of major import and export commodities, the changes that occurred in the past 10 years are mainly manifested as follows: The export volume of grain decreased, while the import volume increased. The difference between imports and exports reached eight million tons (net import) in 1990; cotton changed from exports to massive imports; large quantities of raw materials such as logs, pulp, sugar, fertilizers, iron ore, and two alkalis were imported; at the same time, exports of energy (coal, petroleum and product oil) substantially decreased. Changes in imports and exports help to alleviate the intensity of development and utilization of domestic resources (Tables 6 and 7).

China has a vast land area, and the amounts of resources in various regions vary greatly. Each region shall also determine its own input and output structure based on its own resource consumption.

8) Regionalization of the resource-saving socio-economic system

There are great differences in the amount and structure of natural resources, the scale of living space, and utilization conditions in various regions in China, and the economic aggregate and structure, population and social characteristics are also different. Therefore, each region shall establish a regionalized resource-saving socio-economic system based on specific conditions. There are roughly the following types:

(1) Regions with a serious imbalance in water and land resources as well as water and mineral resources, mainly referring to North China, Northeast China and Northwest China. The key is to establish a water-saving agricultural planting (structure) system, develop an industrial sector (structural) system that consumes less water and an urban and rural water-saving resident living system, develop a water-saving industrial technology system, adjust the spatial structure, and establish a (fresh) water-saving industry belt in the coastal area.
(2) Regions with a lack of energy and mineral resources and well-developed economy, mainly referring to the coastal areas of East China and Southeastern China, and also include the southern part of Northeast China, Beijing-Tianjin-Tangshan region and other regions to a certain extent. In these regions, it is

Table 6 Changes in China's import and export trade structure

Commodity category		Export				Import			
		1981	1985	1987	1990	1981	1985	1987	1990
Primary products	Food and live animals mainly for eating	29.5	38.0	47.8	66.1	38.0	15.5	24.4	33.0
	Drinks and cigarettes	0.6	1.1	1.8	3.4	3.0	2.1	2.6	1.6
	Non-edible raw materials	19.0	26.5	36.5	35.4	43.1	32.4	33.2	41.0
	Mineral raw materials and related raw materials	52.0	71.3	45.4	52.4	0.9	1.7	5.4	12.1
	Animal and vegetable oils, fats, etc	0.9	1.4	0.8	1.6	1.0	1.2		9.8
	Subtotal	102.0	138.3	132.3	158.9	86.0	52.9	69.2	98.5
Manufactured products	Chemicals and related products	13.0	13.6	22.4	37.3	27.0	44.7	50.1	66.5
	Light textile, rubber, mining and metallurgical products and manufactured products	47.0	44.9	85.7	125.8	43.0	119.0	97.3	89.1
	Machinery and transportation equipment	11.0	7.7	17.4	55.9	63.0	162.4	146.1	165.8
	Miscellaneous products	37.0	34.9	62.7	126.9	6.0	19.0	18.8	21.0
	Subtotal (including uncategorized products)	108.0	135.2	262.1	462.1	149.0	369.6	363.0	434.9

Unit: 100 million dollars
Source China Statistical Yearbook, 1983 to 1991

necessary to gradually adjust industries that consume a lot of resources, develop deep-processing production and high-tech industries, increase exports of manufactured products, and obtain foreign resources through foreign trade. That is to say, the new industrial structure system and foreign trade product system shall be established.

(3) Regions with narrow land and living space and fairly developed social economy, including many large and medium-sized urban agglomeration areas, coastal plains, hilly areas, and some river valley plains in the mainland. In order to

Table 7 Changes in imports and exports of main commodities in China

Commodity name	Export			Import		
	1981	1985	1990	1981	1985	1990
Coal (10,000 t)	694	777	1729			
Crude oil (10,000 t)	1387	3003	2399			
Product oil (10,000 t)	467	621	526			
Grain (10,000 t)	99	923	583	1447.7	600	1372
Cotton (10,000 t)		34.7	16.7	80.1	0.02	41.7
Cement (10,000 t)	0.7	14.2	682.9			
Aquatic products (10,000 t)	11.6	12.0	35.8			
Fruit (10,000 t)	19.9	21.4	22.6			
Cotton cloth (10,000 t)	12.78	16.73	22.22			
Cotton yarn (10,000 t)	4.77	15.48	17.62			
Bicycle (10,000 sets)	72	54	378			
Camera (10,000 sets)	5.0	8.4	419.2			
Low-grade paper board (10,000 t)	21.5	18.5	13.0			
Steel material (10,000 t)	61.4	15.5	199.2	354.4	2005	419
Log (10,000 m^3)				187	969	415
Pulp (10,000 t)				6.4	54.7	34.2
Fertilizer (10,000 t)				555.1	761	1627
Two alkalis (10,000 t)				25.5	128.0	25.1
Iron ore (10,000 t)				335.5	1006	1434
Sugar (10,000 t)	12.54			99.1	191	113
Wool (10,000 t)				5.9	11.3	3.3
Machine tool (10,000 t)	3.33			0.3	1.3	4.4
TV (10,000 sets)				398.4	509	67
Automobile and chassis (10,000 sets)				4.2	35.3	6.3

Source Same with Tables 5 and 6

solve the problems concerning population, resources and environment in these regions, it is necessary to focus on saving land resources, restrict the industrial production that occupies a large amount of land and generates large-scale transportation, and establish a scientific urban scale structure. It is necessary to improve the facilities for transportation and communication with the surrounding regions, and increase the accessibility of these regions, so as to guide the population and economy of the agglomeration areas.

(4) Regions where land resources and some other resources are relatively abundant, but the economic development level is low due to some poor natural conditions and other social and historical factors. At present, most of the old revolutionary base areas, areas with large ethnic minority populations, border areas, and poor areas belong to this type. The industrial groups based on advantageous resources

shall be progressively established in these regions, and the pursuit of a complete production system shall be prevented.

The above regions are roughly divided according to resource characteristics and resource types. Taking into account the development and implementation of the plan for a resource-saving society, the structure, index system and measures of the resource-saving socio-economic system can also be researched according to each administrative region.

9) Build a developed transportation, post, and telecommunications system

Certain resources and living spaces have different economic aggregates of the tolerable population under different levels of transportation and communication systems. The bases are as follows: First, the accessibility of any location in the space is improved, so that its resources and space can be fully utilized (that is to say, in the case of a homogeneous surface, the degree of utilization of resources and space is related to the accessibility of the region or point), which attracts local resources and space into the socio-economic cycle to a greater extent; second, relying on a developed communication network, the flow of some people and goods can be reduced, thus saving resources and space. According to analysis and calculations, the energy consumption per unit of GNP in China was 3.5 times the world average, the freight turnover consumed was 4.2 times the world average, but the information flow was less than 1/10 of the world average from 1952 to 1987. Due to the undeveloped information connection among regions, repeated construction among regions and even adjacent regions has caused a lot of waste. In recent years, in-transit funds have reached 30 billion yuan in China, while they are only 500 million dollars in the United States; in the existing passenger traffic volume, information carriers account for about 60%, of which 35.1% can be replaced by communication [12].

Satellite communications and video conferences that have emerged in recent years and can advance the industrial revolution and social progress in the future can further shorten the spatial distance across the country and in the world. By holding the video conference, relevant personnel can participate in the conference without a long journey, and can also check goods and samples mutually, which can save the space for transportation of personnel and goods, greatly save time, expand the ability of people, shorten the decision-making process, and improve the efficiency of the meeting. The economic effect of the video conference is also obvious. According to the operating practice and analysis in Western Europe, assuming that the travel cost per person per time is 1000 dollars and the monthly maintenance cost of the video conference room with an investment of 500,000 dollars is 20,000 dollars, if there are four persons in each official contact, personnel travel is suitable in the circumstances where there is one meeting and conference each month, and the video conference will be economical in the circumstances where there are more than one meeting and conference each month.

In Western developed countries, with the increase in per capita national income, attendance rises substantially. Will China follow this development trend, and how can this trend be embodied and controlled specifically? According to China's population and economic strength, we cannot reach the welfare level of some Western industrial countries for a long period of time, and China's infrastructure level and living space do not allow travel in large numbers. For this reason, the development of post, telecommunications, and communication facilities is indeed an important way to save resources.

When the post and telecommunications industries are highly developed, in addition to video conference, office work, production of some industrial parts, work of certain processing industries and many tertiary industries can also be done by individuals at home, which can be linked together through wide cables, terminals, and control rooms. This can greatly reduce the transportation of personnel and save resources and space.

5. **Concept change and countermeasures**

(1) Values. The determining factor that affects the social structure and economic structure is the consumption structure of the society. From the perspective of long-term development, the consumption structure that Chinese people seek will fundamentally determine the resources and environmental conditions of China and various regions. However, the consumption structure is directly affected by people's value orientation. It is necessary to seek the food, housing, transportation, and social communication conditions required to meet the people's living and development needs for every member of society, but the possession of resources and wealth beyond actual needs shall not be pursued. Of course, this value is set for the whole of society. Consumption that greatly exceeds actual needs is a waste of resources and space. People's values refer to people's evaluation of various social practices, including methods of course. People need to occupy and use certain materials, resources and space to survive and develop. With the improvement of the economic, scientific and technological levels, people need not only to satisfy material life, but also to enjoy spiritual life. Physiologically speaking, human can have different standards for material needs, for which there are also limits. Values are affected by geographical environment and sociocultural history. The Chinese nation has a fine tradition of diligence, frugality, and simplicity. We can consider: What kind of consumption and level are not worthwhile and do not deserve our efforts. For example, whether to pursue a car for every two people, whether one person should occupy more than one hundred square meters of housing, whether not the upgrading of durable consumer goods should be this rapid when comparing to our Western counterpart the United States.

(2) Sense of responsibility and sense of crisis. The Chinese nation will mainly rely on the current land for survival and development forever. However, today's achievements are very likely to mean tomorrow's difficulties. We must not be short-sighted in any case but shall think about our future generations. If the natural basis for the survival and development of the Chinese people is destroyed, it means that national security is threatened. Once important resources are

exhausted, the ecosystem will be under excessive pressure, and it will be unable to ensure stable economic development or national security. This is a necessary sense of crisis.

(3) Measure socio-economic development with a new indicator system, and adopt an accounting system that controls the waste of resources. For a long time, China has used indicators, including total industrial and agricultural output value, labor productivity, and capital profit and tax rate, to measure the scale and level of economic development. Its one-sidedness and adverse effects are obvious. In recent years, common international practices have been adopted, that is, using the gross national product, national income and total product of society as the main indicators for measuring socio-economic development. However, these indicators still do not meet the requirements in terms of saving resource consumption and establishing a resource-saving socio-economic system. They mainly manifest the level, scale, speed, benefits, etc. of socio-economic development, and are an aspect of the "human-land" relationship. If such indicators are used to guide development, the government and public opinion will make every attempt to increase GNP to achieve affluence and prosperity. However, no attention will be paid to the sustainable use and protection of resources, and the overall development of people will be ignored. If the goal of meeting basic human needs is followed, it is necessary to add some indicators that reflect the overall development of the society, such as employment, life expectancy, infant mortality rate, the average growth rate of GNP, etc. In addition, it is necessary to gradually adopt an accounting system that controls the waste of resources. That is to say when calculating the output value and GNP, it is necessary to calculate the future loss equivalent to the resource consumption and the fund required to compensate for ecological damages, and include this in the cost accounting of output value.

References

1. B. Zhu, *Research on Policies of Resources, Environment and Development* (1992)
2. L. Fang (ed.), *Research on China's Industrial Regional Distribution in 2000* (China Ocean Press, 1993), p. 82
3. B. Li (ed.), *Industrial Policy and National Economy* (Shanghai Scientific and Technological Literature Press, 1990), p. 402
4. D. Lu, et al., *Research on Liaoning Resource Development and Industrial Transportation Layout* (China Planning Press, 1990), p. 12
5. M. Wang, Q. Zeng, *Global Atmospheric Chemistry and Future Climate Change Trends, Collection of the Second Faculty Committee Conference of the Division of Earth Sciences* (Chinese Academy of Sciences, 1988), pp. 236–238
6. Z. Zhu, *China's Desertification Disasters and Countermeasures, Analysis on Natural Disasters and Disaster Mitigation Countermeasures in China* (Hubei Science and Technology Press, 1992), p. 346
7. D. Lu, et al., *Theory and Practice of China's Industrial Distribution* (Science Press, 1990), p. 60

8. World Bank, *Issues of China's Long-Term Development and Solutions, Attachment 3, Energy Issues* (Chinese Financial & Economic Publishing House, 1987), p. 15
9. National Bureau of Statistics, *China Statistical Yearbook* (China Statistics Press, 1991)
10. W. Chen, Issues of urban land use in China and countermeasures. Econ. Geogr. **8**(4) (1991)
11. World Bank, *Issues of China's Long-Term Development and Solutions, Attachment 2, Agriculture from Now to 2000* (Chinese Financial & Economic Publishing House, 1987)
12. State Development Planning Commission, *Outline of Overall Planning of National Land (Draft)* (1987), p. 63

Chapter 6
Spatial Structure and Regional Development

1 Spatial Structure Theory and Its Practical Significance

From the middle of the last century to the 1940s, that is, within a century, German scholars successively proposed agricultural location theory, industrial location theory, urban location theory and market location theory [1], making major contributions to theoretical research and model research of modern regional development. However, their research objects are basically single socio-economic objects, that is, industrial enterprise, agricultural management mode, city, market area, etc., and the spatial motion and space positioning laws of these individual things are summarized.

After World War II, the social economy of the developed industrial countries and a number of third world countries entered a long period of rapid development. This promoted the development of location theory to a great extent. The objective requirement for further development of the theory was also proposed to achieve regional economic development. Due to the enhancement of the role of factors affecting the socio-economic development of countries and regions, the forms and results of the interaction among factors became more complicated, making it difficult for the location theory before the 1940s to explain the increasingly complex socio-economic phenomena fully. Therefore, it was difficult to apply the theory in regional development planning and management practice. Under the impact of this new socio-economic development situation, the location theory has continued to develop, gotten rid of some limitations of classical location theory, and played a huge role in actual regional development planning and research.

1. **Two directions for the development of location theory**
1) Regional analysis direction

After World War II, American scholars, mainly regional economists Isaard, I. Burton, W.L. Garrison, Bailey, etc., applied mathematics to regional development research and planning, setting off a wave of "quantitative revolution". From the second half

D. Lu, *Regional Development and Its Spatial Structure*,
https://doi.org/10.1007/978-981-99-7683-6_6

of the 1950s to the 1970s, when studying regional issues, scholars widely used mathematical statistics and linear programming to analyze the regional social-economic structure and predict future development; applied the input–output method, analyzed the interaction among various sectors, planned the economic structure in the future, or predicted the future development and change trends. Most of these scholars, i.e., the main founders of "regional science", inherited the method of investigating problems and the rigorous spirit of German scholars when studying location theory, and were also required to get rid of the inherent impracticality of the single-element professional pure theoretical derivation of location theory scholars. They published a large number of books, which had a great impact on geography and economics during that period. One of the important results of their exploration is to lay the foundation for the large-scale application of electronic computers and mathematical models in regional development and planning today.

In the *Location Theory and Regional Analysis Methods*, the author once proposed that the location theory further developed after World War II, forming two main interrelated but different directions, that is, the direction of regional science and the direction of regional analysis of human geography. This was mentioned in consideration of the disciplinary characteristics of regional economists and geographers in developing location theory and the degree of concern for regional policy issues. In fact, their common point, focusing on the application of mathematical methods in the analysis of regional issues, was the primary. Most regional scientists came from the field of economics. In addition to applied mathematics and computers, recommendations on regional policy (which eventually led to the emergence of an independent discipline, regional policy) were regarded as important research goals. After introducing mathematics and computers, most geographers still laid emphasis on the study of spatial differences.

2) Direction of spatial structure

Spatial structure refers to the interaction of socio-economic objects in space and the degree and form of spatial agglomeration formed. Because this theory does not summarize the law of spatial distribution of single elements but integrates almost all socio-economic objects, it can also be called the overall location theory. The prototype of the spatial structure theory was first proposed by the German human geographer A. Schluter in 1906 in the "landscape" theory of human geography. After that, W. Christaller and Losch further developed the content of the "landscape" concept. Böventer [2], a German scholar, carried out the systematic theoretical analysis and model derivation of the spatial structure. He tried to integrate the location theories of Weber, Thunen and Losch, and believed that location theories should examine and clarify not only production and goods as deeply as possible but also the geographical distribution of production factors such as place of residence, employment and mobility. He combined the location theory with the development theory, analyzed and demonstrated the general characteristics of the spatial structure at each stage of social economy, and analyzed in detail the most important factors that determined the spatial structure and its differences: agglomeration, freight and the dependence of the economy on the land of local production factors; and believed that freight is the decisive factor for the characteristics of the input–output relationship and the spatial flow of production factors; and that the impact of land on the spatial structure mainly depends on how it is used. Among them, productive use includes the land for agriculture, mines, industry and infrastructure, and consumption use includes residential and recreational land. The interaction of these factors determines the characteristics of the spatial structure. In addition to the above three main factors, Böventer believed that objects of the same nature in space, such as factories that manufacture the same product, tertiary industries that provide the same kind of service, and markets that sell the same goods, always had a competitive relationship. As a result of competition, a landscape structure model of the location at the hexagonal apex was generated, that is, the theoretically optimal arrangement. He also further deduced how the production scale and consumption scale of each location depend on the production and demand mechanism of various products, especially the relationship with the scale economy of production and the advantages of inter-sectoral agglomeration. Böventer's development of the location theory and spatial structure was also manifested as follows: The scale of industrial and agricultural enterprises, urban scale as well as urban scale structure was theoretically demonstrated from the perspective of optimal land use and reasonable agglomeration.

2. **Connotation of spatial structure theory**

According to the scholars' actual connotation of the spatial structure concept and the definition of the spatial structure concept summarized by the author, we can believe that the spatial combination status of the regional social economy can be observed from different spatial scales, levels and angles, and different questions can be raised. In response to these different questions, their own development practices and theories have been produced. However, anatomical study and integrated overall study shall be performed for these basic questions, so as to scientifically understand and plan the spatial structure of a region.

(1) For a spatial structure composed of different socio-economic "sparse" (so-called structurally "thin") and "dense" (so-called structurally "thick") belt-like or planar regions, it generally refers to the imbalance of socio-economic development among large regions (within a country). This structural problem mainly comes from the zonal or regional differences of nature, the relationship between different positions relative to the ocean and the political center in history, as well as the location relationship with international economic agglomeration areas. The imbalance of large zonal development has its objective law: the inverted U-shaped correlation law between economic growth and development imbalance. People should use this law to develop the regional economy by stage, so as to achieve regional development from being unbalanced to being more balanced.
(2) Frame or context of socio-economic spatial organization. For example, the growth pole model proposed by Western scholars, the development axis model proposed by the German planning circles, and the "point-axis system" model proposed by the author based on the development axis and central place theory (1984), all these different structural frame models reflect the effective forms of

socio-economic spatial organization in different degrees, which are important structural models used to formulate the rational distribution of productivity in large regions and key urban development strategies, and are also operational models to scientifically solve the problems of "sparse" and "dense".

(3) Optimal enterprise scale, residential area scale, urban scale, and central place hierarchy system. The bases of theoretical derivation are as follows: One is the agricultural location theory, and the other is the combined effect theory. The derivation of the optimal enterprise scale is combined with that of the reasonable scale of the urban residential area, and the city is regarded as an enterprise and understood as a production process. The "threshold" theory is used to apply the central place level system in the reality of regional planning.

(4) Spatial structure of land use centered on urban residential areas (markets). This is the development of Thunen's theoretical model and the theory of location differential rent. It uses the concept of production and consumption functions to derive the net income function of each operation mode of suburban agriculture, and the division of operation zones is obtained therefrom.

(5) Spacial interaction. It includes the flow of goods, people, and finances among regions, the attraction scope of central cities at all levels, the diffusion process of innovation, information and technological knowledge, etc. These aspects are important reflections of spatial structure characteristics.

(6) Characteristics and evolution of the spatial structure at each stage of socio-economic development. The general trends and types of dynamics and evolution of spatial structure changes are revealed through the analysis of the general mechanism of action.

3. **Spatial structure—an indicator of regional development status**

1) Three-dimensionality of spatial structure

The "space" in the spatial structure concept of social economy is not equal to the "absolute space" in physics or the static "pure space" in geometry, but refers to many types of objects such as agriculture, industry, urban residential areas, road and communication facilities, cultural and commercial supply facilities, etc., where the following phenomena are constantly taking place, such as commodity production, transportation of raw materials and finished products, information transmission, commodity sales, and other reproduction processes, development of new districts, population flow (dispersion and agglomeration), town expansion, generation of new residential areas, diffusion of new technologies, etc. Each thing and object as well as the movement between them will form a spatial situation, and their combined relationship in a whole will create multiple spaces. It corresponds to a comprehensive and holistic space system, including a system of various elements that exist in space. The investigation and analysis of the space system emphasize the interconnectedness of various components in the system, that is, the subsystems, and also emphasize the process of their connection and change, that is, the instant reflection of various dynamic sequences. At different moments of development, the relationships between various things mentioned above, including the positional relationship, will change to different degrees. Therefore, things and phenomena have two-dimensional meanings of length and width and are also given the vertical (time) concept. The three-dimensionality of this space is an important characteristic of space. On the other hand, the three-dimensionality of space also refers to the positional relationship of related things and phenomena in space. To determine the location of things, three mutually perpendicular lines can be drawn from any point in space in terms of coordinates, just like the three lines of length, width and height in geometry. In this way, focusing on the positional relationship and location structure in space, the three-dimensional term "spatial structure" is undoubtedly more appropriate than "territorial structure".

2) System and structure

Regardless of a country, a region, or a city, the socio-economic components have different degrees of interaction and interrelationship. After long-term planning and development, the social economy of the region often forms a system with certain structures and functions, that is, the social system and economic system of the region (or country). Once a "system" is formed, it is an organic whole, like an organism. If a certain part is affected by the outside, it may affect the entire organism. The characteristics of this organism make it possible to regulate and control the state of regional socio-economic development through laws, government behaviors, and enterprise behaviors. Any system has a certain function or is manifested as a certain characteristic, which is mainly determined by the structure of the organism. The structure of the regional socio-economic organism is constantly changing. Changes in quantity and population composition, the expansion of production, the generation of new industries, the construction of transportation hubs and transportation routes, etc., have caused changes in the capabilities, levels, and external radiation effects of the regional social economy. Therefore, it can be concluded that:

(1) The characteristics and problems of the socio-economic system of this region can be grasped by analyzing the functions and characteristics of the regional socio-economic system;
(2) The characteristics and problems of the socio-economic system of this region can be grasped by analyzing the internal structure of the regional socio-economic system, that is, the components of the system and the interaction among the components. This is to study the "system" through the "structure". This is an important methodology for the research on the issues of regional development.

For example, it is necessary to dissect all main structures of regional (or national) economic development to analyze the economic system of a region or a country, thus trying to understand the development level, direction, trend and strength of the national economy of this region or country, regional economic cooperation, the relationship between economic development and resources and environment, etc., as shown in Fig. 1.

Since it is necessary to research the "system" through the "structure", the "structure" is considered a substitute for "system". Here, the following specific goals are mainly achieved through the research on the spatial structure of the regional social economy:

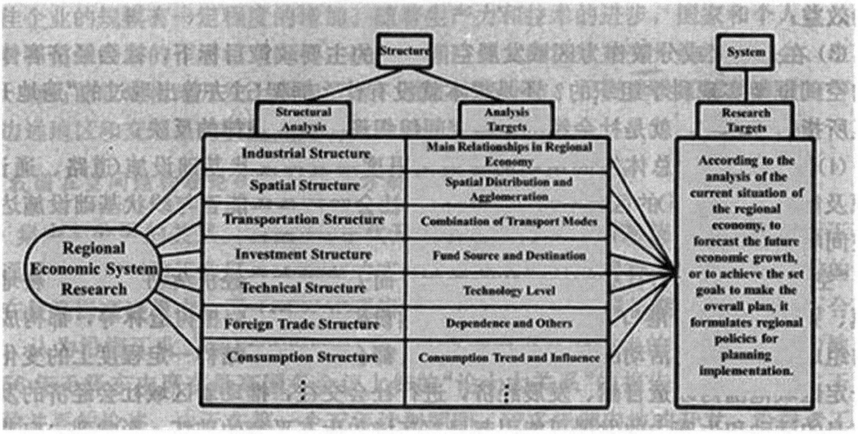

Fig. 1 Schematic diagram for grasping the "system" through the analysis of "structure" (taking the development of the regional economic system as an example)

(1) Judge the stage of regional economic development, because the socio-economic spatial structures at different development stages have different characteristics;
(2) Reveal the characteristics of the regional economic spatial distribution, the relationship between agglomeration and dispersion, and their benefits;
(3) Forecast the evolution trend of the regional economic spatial structure status;
(4) Propose the adjustment direction for the specific regional spatial structure and the approach to adjusting the direction according to the evolution law and actual trend of the spatial structure.

3) Spatial structure characteristics are important indicators of regional development status

Since the issues of regional development were raised after World War II, academic circles, business circles, and politicians have believed that the characteristics and level of the industrial structure of a country or a region are crucial for the development of the entire national economy and for participating in international and domestic interregional development. Whether the industrial structure is advanced, whether the industrial structure can fully reflect its own potential and advantages, whether the measures can be taken in time to promote its structural transformation, etc. are even a matter of life and death. People do not have such a clear understanding of the importance of the socio-economic spatial structure. However, the laws in this aspect have already made China's economic construction pay a heavy price.

From a large number of instances of regional exploitation and regional development, it can be concluded that the spatial structure of the social economy, similar to that of a region (country), is an important aspect reflected by the essence of the regional development status, and a compass for investigating and identifying regional development status and regional socio-economic organisms from the perspective of spatial distribution and spatial organization. Whether the regional development status is healthy, whether the relationship with the outside and the organization of the internal parts are orderly, whether the budding and vigorous factors are placed in a favorable position (spatial location and structural location), etc. The definite judging criterion can be given from an important aspect by analyzing the spatial structure of the social economy. Taking the evaluation of productivity distribution and objective policies for regional exploitation and regional development in several major stages in China in the past half a century as an example, the following questions can be raised for each major stage:

(1) Whether the overall framework (spatial strategy) of territorial development and productivity distribution conforms to the inverted U-shaped correlation law between economic growth and spatial imbalance, and whether it conforms to the objective needs of the development process from spatial imbalance to a more balanced one;
(2) At different stages, which strategy may exert the best economic and social benefits, spatial agglomeration or spatial dispersion?
(3) What kind of spatial frame is used to achieve scientific organization of socio-economic objects when agglomeration or dispersion is taken as the main policy objective of the regional development space strategy? Or is there no frame at all (the "blossom everywhere" that appeared in the past)? The "frame" referred to here is a reflection of the evolution law of the spatial organization of the socio-economic object;
(4) Whether the overall spatial organization of the social economy can maximize the function of the regional economic organization of linear infrastructure (roads, communication lines, water sources, energy supply lines, etc.), and whether the socio-economic object and the linear infrastructure can achieve the best spatial fit?

"Space" is a place where human beings carry out social-economic activities. All social-economic activities of human beings, such as opening up arable land, building villages, towns, railways, highways and ports, constructing water conservancy projects, building mines and factories, and planting trees, constitute an integral part of the spatial structure. Every location decision of such activities will bring changes to the spatial structure to a certain degree. Human beings transform nature, develop the economy, and conduct social interactions within a certain area, thus promoting regional socio-economic development. The development of human activities and productivity may cause damage to the local environment and ecological balance, and affect socio-economic development within a certain range. Every development and change of the social economy will change the position and interrelationship of the original socio-economic objects within a certain range, the strength of the regional socio-economic "power", the degree of spatial agglomeration, the imbalance of development, etc. Therefore, the regional development status and the spatial structure state are closely related to each other.

In practice, the spatial structure state is affected by many factors. In addition to three basic factors, such as freight, rent and agglomeration, there are also factors such as the distribution of resources, topography and climate, historical characteristics, social structure, the relationship with the surrounding area, the decision-making standards and decision-making level of the decision maker, etc. Among them, some factors ultimately work through the above three basic factors (all economic factors). For example, mines are opened up in the resource-producing areas, and related industries, towns and populations gather in their vicinity. The form is determined by the resource distribution and is actually formed due to the impact of transportation costs and agglomeration effects. In addition, there are some factors that do not play a role independently through the above three basic economic factors, such as decision-making standards and levels and national characteristics (production, consumption, housing, etc.) that have been formed for a long time in history. If you want to find a dominant power, the decision-making standard and level of the decision-maker are the most critical. It reflects how people recognize and deal with the factors above. If the relationship and requirements between them can be handled scientifically from a holistic perspective, and the contradictions and conflicts can be resolved correctly, various components in space will be coordinated with each other, so that the socioeconomic objects can be adapted to the natural economic foundation of the region. In many cases, decision-makers represent the interests of various classes and groups. Many of their decision-making standards are the same, which are often rational and in line with economic principles in terms of local and individual location decision-making and decision-making of inter-regional flow of personnel, materials and finance. However, it is often blind and passive to a considerable degree to determine the overall spatial structure. After World War II, some Western countries developed and implemented a series of regional policies and measures to adjust the spatial structure not meeting the requirements, and meanwhile, established a new spatial structure to avoid passivity to a certain degree. The state guides the location decision-making of various regions and departments by developing a unified socioeconomic development plan and productivity layout plan, which objectively creates the preconditions for establishing a reasonable spatial structure. However, in the practice of regional exploitation and development, various departments and regions will also have contradictions in the development goals and the occupation and utilization of the space and resources required to achieve the goals. The decision-making process is often extremely complicated. Therefore, it is hard to avoid mistakes, and unreasonable spatial structures may also generate.

The spatial structure characteristics are closely related to the level of social and economic development and welfare standards. Firstly, the same factor functions in different scales and ways at different development levels. In addition, new factors will affect the characteristics of the spatial structure. For example, with the development of society, the use of land for agriculture is relatively reduced, and the scale of development of industry, transportation and towns is less affected by the land quality and area than agriculture. In general, due to the expansion of the spatial flow and communication volume of goods and all means of production (except for land), and the increase in frequency and speed, the role of freight is relatively reduced, and the scale of the best enterprise increases to a certain degree. With the advancement of productivity and technology and the increase in national and personal income, the role of freight and agglomeration factors is relatively reduced. The decision-making standard changes from maximum economic benefits to taking into account economic, social and ecological benefits, and it is possible to develop land and resources in a more balanced way. Through government subsidy, investment in the construction of regional infrastructure, etc., the social economy in remote areas and areas with poor traffic conditions is developed.

4) China's basic categories in the research and practice of spatial structure theory

(1) Relationship between concentration and decentralization. China began to introduce technology and equipment from the former Soviet Union for industrialization in the 1950s. Through the practice of regional planning for industrial areas carried out at that time, the planning and academic circles raised the issue of the relationship between concentration and decentralization. In the highest national decision, the orientation of the macro pattern of territorial development is to change the unreasonable industrial layout framework formed in old China for a long time. It was believed that coastal industries and cities were too concentrated. The national industry shall be appropriately dispersed, and the construction of the inland shall be strengthened. In 1956, the report "On the Ten Major Relationships" made by Chairman Mao Zedong at the Supreme State Conference included an exposition on the relationship between concentration and decentralization. Due to the emphasis on inland development during the "first five-year plan" period, the role of the coastal old industrial cities and economic centers was not fully played. In Chairman Mao Zedong's report, it was emphasized that the role of the old industrial cities along the eastern coast should be fully played, that is, "the industrial bases along the coast must be fully utilized and developed, so as to have more strength to develop and support inland industries" [3]. However, due to the influence of ultra-left trends of thought such as the "Great Leap Forward", this correct macro strategy was not implemented. In the 1960s and 1970s, national leaders made the decision to carry out large-scale "third-front" construction due to the international situation at the time. Widespread decentralization was shown in terms of national construction, territorial development and regional development. This situation was not changed until the early 1980s. Since the 1980s, the space strategy of China's regional development showed a trend of centralization to a considerable degree again. In academic circles, there had been "coastal strategies" that leaned towards the coastal areas, "gradient transition strategies" for gradual spreading from the coast to the central and western regions, and "leapfrog strategies" that greatly moved the development center to the west (see Chap. 3 of this book).

(2) Coast and inland. The issue of the relationship between the coast and the inland is one of the core issues that have always affected decision-making at the highest level in China. If the emphasis is put on the development of coastal areas, agglomeration will be the dominant trend in the social economy; if the emphasis is put on the development and utilization of western resources and the large-scale development of the central and western regions, decentralization will be objectively taken as the main goal for the national productivity distribution. In

academic circles, the academic viewpoint on the relationship between the coast and the inland is basically the same as that on centralization and decentralization for a specific person.

(3) Reasonable urban scale and urban system. The reasonable urban scale and the urban system is an important aspect of the socio-economic spatial structure within the meso scope, and has always been an issue emphasized by the regional planning and urban planning circles; since the mid-1970s, the academic circles in China have begun to extensively study the rational scale of cities, the urban scale and structure in China and the policy issues thereof. The starting points of the research on the reasonable scale of cities are generally as follows: how to make the internal structure of the city "facilitate production and life", how to reasonably determine the urban scale according to the nature of the city, etc. However, no one has conducted an in-depth exploration of how to quantify a specific evaluation in practice or has proposed a more recognized evaluation target and evaluation indicator system. Internationally, very few geographers and urban planning scholars in Western industrialized countries believe that it is necessary to plan the best urban scale. Only the location theory scholars and spatial structure theory scholars, starting from their respective theories and assumptions, have deduced the optimal urban scale (curve) from a purely theoretical perspective. In the former CMEA countries, it was generally proposed that urban planning and construction shall be carried out according to the best scale in practice, which was mainly based on economic standards. For example, the Polish economist B. Maresz used the "threshold theory" as the limit of the reasonable scale of urban development in his book *Urban Construction Economy* [4], and scholars of the former Democratic Republic of Germany proposed to find the best urban scale and structure in a certain area based on the most economical standard for the total regional infrastructure construction in the 1980s.

(4) The Third-front and the Three Regions Construction was arranged according to Front 1, 2 and 3 from the 1960s to the mid-1970s. As a result, the original long-term socio-economic spatial structure was changed to a certain degree, so that the national spatial structure, which was originally quite unbalanced, became more balanced. However, a heavy price was paid for this, even if the national economic growth rate declined at this stage. The "Development Strategy for Coastal Areas" implemented in the Sixth Five-Year Plan and the 'Three Regions" strategy implemented in the Seventh Five-Year Plan were also the same in essence. Moreover, the "Three Regions" policy was developed from

the "Development Strategy for Coastal Areas"[1] during the "Sixth Five-Year Plan" period, which was standardized and operable.

(5) Point-axis system and the T-shaped structure of China's regional economic and territorial development. At the end of September 1984, the author made a report themed "The Scientific Basis of the General Map of China's Industrial Productivity Distribution in 2000" at the national symposium on economic geography convened in Urumqi. He proposed the "point-axis system" theory of socio-economic spatial structure for the first time, and also proposed that China's regional economic development and regional economic development in the next few decades would take the coastal areas and areas along the Yangtze River as the primary axis, both forming a T-shaped macroscopic framework. In the past decade, this theory and model have been widely used throughout the country and in various regions.

2 Evolution Stage and Control of the Socio-Economic Spatial Structure

1. Spatial structure is a function of historical development

The spatial structure of the social economy is constantly changing and developing. The reason is that people often solve the problem of location selection in different ways at different stages, that is, countless individuals determine the best location from different viewpoints and angles. As a result, all socio-economic objects are gathered (or dispersed) in roughly the same direction. In this way, the social economy of different regions at the same development stage generally shows basically the same pattern or framework in terms of spatial structure.

Human beings have a common stage of socio-economic development. In the process of social development from a low level to a high level, the social structure as well as economic and technical levels are gradually evolving and improving. Even if the social systems of countries and regions are different, they still have the same rules for productivity distribution and the same characteristics of spatial structure. For a long time, there has been considerable disagreement on whether this law is universal in socialist countries and capitalist countries. It should be said that both

[1] The "Development Strategy for Coastal Areas" implemented during the Sixth Five-Year Plan period (1981-1985) was developed based on the fact that the emphasis of China's economic construction was put on the "third-linr areas" in the central and western regions in the 1960s and 1970s. According to this strategy, the state first implemented an opening-up policy in some coastal areas, increased investment in ports, power stations and other transportation, communication and energy infrastructures, strengthened the basic raw material industry, accelerated the pace of urban modernization, and adjusted the industrial structure, etc.. During this period, the state's investment in fixed assets in coastal areas accounted for approximately 51% of the total in China (not including investment regardless of regions), but the net value added of the gross national product accounted for 62% of the total in China. Therefore, the national economic growth rate was accelerated, and huge social and economic benefits were obtained. As a result, the development strategy of Three Regions was formally proposed in the "Seventh Five-Year Plan".

capitalism and socialism are based on modern mass production, especially in terms of industry, transportation, and urban construction. Their development and distribution are affected by natural and technical factors, and the economic benefits and convenience for people's production and life shall be considered. Therefore, there is no difference in the socio-economic spatial organization formed under normal circumstances. For example, the formation and deepening of the division of labor in society is the product of social progress, and one of the results is the formation and development of cities. There are many location factors. Among them, cities are often formed in favorable positions at the intersections of traffic arteries. Cities radiate outwards along the traffic line they emit, and have hinterland (attraction ranges) of different sizes according to their economic strength and service functions; industrial areas based on import and export trade often appear near seaports; the agglomeration areas of different scales and degrees have been formed in many countries today. Some countries also have the problem of over-density or over-sparseness and a series of socio-economic and environmental problems caused therefrom; land utilization in urban and suburban areas has almost the same categories and layout characteristics. However, different countries and regions have different overall levels of productivity development, different abundances of resources such as land and minerals, and social structures and consumption characteristics that are not exactly the same. In countries or regions with abundant land resources but low productivity levels, the degree of population and economic agglomeration is generally small, and the population density in urban areas is also small. Now, there are no spatial structures that are not exactly the same even in countries with the same social system, including the scale of agglomeration, the series of urban levels, the categories of land utilization and their distribution characteristics. Here, the characteristics of the country or region itself are one of the important factors leading to differences in spatial structure.

In terms of the level of socio-economic development, human society can be divided into a series of stages. In the long-term primitive society, most human beings lived on hunting. Feudal society was dominated by agriculture and handicraft industry, and modern industry started from the capitalist stage. The social economy and its various structural states have been constantly changing by the end of the twentieth century. The previous stage was always injected into the later development as a historical feature. This is the view of structural evolution that we should have.

2. **Four stages of socio-economic development and their spatial structure characteristics**

In order to systematically reveal the general characteristics of the spatial structure at different development stages, we exclude the influence of factors such as social systems and the relative position of land and sea, and the premise is basically the same as that assumed by location theorists, that is, the differences in natural conditions within the region are not considered.

1) Stage where the agricultural economy is dominant

This stage has gone through a long historical period, during which the level of productivity development is low, and the production of each member of society and the social grass-root organization can only basically meet its own needs. The vast majority of people are engaged in agriculture in a broad sense. Within a certain geographical area, residential areas emerge at different intervals, which are mainly taken as the places of residence for farmers who cultivate the surrounding land and basically do not undertake the service function. Of course, the commodity economy gradually develops, and various types of small cities slowly emerge, which, however, are mainly taken as the place for commodity exchange and a small-scale management center, with a small scope of influence. The hinterland between cities is also unconnected, that is, there is no hierarchical relationship. There is no affiliation among cities in terms of nature and scale. This situation can be explained as a low-level residential area distribution system. There are few people, materials, and information exchanges between urban and rural areas, the level of regional infrastructure such as roads is low, and a distribution network cannot be formed. Therefore, the economy is stagnant, and social development is slow. Among the basic factors that lead to rapid changes and differences in spatial structure, some have not yet appeared, and some have a weak influence so that the spatial structure state is of great stability, that is, in a "balanced" state. Therefore, the basic characteristics of the socio-economic spatial structure at this stage can be summarized as follows:

(1) The problem of unbalanced economic development among large regions, that is, density, is not formed;
(2) The framework of socio-economic spatial organization is in its original state, the hierarchical characteristics of the road network are not obvious, and the function of the development axis cannot be formed;
(3) The urban residential areas in the region do not form a hierarchy-scale system, and the relationship between them, that is, the distribution of hierarchy-scale, presents a gently sloping curve (an in Fig. 2).

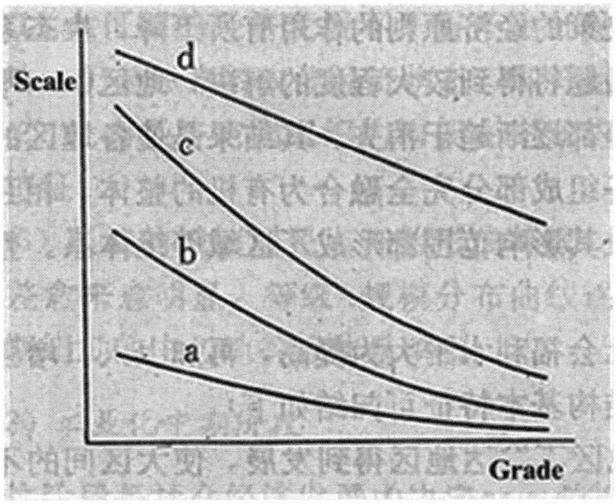

Fig. 2 Hierarchy-scale curve of urban residential area at different development stages of the region

This structure state conforms to the spatial structure model derived from Thunen's agricultural location theory.

2) Stage of transition from an agricultural economy to industrialization

The social economy develops rapidly due to the reform within the society and the changes in external conditions: The social division of labor becomes apparent, agriculture develops, prosperous handicraft industry and mining, as well as small-scale raw material industry and manufacturing industries, emerge, water transportation is opened up, and railway network and road transportation begin to appear. In this way, the scale of commodity production and exchange is greatly expanded compared with that in the previous stage. As a result, a large number of rural people flow into cities, and urban–rural exchanges become frequent. Cities that used to develop extremely slowly are expanding, and new towns are formed near the mining industry and ports. Regional economic growth is concentrated in the center of a region of the city. The development of the center takes in the nutrition (including resources and manpower) within the surrounding area and the influence scope and exerts its economic and social functions (exporting commodities and exerting political and economic leadership and organizational functions). The imbalance in the degree of spatial agglomeration appears, and the development in the fringe areas far away from cities and countries (regions) is still backward, basically the same as that in the previous stage. Therefore, the spatial structure is unstable.

During the transition from the stage where the agricultural economy is dominant to the initial stage of industrialization, there are very different durations in different countries, which mainly depends on the political stability and the advancement of the political system. The basic characteristics of the socio-economic spatial structure at this stage can be summarized as follows:

(1) Unbalanced economic development among regions began to appear. In other words, development in some areas begins to accelerate, while other areas are still in the original state of the agricultural economy, that is, a certain spatial economic gradient begins to form.
(2) The framework of the socio-economic spatial organization starts to be in a point-axis state in the regions that developed first, forming the hierarchical characteristics.
(3) The urban residential areas in the region begin to form a hierarchy-scale system. Generally, the scale differences among cities with adjacent hierarchies become more and more obvious, and the hierarchy-scale distribution curve gradually becomes steeper (b in Fig. 2). This structure state is in line with W. Christaller's model of the central place hierarchy system.

3) Middle stage of industrialization

This stage is a decisive turning point in socioeconomic development. Due to the preparation in the previous stage and social transformation, productivity is further liberated. The social and private investment capability is expanded, and national income increases substantially. Through the redistribution of input capital and national income, 10–20% or more of the funds are used to expand reproduction.

Therefore, the national economy enters a period of strong dynamic growth. The main sectors that develop are iron and steel, machinery, chemicals, power, textiles, etc. At the same time, science and technology develop rapidly, the tertiary industries start to appear in large numbers, a dense transportation network penetrates into all parts of the region, and a variety of transportation methods constitute a comprehensive system. While large cities and agglomeration areas continue to develop, the regional second-level and third-level centers are strengthened along with resource development and economic development, and the pure "center-periphery" structure at the previous stage gradually becomes a multi-core structure. Exchanges and communication among cities are extremely frequent. The resources and potential of backward areas are increasingly attracted to the economic cycle and allocated to the original centers, thus forming new centers. At this time, you can see the initial state of the functional division and hierarchy of cities: Generally, the larger the city, the stronger its comprehensiveness, and the larger its attraction range; the second-level cities in the region are within the attraction range of the first-level cities, and the level and type of service industry are inevitably lower than those of the higher-level cities. Since the economic strength is not very strong, not all the fringe areas are fully developed. The spatial structure is still changing.

In general, the economic growth rate at the middle stage of industrialization is relatively high. The duration of this stage in various countries and regions varies greatly due to various reasons. Although this stage of development is quite short in its long history (30–100 years), the changes in socio-economic development and human changes are amazing, and the most prominent manifestation is the intense economic industrialization and social urbanization. Accordantly, changes in the spatial structure are also rapid. The basic characteristics of the socio-economic spatial structure at this stage can be summarized as follows:

(1) The degree of imbalance of development (exploitation) among large regions is getting higher and higher, and the principle of agglomeration economies is dominant in the decision-making of socio-economic location. In this case, the research on regional development and regional policies aiming at eliminating the imbalance among large regions is taken seriously in general.
(2) The framework of the socio-economic spatial organization is formed, and the development axis and urban residential areas form a "point-axis system". Society and industries are mainly concentrated in high-level cities and axes.
(3) Due to the strong effect of factors of agglomeration economies, big cities are increasingly developing, and the included angle between the scale-hierarchy curve and the horizontal ordinate reaches the maximum (c in Fig. 2).
4) Later stage of industrialization and post-industrialization stage

Science and technology are highly developed, and are widely used in various sectors of the national economy, urban–rural construction, the development and protection of territorial resources, etc. The growth rate of production greatly exceeds that of the population. The income of the society and its members reaches a very high level, and there is a modern and developed transportation and communication system. Computers and information systems penetrate into various fields of people's work and life. At this stage, the role of economic principles in location decision-making of socio-economic objects declines, and more attention is paid to the principle of ecological system balance. Therefore, the problems of over-density and over-sparseness will be solved to a greater degree, and the imbalance among regions and differences in employment, income, consumption levels and choice opportunities will gradually disappear. The result is that the space and resources of various regions are utilized in a more sufficient and rational way, and the components of the spatial structure are completely integrated into an organic whole, interacting with and depending on each other. Accordantly, urban residential areas, service facilities and their influence scope form a regional hierarchy system respectively. The entire spatial structure system is able to restore "balance".

At this stage, the industrial structure tends to be more advanced, the social welfare standards are greatly improved, and the population growth slows down, thus making the economic growth rate decline. The basic characteristics of its socio-economic spatial structure can be summarized as follows:

(1) As the problems of over-sparseness and over-density are gradually solved, the underdeveloped areas and undeveloped areas develop, so the imbalance among large regions is less and less obvious. The role of agglomeration economies in location decision-making declines, while more and more attention is paid to balancing distribution and exerting social and ecological benefits.
(2) A perfect "point-axis" spatial structure system is formed, but the difference in the hierarchy is smaller and smaller.
(3) As the effect of agglomeration factors is not so strong, while the effect of decentralization is more and more important, the hierarchy-scale curve of urban residential areas in the region becomes flat again (d in Fig. 2).

3. **Regulation of the conversion of socio-economic spatial structure**

The above four stages of spatial structure evolution reflect the general law of socio-economic spatial agglomeration or dispersion trends. According to the objective laws of the market and the behavior standards of social-economic activities of human beings, the spatial structure always transitions from a lower level to a higher level, thus promoting the sustainable development of the social economy, which is a matter of common sense and is the interaction result of economic "power" and social "power". However, the policies and the society are required for necessary macro-control of the time for evolution between stages, the corresponding regional development direction, control measures and other issues.

From the evolution of spatial structure at the above four stages, it can be seen that: The spatial structure of the social economy is "balanced" theoretically in the long agricultural society. Of course, this original balance state is manifested on the basis of villages or small towns. With the development of the social economy, agglomeration becomes the main trend. At the later stage of industrialization or the post-industrialization stage, the spatial structure of the social economy returns to a "balanced" state. Now we want to ask: What factors cause the change (or conversion) of the spatial structure at different development stages? Can this conversion process be artificially controlled and policy-oriented?

The root cause for the evolution at four stages of the socio-economic spatial structure is economic factors, that is, the agglomeration economies lead to the conversion from the first stage to the second and third stages. Afterward, the allowed boundary is exceeded due to the development of the "agglomeration economies", forming the so-called "diseconomies of agglomeration", and a high-level "balance" state appears.

1) Standards for judging the need to adjust the socio-economic spatial structure
(1) Judgment of the stage of socio-economic development. Generally speaking, the spatial structure characteristics of the above four stages are universally representative. For an economically developed region or country at the later stage of industrialization, it is generally required to make macro-adjustment to the spatial structure, that is, to develop new areas, and gradually achieve more balanced regional development. When a region is at the early stage of industrialization, if it is required to continue industrialization and urbanization in a decentralized way like an agricultural society at this time, the industrialization process will not proceed smoothly. Because the limited capital and material resources are dispersed to an excessive geographic range, it is impossible to concentrate the capital and material resources required for industry and urban construction in a smaller range.
(2) Judgment of the degree of spatial agglomeration and agglomeration economies. To judge the relative concentration of industry, cities and transportation within the region, the following indicators can be considered:

   a. Hierarchy-scale structure relationship of cities in the region. If the gap in scale among various levels is large, it indicates that cities are relatively agglomerated. Otherwise, they are relatively balanced. Also taking into account other indicators, one can judge whether it is necessary to adjust the spatial structure.
   b. Growth rate of the regional economy in the medium and long term. In the initial stage of rapid growth, agglomeration shall be taken as the main trend, and measures shall be taken to implement centralized development, that is, to start the second stage mentioned above; after decades of rapid growth, it is inevitably required to change to a more balanced growth (layout) and to develop new districts, that is, to start the fourth development stage mentioned above. That is to say, what corresponds to the hierarchy-scale system of the urban residential area of the above four stages is that different development stages have different economic growth rates (Fig. 3).
   c. Is the urban socio-economic development coordinated? In the process of regional socio-economic development in modern times, if the cities are too large and too agglomerated, it will cause many inconveniences in urban production and life and insufficient supply of infrastructure; there will be a serious decline in environmental quality, and social and economic conditions will also be affected to different degrees.
   d. Percentage of the floor area (built-up area) in the total town area for Front 1 and Front 2 cities. If the proportion is very large, it indicates that the overdense state shall be changed.

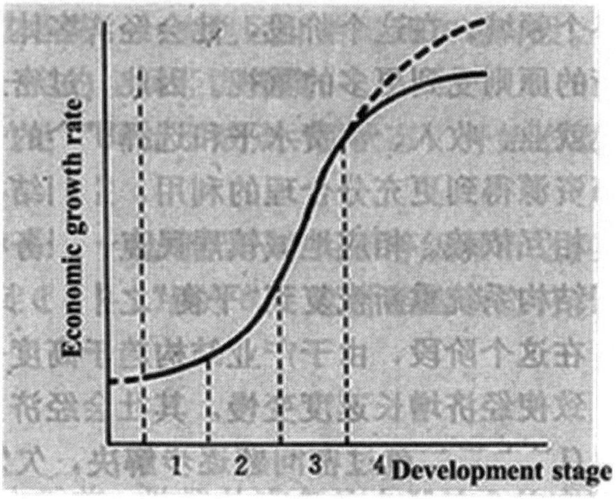

Fig. 3 Economic growth rate at different stages of socio-economic development

2) Possible measures to adjust the socio-economic spatial structure

To adjust the spatial structure of the social economy, regardless of the transition from the first stage to the second and third stages, or the transition from the third stage to the fourth stage, the measures taken are manifested as follows:

(1) Orientation of major industrial construction projects. Whether such large or extra-large projects are built in urban areas that are constantly agglomerating or in new areas in the process of decentralization, they will attract some enterprises, cities and public facilities to the area where they are located, which is strongly oriented. If one or even several large projects are arranged in the new area to be developed, the development of the new area can be driven, so as to achieve the purpose of adjusting the spatial structure.
(2) Orientation of infrastructure. In order to promote the development of the new area, some major infrastructure projects, such as railways, ports, power stations, etc., shall be constructed first in accordance with the predetermined overall planning for regional exploitation. The construction of these infrastructures can improve the layout conditions of industries and cities; under the conditions of a market economy, the government is often driven by infrastructure construction to promote the development of new areas. Once the main infrastructure of the new area (generally roads, water supply, communication and energy) can basically meet the needs of large-scale construction, the specific development depends on the company and the competent economic department.
(3) Policy orientation. It refers to the government developing and implementing several policies that are favorable for or encourage the development of new areas, so as to attract enterprises and companies to invest. These policies involve the lease of land by lot and its price, taxation, import, and export trade operation rights, import and export quota freight rates, tariff policies and so on.

3) Instances—adjustment of spatial structure in the Beijing–Tianjin–Tangshan region

A. Two different schemes for the development of spatial structure

The Beijing–Tianjin–Tangshan region stated here follows the name before the mid-1980s. In addition to Beijing, Tianjin and Tangshan in Hebei, it also includes Qinhuangdao and Langfang of Hebei. The land area is more than 50,000 km^2. The gross national product in 1992 was 146 billion yuan, accounting for 6.1% of the total in China; the total industrial and agricultural output value was 245.4 billion yuan, accounting for 5.8% of the total in China. This area is the second largest agglomeration area in China, second only to the Yangtze River Delta.

The strategic concept of adjusting the spatial structure of the Beijing–Tianjin–Tangshan region was put forward when the State Planning Commission organized the territorial planning of the Beijing–Tianjin–Tangshan region in the early 1980s.

In 1980, the total industrial output value of the Beijing–Tianjin–Tangshan region was 47.5 billion yuan, accounting for 9.5% of the total of China. The urban population was 10.58 million, accounting for 40% of the total population. Among the urban population of 10.58 million, 86% were concentrated in Beijing and Tianjin.

The total population in the urban area of Beijing reached 5.97 million in 1980 (including the non-agricultural population of 4.77 million), with a population of 2181 per square kilometers. The total population in the urban area of Tianjin reached 5.33 million (including the non-agricultural population of 3.92 million), and the average density is 1247. In terms of industrial distribution in Beijing (judging from the number of enterprises), the urban area was the densest, followed by the third ring, the inner suburban area and the outer suburban area in sequence. The urban and inner suburban areas, accounting for 8.5% of the area of the city, occupied 69.2% of the industrial enterprises, 76.6% of industrial workers, 71.6% of industrial fixed assets, and 76.2% of industrial output value; while the outer suburban area, accounting for 91.5% of the area of the city, only occupied 31.8% of the industrial enterprises, 23.4% of industrial workers, 28.4% of industrial fixed assets, and 23.8% of industrial output value. There were 12.8 industrial enterprises and 5032 industrial workers per square kilometer in the urban area of Beijing, and the industrial land area accounted for 19% of the land area. This phenomenon of over-concentration and crowding was especially severe in Xuanwu and Chongwen districts. Within the area of 32.4 km^2 of the two districts, there were 605 enterprises, with an average of 19 enterprises per square kilometer. The land area for enterprises in Xuanwu District accounted for 56% of the total area, which had almost become a concentrated industrial area. Such a situation where the industries were highly concentrated in the downtown was also rare in many large cities in foreign countries. 83% of the urban population in the whole city was concentrated within the urban area. The population density in the urban built-up area with an area of only 346 km^2 had reached 13,600 persons per square kilometer, surpassing that of London (9000 persons per square kilometer), Paris (8000 persons per square kilometer), Washington (more than 4000 persons per square kilometer) and Moscow (9500 persons per square kilometer).

The over-concentration of industries and population in the urban area of Tianjin went even further than that of Beijing. Among the 4171 factories and enterprises in the city, those distributed in the downtown area accounted for 68.3%, those distributed in the inner suburban area accounted for 16.7%, and those distributed in five counties in the outer suburban area only accounted for 15.0%. Within the area of 154 km^2 of the downtown area, there were 2844 factories and enterprises, with an average of 18.5 enterprises per square kilometer. 80% of the industrial output value was concentrated in the downtown area and the inner suburban area. Due to the large population in the urban area and the over-concentration of industries in the two metropolises, Beijing and Tianjin, a series of knotty problems had arisen, such as environmental deterioration, shortage of land, lack of water sources, energy shortage, traffic congestion, and large debts in housing and municipal facilities, which became the focus of various contradictions encountered during the regional exploitation.

In sharp contrast to the high concentration of industries and population in Beijing and Tianjin, the development of industries and population in small and medium-sized towns in the Beijing–Tianjin–Tangshan region was relatively slow. Due to the lack of investment in small and medium-sized towns, the construction standards of urban life service facilities were low, and the attractiveness was poor. Some enterprises were reluctant to take root in small and medium-sized cities and towns, making the rational distribution of industries unable to be carried out smoothly, and there was great resistance to dredging the population of large cities. The seriousness of the problems lies in the fact that although Beijing and Tianjin had clearly shown many disadvantages of the over-concentration of industries and population, there was still a tendency for continuous concentration. During the period from 1976 to 1982, the population in the urban area of Beijing increased by 980,000, with an annual growth rate as high as 3.9%. The population in the urban area of Tianjin increased by 590,000, with an annual growth rate of 2.6%. Large cities have better infrastructure and social, cultural, and life service facilities, as well as more opportunities for cooperation and employment, which are more attractive to enterprises and employees. Therefore, some enterprises will carry out large-scale expansion and new construction in urban areas in the name of transforming old enterprises and taking the road of connotative development, and some rural and non-local people will also try to squeeze into urban areas. If uniform planning is not performed based on long-term interests and from an overall perspective, a series of problems caused by the over-concentration of industries and population in the Beijing–Tianjin–Tangshan region will be further aggravated in 20 years.

At that time, we made a specific assumption: If the spatial structure is not adjusted, social and economic facilities will continue to agglomerate in the urban and inner suburb areas of Beijing, Tianjin, and Tangshan. From the mid-1980s to 2000, the following will be increased in the urban and inner suburban areas of the three cities on the original basis: the urban population of two to three million, the annual consumption of freshwater of about five billion cubic meters, high-quality arable land of about 250 km^2, and railway freight volume (arrival and departure) of 20–30 million tons. The emissions of others such as "three wastes" will also increase substantially. This is an extremely heavy burden on the socio-economic development of the urbanized areas that have been quite agglomerated and over-expanded, and the cost of solving these problems is huge.

B. Main direction of spatial structure adjustment

At that time, the author believed that the spatial structure must be adjusted from the perspective of rational development and utilization of territorial resources in the medium and long term. Therefore, the author proposed another scheme at that time, that is, to adjust the spatial structure of the social economy in the Beijing–Tianjin–Tangshan region at three different levels and regional categories: With regard to the relationship between the coastal zone and the inland, the focus of territorial development and construction layout shall be promoted to the coastal zone in the future; in terms of the relationship between Beijing, Tianjin and Eastern Hebei (Tangshan and Tianjin at present), it is necessary to focus on the development of Eastern Hebei in the future; in terms of the development focus in the two metropolises, Beijing and Tianjin, it is necessary to focus on the development of small towns in the outer suburbs.

These three basic formulations and specific elaborations were written in the State Development Planning Commission's research report on the comprehensive planning of territorial development in the Beijing–Tianjin–Tangshan region. According to this scheme, it is necessary to make full use of the relatively vast and rich land resources, water (seawater, for industrial cooling, etc.) resources, port transportation conditions, as well as more suitable pollutant control and dilution conditions in the coastal zone. The development facts since the 1980s have also proved that this is one of the important ways to bring the economy of North China and even some regions in Northwest China to the world market.

The large-scale planning research organized by the State Development Planning Commission, involving more than 500 people from many departments of the State Council, Beijing, Tianjin, Hebei, scientific research, and university departments, had a great impact. Since then, Tianjin Binhai and Eastern Hebei have accelerated the pace of construction. As of 1992, the total industrial output value of Tianjin Binhai accounted for 40% of that in the whole city. The economic strength of Tangshan and Qinhuangdao had also been substantially improved.

The main directions and specific requirements proposed at that time for the territorial development and construction layout were as follows:

(1) Advance to the coastal zone. The coastline of the Beijing–Tianjin–Tangshan region is 564 km long, there are more than one million mu of saline-alkali wasteland in the coastal zone, with large areas of low-lying land, reed fields and salt fields, and the petroleum resources have good prospects. In addition to Tianjin, Tanggu New Port and Qinhuangdao Port, Laoting Yutan also has the conditions to build a deep-water port. Dagang Power Plant has built large-scale hydraulic facilities that use seawater for cooling. The industrial and urban layout of the Beijing–Tianjin–Tangshan region is advancing towards the coastal zone. The advantages of developing steel, electric power, petrochemical and marine chemical industries that require a large amount of land, water consumption and

transportation and cause huge pollution in the coastal zone are very obvious, that is we can save the precious arable land and freshwater resources to the maximum limit, and use cheap water transportation at the port nearby, so as to reduce the huge pressure on trunk railway transportation. The coastal zone has good atmospheric diffusion conditions, strong sewage dilution ability, large environmental capacity, and much lower sensitivity to environmental pollution compared with the urban areas of the three major cities. The impact of human activities on the environment is relatively small. In a word, the relationship between population, economy, resources and ecology can be easier to achieve coordination by developing industries and building cities in the coastal zone.

(2) Focus on the development of Eastern Hebei. The rich mineral resources such as iron ore, coal, limestone and refractory clay in the Beijing–Tianjin–Tangshan region are mainly concentrated in the Eastern Hebei, especially in the middle and lower reaches of the Luanhe River. The development potential of water resources in the Luanhe River Basin is also greater than that in other regions of Beijing, Tianjin and Tangshan. In the early 1980s, power stations, reservoirs, water diversion projects and several trunk railways that have been built or are under construction will further improve the construction conditions in Eastern Hebei. In 1980, the total industrial output value of Eastern Hebei accounted for only 8.7% of that of the Beijing–Tianjin–Tangshan, which was mainly concentrated in the urban area of Tangshan. There were not many industries in the regions beyond Tangshan. In order to develop steel, building materials and other raw material industries in the Beijing–Tianjin–Tangshan region, it is necessary to focus on the development of mineral resources in the middle and lower reaches of the Luanhe River, and drive the development of processing industries and towns on the basis of resource development and utilization. However, the raw material industry established on the basis of the development of mineral resources has a large transportation volume, large floor area, high water consumption and high pollution degree. Therefore, new large enterprises, including steel, electric power, building materials, chemical industry, etc., shall also be built in the coastal area of Eastern Hebei as much as possible.

(3) Develop small towns in outer suburbs. Large cities are more attractive to the development of processing industries, especially technology-intensive and knowledge-intensive industries. It is necessary to move some large and medium-sized industrial enterprises that must be newly built and are highly dependent on big cities in small towns in outer suburbs in combination with the adjustment of the urban layout. However, there must be convenient transport links between these small towns and urban areas. At present, there are four or five satellite cities for Beijing and Tianjin respectively. Various facilities are still very incomplete. During the next two "5-year plan" periods, we shall focus on building two or three satellite cities, so that they can play a role in controlling the dispersion and interception of urban development in big cities.

At that time, it was also proposed that the spatial structure adjustment in the Beijing–Tianjin–Tangshan region shall not be made in three directions at the same time, but shall be made in a step-by-step manner according to the growth of national economic strength. In the 1980s, the emphasis was put on developing satellite towns around big cities and industrial areas in outer suburbs. For the development of the coastal zone, the emphasis shall be put on the Tanggu District of Tianjin Binhai, and the construction of ports, towns and industrial areas shall be carried out. In Eastern Hebei, the existing backbone enterprises can continue to build supporting facilities, and the development of mining areas and infrastructure construction can be strengthened. After entering the 1990s, Eastern Hebei and the entire coastal zone can be developed on a large scale.

3 Imbalance of Regional Development

1. **Universality and severity of unbalanced development**
1) Contribution of unbalanced development strategy implemented in China

In the book *Location Theory and Regional Research Methods* published in 1988, the author initially discussed the issue of economic growth and regional balanced development (pp. 100–101), and cited some facts to prove the inverted U-shaped correlation law between economic growth and regional unbalanced development, which was proposed by J. Williamson. These statements have been confirmed by the fact of the spatial distribution of the national regional economic growth in China for decades, especially since the 1980s. In the early 1980s, China proposed the "Development Strategy for Coastal Areas" after summarizing the inappropriateness of the macro layout and the requirements for implementing the policy of opening to the outside world during the "third-front" construction period. Afterward, three major zones, that is, east, middle and west, were divided, and the regional inclination strategies were developed accordingly. The correctness of the national macro-regional policies became one of the important factors for China's long-term economic growth since the 1980s. From 1981 to 1990, the average annual growth rate of GNP in coastal areas (12 provinces, municipalities and autonomous regions) was about 1% higher than that of the whole country. Since the GNP in the coastal areas accounted for more than half of that of the whole country, its contribution to the national economic growth was very great. It can be said to be the backbone of the sustained and rapid growth of the national economy as well as the rapid improvement of comprehensive national strength in the 1980s.

2) Widening trend of the gap among regions in China

In the process of advocating to give full play to the advantages and realizing the rapid growth of the national economy, the gap among large regions, especially between the coastal regions and the central and western regions, is widening. In terms of the per capita GDP of provinces and autonomous regions, the gap between Guizhou in the western region and Liaoning and Guangdong was 502 and 192 yuan respectively in 1978, which expanded to 1817 and 1933 yuan respectively in 1991. For another example, the gap in the per capita GDP between the Ningxia Autonomous Region and Liaoning and Guangdong was 323 and 13 yuan respectively in 1978, which expanded to 1256 and 1372 yuan respectively in 1991 (Table 1). The absolute difference in per capita GDP was so large! If the central government fails to implement policies to narrow regional disparities, it may endanger social stability in the future.

Table 1 shows that there are great differences among regions in China in terms of income per capita. The per capita GDP of low-income regions is only 64% of the national level, while that of high-income regions is 190% of the national level, three times that of low-income regions.

The same is true when measured by the economic growth rate. From 1979 to 1992, the average annual growth rate of the national GDP was 9.0%. Guangdong had the highest growth rate, 13.3%, followed by Zhejiang (12.5%), Fujian (11.6%), Shandong and Jiangsu (11.1%).

3) Basic reasons for development imbalance

(1) Natural geographical environment. "Determinism of geographical environments" has become popular since the last century. This theory has been criticized by some people in society and academic circles since this century. This theory is of course correct from the following perspective: High mountains, topography, particularly unfavorable climates, etc. hinder the rapid development of the social economy, causing the gap in the development level of the region with fertile soil and mild climate in the plain. However, it is wrong to generally believe that differences in natural conditions will inevitably lead to a gap in the level of socio-economic development

The most fundamental reason for the differences in population density, economic density and economic strength among regions in China at this stage is that China has three natural regions and three gradient terrains [5]. The three comprehensive natural regions, that is, the eastern monsoon region, the northwest arid region and the Qinghai–Tibet alpine region, fundamentally affect and even determine the gaps in the regional socio-economic development in China. Among them, the eastern monsoon region accounts for 45% of the land area in China, with a population accounting for 96% of the total population in China and the total industrial and agricultural output value accounting for about 97% of that in China in 1990.

In terms of the natural geographical environment, regional differences in natural systems (mainly mineral resources, energy, water sources, land, etc.) had a huge impact on regional development in the early stage of industrialization. With the development of inter-regional trade, most of the mineral resources can be transported or imported from other regions, which is basically no longer the reason for the booming of regional differences. However, when other conditions are the same, regions with rich mineral resources may still promote faster development of the social economy and increase in strength. The lack of water sources in northern China has certainly affected the economic growth rate and the growth of per capita income.

Table 1 Changes and ranking of per capita GDP in provinces, municipalities and autonomous regions from 1978 to 1991

Serial no	Region	Per capita GDP (%)				Equivalent to the national average (%)			Growth rate of per capita GDP 1979–1992 (%)
		1978	1980	1985	1991	1978	1985	1991	
Low-income regions									
1	Guizhou	175	217	418	890	46.7	51.4	50.6	7.7
2	Anhui	242	286	618	1052	64.5	75.9	59.8	6.7 (1979–1991)
3	Guangxi	225	278	471	1058	60.0	57.9	60.2	6.5
4	Gansu	348	387	608	1138	92.8	74.7	64.7	6.7
5	Henan	232	317	580	1141	61.9	71.3	64.9	8.4
6	Yunnan	223		483	1147	59.5	59.3	65.2	8.1
7	Sichuan	253	329	598	1180	67.5	73.5	67.1	7.9
8	Jiangxi	276	342	595	1212	73.6	73.1	68.9	7.8
9	Hunan	285	365	622	1280	76.0	76.4	72.8	6.6
10	Shaanxi	294	338	608	1292	78.4	74.7	73.5	7.3
Lower-middle-income regions									
11	Tibet			894	1388		109.8	79.0	6.7
12	Ningxia	354	409	703	1451	94.4	86.4	82.5	6.4
13	Inner Mongolia	307	345	717	1466	81.9	88.1	83.4	6.7 (1979–1991)
14	Shanxi	365	434	809	1467	97.3	99.4	83.4	6.1
15	Hebei	364		719	1545	97.1	88.3	87.9	7.0
16	Hubei	332	428	808	1584	88.5	99.3	90.1	7.7
17	Qinghai	431		808	1592	114.9	99.3	90.6	4.5
18	Hainan			712	1645		87.5	93.6	9.1
19	Jilin	391	450	838	1718	104.3	102.9	97.7	7.2
Higher-middle-income regions									
20	Fujian	273	345	712	1803	72.8	87.5	102.6	9.8
21	Shandong	327	413	839	1876	87.2	103.1	106.7	9.2
22	Xinjiang	313	410	820	2047	83.5	100.7	116.4	9.4
23	Heilongjiang	563	694	1042	2099	150.1	128.1	119.4	5.1
24	Jiangsu	430	541	1053	2143	114.7	129.4	121.9	10.1
25	Zhejiang	470		1029	2310	125.3	126.4	131.4	11.5
High-income regions									
26	Liaoning	677	812	1378	2707	180.5	169.3	154.0	6.7
27	Guangdong	367	473	982	2823	97.9	120.6	160.6	11.5
28	Tianjin	1150	1391	2192	3944	307.0	269.8	224.3	6.5
29	Beijing				5781			328.8	

(continued)

Table 1 (continued)

Serial no	Region	Per capita GDP (%)				Equivalent to the national average (%)			Growth rate of per capita GDP 1979–1992 (%)
		1978	1980	1985	1991	1978	1985	1991	
30	Shanghai	2498	2738	3855	6675	666.1	473.6	379.7	6.9 (1981–1992)
National average	375	456	814	1758	100.0	100.0	100.0	7.6	

① This table is from Hu An'gang's report on national conditions, *Research Report on Development Issues in Underdeveloped Regions,* February 1992. The statistical data is from the statistical yearbooks of China and various provinces, municipalities and districts in 1993
② The per capita GDP in the table is the current price, and the per capita GDP growth rate is calculated according to the comparable price

(2) Geographical location. Geographical location can be regarded as both natural conditions and acquired geopolitical and geoeconomic conditions. Its main connotations include the following aspects:

a. Sea-land relationship. If the region is close to or near the ocean, it is easier to participate in a large range of social-economic activities and to enter a large range (even on a global scale) of the economic core areas. Since it is a long-term trend that modern economies are attracted by the ocean, the economies of such regions and countries are more likely to be more developed than those of landlocked countries and regions.
b. Relative position to the economic core area and big city. Generally speaking, a region (or country) close to a large range of economic core areas or large cities has more development opportunities (investment, business activities, information acquisition, etc.). Therefore, it develops faster than the fringe areas far from the core area. However, in the initial development stage of industrialization, economic resources, human resources, etc. tend to be concentrated, and resources (economic, human and natural resources) located in the vicinity of the core area are attracted to the core area. As a result, the region may not be developed as it should be.
c. Geopolitical alliances and conflicts. Due to international political and military struggles, for the "broken zone" and "conflict zone" for which two or more groups compete [6], the long-term political and economic instability will hinder the development of the social economy.

(3) Historical factors. Strictly speaking, historical factors are also the interaction results of natural environmental factors and social economy in the historical period. Due to inertia, they still have an impact on development in the future. There are roughly the following aspects:

a. Regional differences in the socio-economic basis, which are formed in history. Among them, the more developed regions often have more advanced industries, perfect infrastructure as well as higher cultural literacy and level of science and technology. In the future development, these factors will generally continue to play a role in promoting the rapid development of these regions.

b. National psychological characteristics, values for development, entrepreneurial spirit, etc. formed in the historical period. Generally speaking, nationalities that conquered other nationalities (regions) many times in history have the spirit of stamina and diligence. Once a nationality with a strong sense of nationality and national spirit becomes independent, it will have the intention of striving for greater development. A nationality that once occupied the territories of other countries (nationalities) by force often tends to use its own commodities to occupy the markets of other countries and regions. However, people in underdeveloped regions or countries are often more conservative in concept and lack a sense of crisis and urgency in the face of intense economic competition among regions and countries.

(4) Systems and other economic factors that lead to vicious circles. This is an extremely important reason for the backward economic development in many undeveloped countries and regions in the world today, which may include the following aspects:

a. Over the past half century, great changes have taken place in the world's political economy. The traditional socialist system has proved to have many shortcomings, which are mainly manifested in excessive authority centralization in terms of economic management as well as centralized control over the revenues and expenditures, thus making enterprises and entrepreneurs lack market concepts and a sense of competition. Under this system, the driving force for the technological progress of enterprises is weak, resulting in low social labor productivity. Since the 1980s, China has carried out a large number of reforms, which are reflected in a centralized way in the organization of economic life according to the concept of the market economy. The actual growth in various regions shows that all provinces, municipalities and districts with a high degree of marketization and a low degree of nationalization have faster economic growth. These regions include Guangdong, Zhejiang, Fujian, Shandong, etc. However, some underdeveloped regions, such as Qinghai, Ningxia, Gansu, Yunnan, Guizhou, etc., have not yet gotten rid of the planned economy and the traditional ownership model dominated by the state-owned economy, which are the reasons for the relatively backward economic system in these regions.

b. Poor financial conditions. Underdeveloped regions and countries generally have a low self-financing capacity. Due to the low fiscal absorbability, the revenue is less than the expenditure, so the deficit is more and more severe. Underdeveloped countries mainly rely on debts to survive. An underdeveloped area must rely on financial subsidies from the central government. For example, in 1992, the self-financing rates (the proportion of local fiscal revenue to fiscal expenditure) of the following regions in China were as follows: 6.5% for Tibet, 48.3% for Ningxia, 59.1% for Inner

Mongolia (figure in 1991), 68.5% for Anhui (figure in 1993) and 74.7% for Gansu. The "fiscal gap" that is difficult to fill is a common phenomenon in underdeveloped regions.

2. **Re-demonstration of the law of interaction between economic growth and regional unbalanced development**

1) Role of inverted U-shaped correlation law in different geographical categories

In the past decade, Chinese authors have published many treatises on the inverted U-shaped correlation law between economic growth and regional unbalanced development, that is, the capability of J. Williamson's "theorem" to explain the evolution of regional economic development as well as spatial agglomeration and dispersion. Most scholars believe that this doctrine summarizes the basic characteristics of the change process of socio-economic spatial agglomeration, which has macroscopic guiding significance for territorial development and economic distribution in China. However, some scholars believe that there is no need to follow this rule during the development and construction of a country or even a region. Instead, agglomeration and dispersion shall be adopted as appropriate. Through more than ten years of investigation and analysis of the changing trend of domestic regional (national) economic growth and internal imbalance, the author is more certain that this "theorem" is completely in line with the law, and has an extensive capability to interpret and guiding significance. For this reason, the following points shall be emphasized again in this book:

(1) At the early and middle stages of industrialization in a region, the imbalance of regional development is an inevitable side effect of economic growth;
(2) It is necessary to make judgments according to different levels of regions to plan the overall development layout of the region by using the inverted U-shaped correlation law between economic growth and unbalanced development.

Under normal circumstances, the unbalanced development will last for a long historical period in large regions, especially countries with vast areas; while smaller countries or regions can quickly change from an unbalanced development strategy to a more balanced development strategy.

As far as China is concerned, the balanced development strategy shall not be implemented across the country within 20–30 years. For some large provinces, although the requirement for balanced development is also proposed, the emphasis on investment and development shall not be shifted hastily, such as Xinjiang (between the northern and northern Xinjiang), Shandong (between the east and the west), and other regions. However, for regions below provinces, municipalities and autonomous regions as well as inter-provincial and intercity regions, such as central and southern Liaoning, Beijing–Tianjin–Tangshan, Yangtze River Delta, Pearl River Delta, etc., the population, industry and towns have been excessively concentrated in some areas through the development before the founding of the People's Republic of China, especially over the past half century, and the stage where the spatial structure needs to be adjusted has been reached.

2) Time for China to implement the large-scale shift in spatial structure and its degree

It can be known from Table 1 that the eastern coastal areas have achieved a growth rate that exceeds the national average since the reform and opening up. The eastern region greatly exceeds the western and central regions in terms of economic and social development indicators such as per capita GNP. The gap in overall economic strength between the eastern region and the western and central regions is also wider and wider. Both the central and local governments have paid full attention to gradually narrowing the gap among regions, so as to maintain social stability. According to the instructions of state leaders, a development strategy that "some people and some regions get rich first" shall be implemented in the 1980s and 1990s. The development problem of the western region shall be prominently proposed and solved around 2000 [7]. How to master and actually operate the adjustment of spatial structure?

With the development and growth of the market economy, the initiative of the central government to adjust the spatial structure is increasingly concentrated in the following two aspects, that is, the spatial orientation of major infrastructure and more preferential regional policies. Since around 2000, the central government shall concentrate investment on major infrastructure (railway, highway, aviation and water transfer projects) and other aspects in the central and western regions. Through horizontal coordination, the capital and technology from the eastern developed regions shall be attracted to develop the resources of the central and western regions. In both cases, the stock and increment of economic development cannot be shifted to the central and western regions. With the upgrading of the industrial structure in the eastern region, the production of products with high unit product value will hold an advantage across the country. Therefore, the "shift" that started around 2000 will only be a change in the emphasis of regional investment in some industrial sectors, rather than a major adjustment in the space for national economic development.

From the perspective of the rational development of the eastern region itself, there will also be a need for large-scale overall adjustment of the spatial structure objectively in accordance with the inverted U-shaped correlation law. However, it is very difficult for the conditions for such a shift to emerge in the next half-century. The reason is that: The real agglomeration areas only account for about 1/5 of the area of 12 provinces, cities, and districts along the eastern coast. The trajectory of the relationship between economic growth and unbalanced development in these agglomeration areas may have reached the corner of the inverted "U" shape. To continue to achieve economic growth, there is only a need for the small and medium-scale shift of spatial structure (which has been discussed in other chapters of this book), and no large-scale adjustment is necessary or required.

3) Japan's development experience also proves the scientificalness of inverted U-shaped correlation law
(1) Rapid growth and widening of the gap among regions

From the early 1950s to the mid-1980s, Japan's economic growth can be divided into three stages, that is, economic recovery, high-speed growth and low-speed growth stages. The first stage is to recover and rebuild the economy, restoring it to the pre-war level; the second stage is a period of high-speed growth from the mid-1950s to 1973. Among them, the average growth rate was 10.5% from 1960 to 1968 and 8.8% from 1968 to 1971. The growth rate was reduced due to the first oil crisis.

In the process of high-speed growth, two regional problems emerged: the high concentration of economic activities in big cities and the widening of the gap in per capita income levels among regions. Table 2 reflects the concentration of population, employment and income in the three major megalopolises, that is, Tokyo, Osaka and Nagoya, at the final stage of rapid economic growth. The gap in the regional per capita income reached the highest level in 1960, as shown in Table 3. The data in the table represents the ratio of the per capita income of each region to the national average. Among them, the per capita income of two regions, that is, the capital region (mainly Tokyo) and the Kinki region (mainly Osaka), has been higher than the national average level for 27 years; it also shows that the advantages of these two regions reached the peak in the 1960s and then declined.

Table 2 Change trend of population, employment and income of three major megalopolises in Japan (from 1955 to 1980)

Item	Megalopolis	1955	1960	1965	1970	1975	1980
Population (%)	Tokyo	17.3	19.1	21.4	23.2	24.4	24.8
	Osaka	11.4	12.2	13.3	14.0	14.1	13.9
	Nagoya	5.9	6.1	6.4	6.7	6.8	6.8
	Total	34.6	37.4	41.1	43.9	45.3	45.5
Employed population (%)	Tokyo	16.6	19.1	21.7	22.7	23.7	23.8
	Osaka	10.7	12.1	13.3	13.6	13.4	13.0
	Nagoya	6.2	6.2	6.8	6.9	6.9	6.8
	Total	33.5	37.8	41.8	43.2	44.0	43.6
Income level (%)	Tokyo	24.0	27.4	28.4	30.0	29.9	29.7
	Osaka	14.4	15.6	16.7	16.9	15.6	15.1
	Nagoya	6.3	7.2	6.8	7.5	6.9	7.0
	Total	44.7	50.2	51.9	54.4	52.4	51.8

Tokyo megalopolis: Tokyo, Saitama, Chiba and Kanagawa prefectures, Osaka megalopolis: Osaka, Hyogo and Kyoto prefectures; Nagoya megalopolis: Aichi and Mie prefectures

Table 3 Change trend of per capita income level in various regions of Japan (from 1955 to 1982) (per capita level in Japan = 100)

Region	1955	1960	1965	1970	1975	1980	1982
Capital	122	127	122	121	116	115	116
Chubu	97	101	96	100	96	97	97
Kinki	115	115	115	112	105	104	104
Hokkaido	99	89	87	83	89	93	92
Tohoku	79	73	75	73	83	83	81
Hokuriku	94	92	87	87	92	93	91
Chugoku	86	83	89	91	93	93	91
Shikoku	84	79	80	83	85	84	81
Kyushu	80	73	75	73	84	86	85
Okinawa	64	56	53’	51	74	67	69
Coefficient of variation	0.223	0.270	0.226	0.224	0.154	0.156	0.164

Coefficient of variation = $\frac{\sigma}{\bar{z}} \times 100$.

4 Accessibility and Regional Development

1. Regional accessibility and regional development
1) Geographical location and accessibility

Among the factors affecting the economic development of countries, regions and cities, the importance of geographic location has long been recognized by people. In the process of discussing the impact of geographic location on regional development, the geographic location of a country, region or city is mainly investigated and evaluated from the following aspects:

(1) Relative sea-land location. Is the region located along the coast or inland? Even if the region is located inland, it is necessary to further investigate the possibility and convenience of entering the ocean, such as whether there are convenient transportation channels and ports. Through nearly 100 years of development of the world, proximity to the ocean has proven to be a favorable geographical location.
(2) Spatial relationship with big cities and agglomeration areas. It refers to the distance from important large cities and agglomeration areas within a large range, whether there are barriers from special terrains or natural elements in the middle, or whether there are traffic and information contact tools with sufficient capabilities.
(3) Spatial relationship with competitors. Two or more cities, regions and even countries are adjacent to each other. When they are relatively similar in resource

structure and industrial structure, a competitive relationship may be formed. It is mainly manifested as follows: the same object of raw material source and product output, the same object of resource source, etc. Such a location will have an adverse impact at a certain stage. However, the result of the competition may make the competitors more able to stimulate the motivation and wisdom to seek greater development, which can promote the development of the region in the long run.

(4) Location on the international transportation channel. In the ancient and modern development history of the world, there are many examples of rapid development at a certain stage due to the location on the international transportation channel. The reasons for the development brought about by this location are as follows: the rise of transportation-related industries due to its role as a hub of transit, the increase in the national income due to the development of maintenance, warehousing, processing, tourism, restaurant and recreational facilities that serve international transportation, and inflow of foreign capital and domestic skilled workers due to the location on the international transportation channel.

The importance of geographical location is a historical category. A geographical location, which was not important before, may become extremely important due to changes in certain factors. These factors are as follows: opening of new transportation channels and their gradual prosperity; rise of new cities in a certain location, making the previously unimportant transportation channel important; wars can change the geographical location of a region.

From the above aspects of the geographical location, one of the connotations can be found, that is, countries, regions and cities with favorable geographical locations are easier to participate in the social-economic activities in a large range, that is, easier to get in touch with agglomeration areas and big cities in a large range. That is to say, a favorable geographical location often means greater spatial accessibility.

2) Several categories of spatial accessibility

Spatial accessibility refers to the convenience in exchanges of materials, energy, and personnel between an area (country, region, city, linear and point infrastructure) and other relevant areas (also including country, region, city, linear and point infrastructure).

The level (or degree) of spatial accessibility reflects the opportunities and potential for socio-economic and technological exchanges between the region and other relevant regions. In the process of global and regional economic development, a country or a region cannot develop independently without others. Such contact "opportunity" and "potential" are extremely important development factors.

Although the basic connotation of spatial accessibility is the convenience of social and economic connections among regions, spatial accessibility can still be divided into different categories. The specific meanings expressed and the indicators used by different categories are different. They are also different in terms of the evaluation of the effects on regional development.

(1) Range of the space possibly reached during personnel or goods transportation in various time ranges starting from a port or other transportation hubs, such as the place that the personnel or goods can reach by train along the railway network within 5 or 15 h, started from any coastal port in China. With the port as the center, the points with equal time distance are connected, forming multiple equal accessibility circles. Here, the indicator for measuring accessibility is the time distance. Since the railway lines from the port are unevenly distributed in all directions, and the technical levels and operating speeds of different lines are different, the distances from each equal accessibility circle to the port are different in different directions. From such an accessibility map, we can see the level of difficulty in entering the economic cycle of each region by using water transportation and can also see the convenience of connecting the port to the hinterland.
(2) Regional accessibility from a city. Here, it refers to the access from this city to its attraction range. The unit of accessibility is still time distance, that is, the distance (location) possibly reached in various directions by various vehicles (according to the nearest route and the fastest vehicle) in the unit of time, such as within 0.5, 1.0, 1.5 h, etc. The points with equal accessibility are connected into a curve, forming an equal accessibility circle. The differences with the port-centered accessibility circle are as follows: The city-centered accessibility is usually measured by the combined transport speed, and a large number of points (up to several hundred points) shall be selected. Different vehicles are available from the city center to each point. Due to the unevenness of transportation routes in various aspects, the actual accessibility of points at an equal linear distance from the city center varies greatly. This leads to the nonuniformity of the equal accessibility circle in various directions (Figs. 4 and 5).

The equal accessibility circle shall be connected with the computer software because a large number of discrete points shall be selected during the specific operation.

Regional accessibility from a central city (center) can often explain the city's ability to radiate the economy and technology of surrounding regions in different directions, as well as the degree of response (possibility) of each region to such radiation ability.

(3) The accessibility of a region during the transportation within a large range is expressed in terms of interval travel speed and travel time. Such accessibility can be described as follows:

A large region has several large agglomeration centers (or big cities). The accessibility of each region within this range refers to the convenience of reaching each large agglomeration center from each region (with its center as the starting point). It can be expressed in terms of travel time and travel speed by means of a public or private

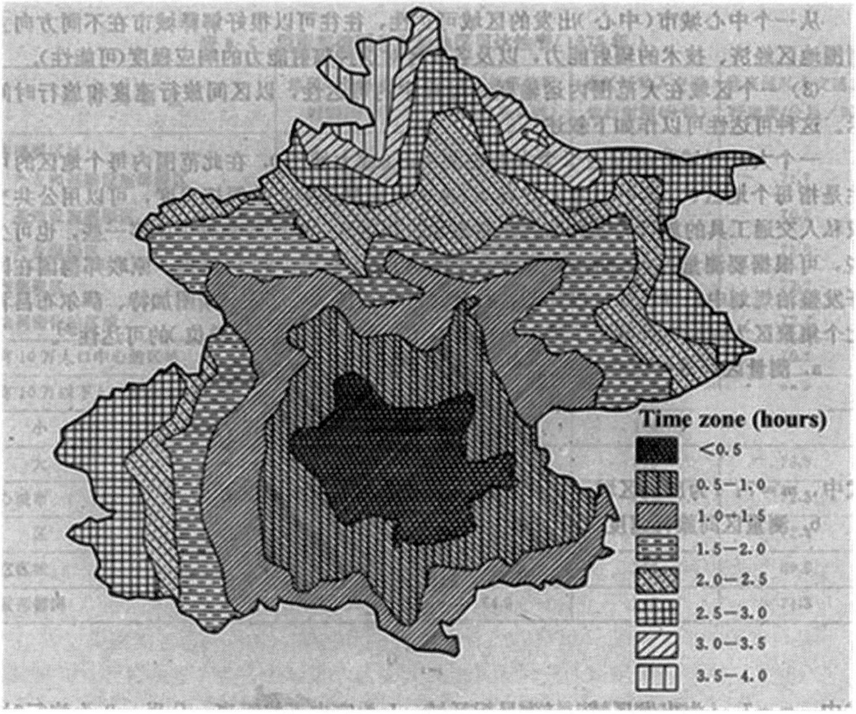

Fig. 4 City-centered equal accessibility circle (taking Beijing as an example)

vehicle. Here, the region can be divided into a large number or a small number, which can be determined according to the size of the region (or urban area) to be measured and evaluated. When planning the territorial development and renovation in the early 1980s, the former Federal Republic of Germany took seven agglomeration areas, that is, Hamburg, Hannover, Cologne, Frankfurt, Munich, Stuttgart and Saarbruecken, as targets, and calculated the accessibility of all cities and counties (administrative units below the state) in the country at that time [8].

a. Formula for measuring the interval travel time:

$$\frac{1}{m}\sum_{j=1}^{m} R_{ij}$$

   where, m = 7, *i* is the departure area; *j* is the target area; R is the travel time, in minutes.

b. Formula for measuring the interval travel speed:

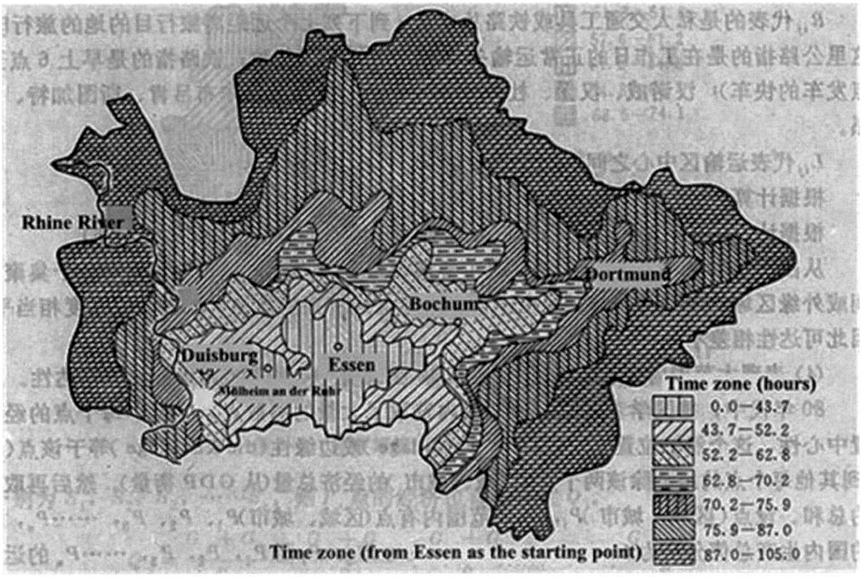

Fig. 5 Equal accessibility circle with Essen as the starting point in the Ruhr region

$$\frac{\sum_{j=1}^{m} L_{ij} \cdot 60}{\sum_{j=1}^{m} R_{ij}}$$

where, m = 7, *i* is the departure area, *j* is the target area, *L* is the straight-line distance in the air, in kilometers, and *R* is the travel time, in minutes.

R_{ij} represents the time of travel by private vehicle or railway from center *i* to the following seven long-distance travel destinations (here, the highway refers to travel by sedan car under normal transportation conditions on weekdays; the railway refers to the express trains departing from 6:00 to 10:00 in the morning): Hannover, Hamburg, Dusseldorf, Frankfurt, Saarbruecken, Stuttgart and Munich.

L_{ij} represents the straight-line distance in the air between the centers of the transportation areas.

According to the calculation results, they were sorted and summarized as shown in Table 4.

According to the calculation results, the accessibility map of cities and counties in the former Federal Republic of Germany at that time was prepared (Fig. 6).

It can be seen from Fig. 6 that the accessibility was high in the major agglomeration areas of the former Federal Republic of Germany and their vicinity; the accessibility was low between the agglomeration areas or in the outer edge areas. However, there was little difference in accessibility due to the highly developed social economy and quite balanced regional development in Germany.

Table 4 Accessibility of various types of regions in the former Federal Republic of Germany (1975)

Structure type area	Railway travel time between regions (minutes)	Railway travel speed between regions (km/h)	Private transportation travel time between regions (minutes)	Private transportation travel speed between regions (km/h)
Region with weak structure: region with weak industries and infrastructure	334	62.9	293	71.7
Region with weak infrastructure	316	62.2	278	70.2
Region with weak industries	283	63.3	248	71.6
Strong agglomeration area	255	67.5	237	72.2
Region that starts to be dense	262	65.6	238	72.7
Region whose central place has a population of 100,000	266	61.3	231	70.2
Region whose central place has a population of < 100,000	350	58.8	299	68.6
Minimum	187	51.3	176	64.5
Maximum	428	74.0	392	78.3
Core city	253	71.4	249	72.3
Suburbs	243	64.1	214	72.4
Other regions	299	60.2	257	69.6
Former Federal Republic of Germany	269	64.9	243	71.3

(4) Accessibility that represents the centrality or marginality of the location of each small area or point (city, etc.) within a large range

In the mid-1980s, German scholars selected a large number of discrete points within the European Community, and calculated the economic location centrality of each point. This economic location centrality (die Zentrallage) or marginality (die Randlage) is equal to the distance from this point (region) to every other point divided by the economic aggregate (measured by GDP) of the two points (regions and cities).

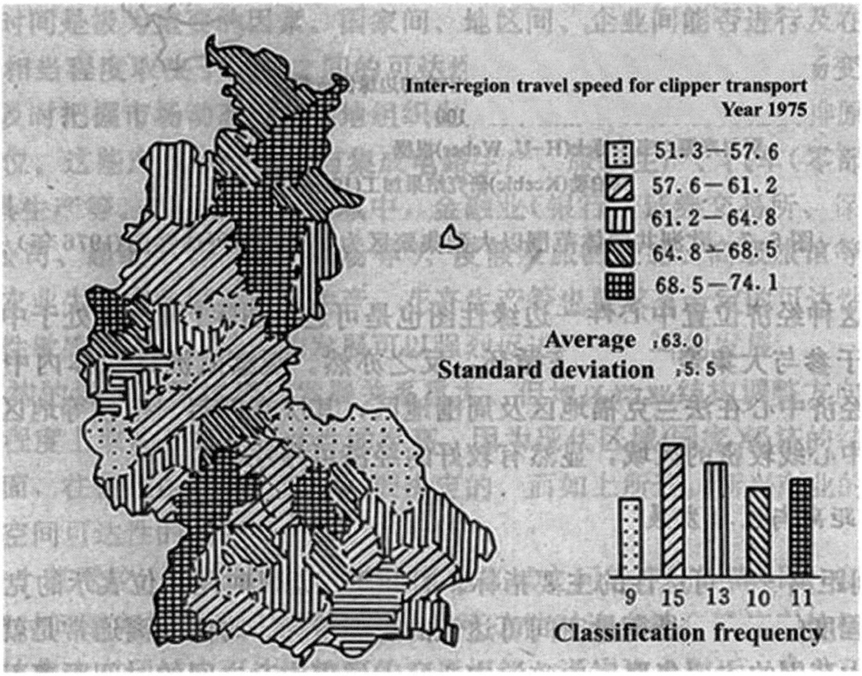

Fig. 6 Accessibility map of various regions in the Former Federal Republic of Germany

Then the values are summed. Set that Point (Region or city) P_i has Point (region or city) P_1, P_2, P_3……P_n within a large range, their GDPs are G_i, G_1, G_2, G_3……G_n, and the haul distance from P_i to P_1, P_2, P_3……P_n is S_1, S_2, S_3……S_n respectively, and then the economic location centrality of Point P_i is Q_i.

$$Q_i = \frac{G_i + G_1}{S_1} + \frac{G_i + G_2}{S_2} + \frac{G_i + G_3}{S_3} + \cdots + \frac{G_i + G_{n1}}{S_n}$$

Here, the unit of Q_i is the number of GNP per kilometer. Through a large number of calculations, and by means of a software system for connecting the isoline, a centrality-marginality map of economic location within the European Community was drawn (Fig. 7).

Why is this economic location centrality-marginality map also an accessibility map? Because a region with high centrality is, of course, easy to participate in the economic cycle of large agglomeration areas, and vice versa. By observing the distribution of the centrality line within the European Community, it can be known that the economic center of Europe is in Frankfurt and its surrounding regions, followed

Fig. 7 European Community-wide map of equal accessibility centered on large agglomeration areas (1976). The isoline value is the marginality index with Rhineland-Pfalz as the center. 8.573 million European currency unit per kilometer = 100. *Source* Processed by Dr. H-U. Weber based on the research results of Keeble (1988)

by Paris, London and other regions; those close to these regions, that is, within the regions with a dense center line, obviously have better economic accessibility.

3) Time distance and regional development
(1) Time distance—main indicator of accessibility. Time distance, expressed in time units, is a concept of the difficulty in overcoming (crossing) spatial distance, and the main indicator for measuring space accessibility. It usually refers to the time distance between a region, country or city and major agglomeration areas, political centers and large urban centers in a large range. Time distance is significantly different from the general concept of distance.

a. Time distance expresses the convenience and difficulty in realizing personnel and material exchanges between two places, which is expressed in time units (hours, minutes, days, etc.); the general space distance is the length of the road or the straight-line distance in the air between two places, which is a pure space concept.

b. The time distance between two places is improved with the construction of new traffic lines and new means of transport as well as the transformation of the original transportation system, that is, being gradually shortened. Therefore, it is a dynamic concept. The general distance will not change unless a new straight line is constructed.

c. The time distance between two places is also related to the reasonable and rapid connection of two or more means of transport. That is to say, when the distance between two places is certain and the transportation routes and means of transport are also certain, the time distance is also different due to the different conditions for connection of different means of transport (for example, whether transfer or exchange is required, and whether there is any conveyor belt in the middle). In urban passenger transportation, for the subways, ground roads (streets) and even ports, long-distance railways, etc., if stations, wharves, large commercial facilities, etc. can be arranged at close locations, the availability between different locations in the urban area can be greatly improved. For this reason, a comprehensive urban transportation hub shall be built.

(2) Impact of time distance factor on regional development. In the long-term economic development process, distance plays an important role in regional exploitation and development, which has been a truth known to all. The time distance, of course, has an important impact on regional exploitation and development since it has the same connotation as the general distance. However, the time distance will have an impact in different aspects since it has more significance than the general distance.

a. The time distance can accurately reflect the potential energy of interaction between two locations or between one region and all other related regions. In the practice of regional exploitation and development, such potential energy is expressed as the strength of social and economic connections among regions and the possibility of impact of one region on other regions (driving, promotion, competition or inhibition). Here, we can assume that: The mileage distance between A and B is greater than that between A and C, but the time distance between A and B is smaller than that between A and C. In this case, if A is a big city with strong radiation ability, then B is more likely to achieve faster development, because it can benefit from the convenient economic and technical exchanges with A.

b. The time distance has different impacts on different industries. Time is an extremely important factor during industrial development in modern times and the economic and technical cooperation among countries and regions. The possibility and scale of cooperation among countries, regions and enterprises depend on the accessibility between them to a considerable extent. The production and markets for many products change rapidly. In order to grasp market dynamics in a timely manner and organize production and sales scientifically, companies and governments must accurately arrange raw materials, semi-finished products and finished products in a timely and proper manner. These industries include semiconductor and integrated circuit production, apparel production, automotive (between parts and assembly) production, toy production, etc. In the field of tertiary industries, the financial industry (banks, stock exchanges, insurance companies, etc.), commerce (commercial companies, supermarkets, futures markets, etc.), vacation and tourism facilities, high-end hotels, etc. require better accessibility. The production of fresh products and flowers in agricultural production also requires certain accessibility. These industries can be called accessibility-sensitive industries, and their development can strongly promote the development of the regional economy.

c. The adjustment of industrial structure has an important relationship with regional economic development. However, the adjustment direction of regional industrial structure and the possibility to achieve effects depend on the improvement of regional accessibility to a certain extent, because whether the structure of the modern regional (national) economy adapts to the market competition is often determined by the development of emerging industries. As mentioned above, the development of the emerging industries depends on the improvement of spatial accessibility to a great extent.
d. There is little difference in the space distance (that is, general distance) from various points in a small region to other regions, but the space distance may be quite different. Since accessibility is extremely important to many socio-economic objects, the result is that these objects are concentrated near transportation hubs and communication centers with higher accessibility in the region. It can be concluded from this that the difference in accessibility within a region determines the spatial structure of the social economy in the region to a certain degree.

4) Accessibility and location differential rent

The basic meaning of location differential rent is that: Different net incomes from land utilization are caused by different distances from the product (or service) consumption center [1]. Under normal circumstances, the differential rent in the urban center is the highest, and accordingly, the land price is also the highest. This law determines the spatial structure of land utilization fundamentally.

Due to the difference in the distribution of the traffic network and the means of transport within the urban area, the accessibility of each location varies greatly. For the location with higher accessibility, the differential rent of the land is bound to be higher. That is to say, the accessibility of the urban area with the urban center as the starting point is consistent with the location differential rent of the urban area. The land rent theory can be used to explain the mechanism of improving regional accessibility to enhance the regional economy.

The consistency between regional accessibility and location differential rent can tell people how to improve the level of land utilization in the region by improving regional accessibility. That is to say, the hilly and rocky land which has a low value and can only be used to plant general crops, grass and fruit trees can be converted to fields for rice, vegetables and flowers with intensive cultivation or the land for urban construction by improving the regional accessibility. Therefore, changing regional accessibility is an important approach to changing the way of land utilization and improving the value of land utilization.

5) An important approach to improving regional accessibility—establish the "point-axis" spatial structure system

As mentioned above, regional accessibility is one of the important factors affecting regional socio-economic development. Therefore, it has become an important goal of regional development to improve regional accessibility, which, of course, is also a precondition for further development of the region.

Of course, there are many ways to improve regional accessibility, such as improving the grades of roads connecting the regional center with other regional centers, increasing the speed of the transportation vehicle, developing direct transportation to reduce the transfer time, etc. However, these measures are very useful to a certain city or a certain defined "point" (such as enterprise, financial unit, etc.). That is to say, once one of these measures is taken, the accessibility of a series of units, cities, etc. will be promoted.

However, the question often encountered in practice is: how to improve the accessibility of a region or a city (with acceptable effects)? That is to say: The accessibility shall be improved for a series of points, rather than a certain point. In this case, the purpose cannot be achieved by implementing direct transportation. The main approach is to build a "point-axis" spatial structure system in the region. With regard to access from other places, the trunk line is used to reach the center of the region. In most cases, a transfer at this point is necessary. With a complete "point-axis" spatial structure system, one or even multiple transfers can be realized in the process of reasonable connection, so as to maximize the accessibility of the whole region. Based on the same truth, the accessibility from any point in one region to the center of other regions is also maximized in the "point-axis" spatial structure system.

References

1. L. Dadao, *Location Theory and Regional Research Methods* (Science Press, Beijing, 1991)
2. E. V. Böventer, *Standortentscheidung und Raumstruktur, S. 50–135* (Hannover, 1979)
3. Edited by Fang Weizhong, Chronicle of the People's Republic of China (1949 to 1980), Page 170, China Social Sciences Press, 1984.
4. S. Jiatai, W. Youren, *Theory and Method for Studying the Rational Scale of Towns* (Nanjing University Press, 1986), p. 9
5. Editorial board of the *Physical Geography of China* of the Chinese Academy of Science, *General Introduction of the Physical Geography of China* (Science Press, 1985)

6. G. Parker, Western Geopolitical Thought in the 20th Century, p. 38, translated by Li Yiming et al., *Chinese People's Liberation Army Publishing House* (1992)
7. "Selected Works of Deng Xiaoping", vol 3 (People's Publishing House, 1993), pp 374
8. Schriftenreihe des Bundesministers fur Raumordnung, *Bauwesen and Stadtesau* (Raumordnurgsbericht, Bonn, 1978), pp.88–90

Chapter 7
Location Differential Rent and Spatial Structure of Urban Land Utilization

1 Land Utilization in Urban Areas Needs to Follows Scientific Laws

1. Urbanization and land utilization in urban areas

Land is one of the critical aspects on which human beings depend for survival and development. Under a certain historical development stage and economic and technological level of a country or region, the amount of land resources that can be used by human beings is limited.

Over the past half-century, China's social economy has achieved great development. Especially since the 1980s, economic industrialization and social urbanization have progressed rapidly. At the same time, land development and utilization have also reached an unprecedented scale. From 1949 to 1993, the national population increased from 542 million to 1.163 billion. Among them, the non-agricultural population increased from 58 to 260.7 million. If a large number of populations that have moved from rural areas to cities since the 1980s are included, the urbanization level of China's population has reached 30.0%. There had been 570 cities in China by the end of 1993. The process has been accelerated due to the following reasons: the increase in the economic strength of industrial and tertiary industries in a large number of industrial and mining cities, open cities, open ports and most counties (banners), a transformation of counties (banners) into cities, division of states to cities and merger of prefectures and cities in China, etc. The main line of the administrative management system below the province in China has gradually changed from rural areas (province → prefecture → county → district → township → village) to an urban system [i.e., province → city (prefecture level) → city (county level) → town → village or street].

D. Lu, *Regional Development and Its Spatial Structure*,
https://doi.org/10.1007/978-981-99-7683-6_7

Large-scale urbanization has rapidly expanded the scope of urban areas. The purposes of many pieces of land are changing, and many pieces of inferior land and even special types of land difficult to use (hills, mountains, tidal flats, shallow seas, sandy land, etc.) are constantly being transformed and used by human beings to expand space for urban development, despite the fact that the cost of this transformation process is often very high.

Due to the rapid development of industrialization and urbanization, China's land utilization is undergoing strong dynamic changes. However, people have noticed that the urban utilization of land is completely different from the agricultural utilization of land, that is, once the utilization state of a certain function is entered (road, industry, residence, commerce, finance, medical care, culture, education and health, science and technology, foreign affairs, non-staple food base, transportation, etc.), it is not easy to change. This requires people to pay special attention to land utilization in urban areas. Therefore, it is necessary to explore the laws of land utilization in urban areas, so as to scientifically arrange the land utilization according to the laws and the needs of socioeconomic development of specific urban areas.

Once the land utilization in urban areas is scheduled and implemented according to the urban planning (law), the plan shall be revised to make changes, which requires a larger investment. The reason is that land utilization in urban areas has different characteristics compared with land utilization in rural areas:

(1) Since cities are areas where modern industry, politics, culture, management and other facilities are concentrated, the value generated per unit land is much larger than that of the rural areas. Accordingly, the investment per unit land is much larger;
(2) Facilities in urban areas are generally permanent;
(3) The land utilization in urban areas is often planned according to smaller statistical units of area (such as square kilometers, mu and square meters). Houses, roads, parking lots, etc. are often built in a small area. Moreover, these facilities cooperate with each other in terms of functions and are indispensable;
(4) Land utilization in urban areas is often linked to space utilization, and the air, ground and underground are utilized in three-dimensional form.

Due to the above features, land utilization in urban areas is highly intensive, interrelated, integral and permanent.

2. Land marketization and urban land utilization

Land can be used, sold and bought by people, which has been common sense. However, what makes land special as a special commodity is not understood by most people. Land has a price and a use value. Therefore, land is also sold and bought by setting a price and with payment by the buyer. In the case of land marketization, it is of significance to study the use value and price of land, especially the land in urban areas. The price of land is determined by its use value. The basic use value of land includes the following aspects: for agricultural farming or grazing, for industrial and transportation facilities, for facilities in residential buildings and ecological environments, for finance, commerce and tourism facilities, for military purposes, etc. The benefits generated by these utilization modes vary greatly. Therefore, the value and the price reflecting the value vary greatly for the same area of land.

The use value of land depends on the following factors under normal circumstances, which can be divided into natural and socio-economic attributes of land:

(1) Size of the area. Land is very precious in regions, cities and even countries deficient in land resources, which can be described as "an inch of land values an inch of gold". In regions, cities and even countries affluent in land resources, small and medium-scale differences in land area often do not significantly impact land utilization and economic development.
(2) Topography as well as soil texture, hydrogeological environment and engineering geological environment of plots and sections. Land evaluation indicators such as terrain slope, soil fertility, ground-bearing pressure, underground aquifer conditions, and these characteristics are reflected by land evaluation indicators. These indicators play an important role in determining the use, engineering measures and price of a specific section.
(3) Geographical location of the land (i.e., specific plot and section). These include natural geographic locations, such as the locations relative to mountains, coastlines, rivers (especially navigable rivers), plains and hills; social and economic geographical locations, such as the locations relative to large and middle-sized cities and their commercial and financial centers, government management agencies, ports and other transportation hubs, highways, railways and other transportation lines, scenic tourist destinations, etc. Here, the locations "relative to…" includes the meanings of spatial distance, time distance, direction, etc. In the long period before the mid-industrialization, the urban center is often the gathering place of advanced socio-economic objects. That is to say, the major commercial, financial, cultural and transportation hubs and other regional and city facilities are often concentrated in the urban center. Therefore, the location of land relative to the urban center becomes an extremely important indicator of land value.
(4) The land value and the price that reflects the value is changing. Especially for land in urban areas, values and prices change rapidly and often great. This is another characterization of land, as a special commodity, different from general commodities. Such changes come from the following: The socio-economic development of the region where the land is located increases the need for land for various purposes, and the result is to change the land utilization mode of a specific section. Generally, a more intensive utilization becomes the utilization type of a substantial increase in the input per unit area.
(5) Land, i.e., specific plot and section, is immovable from the perspective of nature, that is, there is no spatial mobility. However, the specific land is "movable" through evaluation from the perspective of the social economy. Such "movement" refers to spatial displacement in the relative sense, rather than spatial

displacement in the absolute sense, that is, the relative position of the socio-economic object has changed. For example, a piece of land with poor spatial accessibility is far from traffic lines and transportation hubs. However, two or more new traffic lines are passing through or even intersecting here at a certain stage, which greatly improves the spatial accessibility of the place. Such a situation is actually equivalent to moving the land to the vicinity of the original traffic lines. The establishment of new cities, new factories, new scientific research and teaching institutions, etc., can also make the nearby land have the effect of spatial displacement. In order to make the land have the effect of spatial displacement, that is, to increase the value of land utilization, urban areas, industrial and mining areas and resource development areas shall be comprehensively planned.

The above factors always change in strongly industrialized and urbanized countries and regions. Because of the marketization and commercialization of land as a commodity, the land utilization mode, the land price and the spatial structure of urban land utilization formed on this basis are also developing and changing.

Despite rapid changes in the mode of land utilization in urban areas under the background of marketization and commercialization, the spatial structure of land utilization in urban areas has a common pattern and characteristics. Such common patterns and characteristics are governed by a universal law, that is the law of "location differential rent" (Lagerente) (or called "land location rent").

3. **Land utilization in urban areas under the land allocation system**

China experienced a planned economic system for more than 30 years from the 1950s to the 1980s. Under the planned economic system with authority centralization, the system of free use of land, that is, the system of land allocation and use, was implemented. China's land management has changed from free use to paid use since the early 1990s. The principle of paid use of urban land and the theoretical basis of a series of specific policies form the law of location differential rent.

China began implementing the first 5-year plan and carrying out large-scale industrialized construction in 1953. In order to ensure the smooth requisition of land, the Government Administration Council announced the "Measures of the Government Administration Council on Land Requisition for National Construction" in November of the same year, which clearly stated that the land ownership belongs to the state, and the land user only has the right to use; at the same time, it was stipulated that the state-owned land shall be allocated by the government without compensation and free of rent, and shall not be transferred independently. This is the origin of the system of free allocation of land in China.

The system of free allocation and use of land may have played a positive role in the 1950s and 1960s. However, there were also many negative effects even at that time. For a long time, the free use of land had essentially violated the objective requirements of the law of location differential rent, and the land ownership and use rights had been artificially divided. However, the actual implementation result was that: The state lost the ability to dominate and control the land due to lax management and lack of a set of specific policies. Even in the case where the state had the right to regulate, there were two disadvantages in the results due to reliance on the administrative orders of the decision makers:

First, due to the subjectivity of the decision-makers, they often failed to follow the requirements of the law of differential rent when developing the specific policies and plans for land utilization in urban areas, so land utilization was unreasonable, which was mainly manifested as uneconomical.

Second, due to different degrees of excess returns (net returns) in urban land utilization, land users (units and rent seekers) often offered bribes to the personnel of government agencies. As a result, land was rented indiscriminately, resulting in chaos and waste of urban land utilization. At the same time, corruption in the government was also promoted.

Under the above background, the problems of land utilization in urban areas in China have been quite serious for decades. These problems have become clearer since the state implemented a system of paid use of land in 1988.

(1) Unreasonable spatial structure of urban land utilization

Under the guidelines of free allocation of land and overemphasis on the production function of cities, especially the construction of industrial cities, urban land allocation met the industry needs first. Since industrial enterprises did not have to pay the high land rent payable to the state, that is, the location differential rent was artificially removed, and industries were almost unlimitedly concentrated in inner suburban areas, residential areas and even urban areas of large and medium-sized cities. As a result, the land utilization of many large and medium-sized cities was in chaos,[1] and a spatial structure (a ring structure centered on the urban center) conforming to the law of location differential rent cannot be formed. A large number of industrial enterprises, general institutions and organizations, etc. occupied a large area of land. Units with a low profit or no profit per unit area occupied large tracts of land in the downtown and urban areas. For example, the Bund in Shanghai was the place where finance and business were concentrated in Shanghai before the 1950s, and the land price was extremely high. However, after the implementation of free use of land, the purposes of 91% of the original 34 bank buildings were changed. Among them, five were used as factories, 15 were converted into institutions, eight were used as offices for organizations, and two were converted into middle school buildings. One-third of the original large commercial companies in Shanghai were also converted into factories, warehouses, offices, etc. In the entire downtown area of Shanghai, industrial land accounted for 20.5%. In addition, it also led to an imbalance in the proportions of various urban land uses. The proportion of industrial production land was relatively high, while the proportion of living and residential land was relatively low. In foreign urban planning, industrial land was generally required not to exceed 15% of the total urban land. Statistics on 22 cities in the United States in the 1970s

[1] The content of this part is mainly from Yang Lei's "Analysis of the Impact of Free Use of Land on Urban Development", 1992, in "Productivity Distribution and Territorial Planning", 1992, No. 1.

indicated that the urban industrial land only accounted for 10.4% of the total urban land in the United States. The general and reasonable standard for the indicators of living and residential land shall be about 50% of the total urban land. Through the survey of 55 cities in China in 1981, the average urban production land in China accounted for 63%, while the living land including roads, squares, green spaces and public buildings accounted for only 37%.

The result of the chaotic urban spatial structure was that the performance of the main functions of cities was affected. That is to say, the unreasonable occupation of the downtown area deprived the foothold for urban commerce to develop, thus weakening the commercial circulation, the oldest and also the most important function of the city. The reduction of living and residential land directly affected the functions of living, residence and entertainment of cities. It was even more unfortunate that many scenic, historical and cultural cities also developed a lot of industrial land under the policy of free allocation of land, placing the cart before the horse, which destroyed the characteristics of cities, and caused a lot of irreparable damage.

(2) Restriction on the development of small and medium-sized cities, causing a lot of waste of urban land

Under the system of paid use of land, land users must consider the opportunity cost brought about by urban land rent when selecting a site in the city. As a result of selection, low-profit land users tended to prefer urban fringes or suburbs where the land prices were not so high. If such choices were made between cities, land users who were unwilling to bear the high land rent in large cities preferred small and medium-sized cities as a result of the difference in land prices among different cities. As a result, a considerable number of land users could be effectively prevented from selecting sites in cities or large cities due to the regulation effect of the law of location differential rent. For the land users who had already selected the sites in cities, this regulation effect would prompt them to consciously save the land and develop to the air and underground. For enterprises, it could also prompt them to find ways to improve their operating efficiency. Under the system of paid use of land, on the one hand, cities were attractive to land users due to agglomeration benefits. On the other hand, the countervailing power for large-scale agglomeration was also generated due to land rent factors. The two were supplementary to each other, which could promote the coordinated development of cities.

However, many land users that cannot choose sites in cities under the system of paid use of land can easily realize their desire to agglomerate in cities under the system of free use of land. Large cities attracted a large number of industries, organizations and institutions to select sites due to many factors such as wider market, wider service facilities and more skilled workers. As a result, large cities often developed at a speed higher than average. According to the survey of 74 key cities, the built-up area in 1988 increased by an average of 7.9 times compared with that in 1949. Among them, cities with more than two million people increased by an average of 21.9 times in terms of built-up area. This was evidenced by the extraordinary expansion of the scale of large cities. What's more serious was that there were many phenomena of land waste such as excessive expropriation, early expropriation and no use after expropriation in land users due to the free allocation of urban land. The free use of the land had made cities additionally accommodate many land users that should not be accommodated. Coupled with the waste of these land users when using land, there was no doubt that the land use scale of cities was additionally expanded compared with that under the system of paid use of land. For example, Hefei has begun to collect the "development fees for land acquisition" since September of 1983, but only 700 mu of arable land had been acquired in one year by September 1984. During the 5 years from 1979 to 1983, the average annual requisition of arable land in Hefei was about 2000 mu, which was 2.85 times that when the development fees were collected. It can be said that the extra 1400 mu of land each year were caused by unnecessary land acquisition or land waste under the system of free use of land.

(3) Lack of stable funding sources for the construction of urban infrastructure

The city government can increase considerable revenue for the city's finances through land management under the system of paid use of land. Hong Kong is a region where the system of paid use of land is implemented, and the revenue related to land management accounts for an average of 25% of the fiscal revenue of the Hong Kong government every year. The land revenue in Hong Kong reached HK$5.82 billion in 1988, with the highest year accounting for 36% of the government's fiscal revenue. Sufficient fiscal revenue provides a solid capital base for urban construction.

No land use fee was charged due to the free allocation of urban land in a long period of time. As a result, there lacked a stable fund source for the construction of urban infrastructure in China. The investment in the construction of urban public facilities always accounted for about 2% of that in fixed assets in China from the "First 5-Year Plan" to the "Fifth 5-Year Plan". The proportion in the "Fourth 5-Year Plan" period was the lowest, only 0.85%. The proportion in the "Sixth 5-Year Plan" period slightly increased, only 3.4%, while this indicator of developed countries in the world was about 10%. The low level of investment in urban construction was one of the important reasons for many problems concerning urban residence, transportation, water supply and drainage, environmental pollution, etc.

2 Formation of Location Differential Rent

In the long-standing practice of land planning and land development, people have been exploring how to make the most reasonable use of land in urban areas. In early urban construction, people mainly considered the topography, climate, hydrology, and soil characteristics of the specific section, as well as the water conservancy and traffic conditions of the nearby area, so as to determine the purpose of land. However, with the deepening of industrialization and urbanization, people have realized that it cannot be taken as one of the criteria for determining urban land utilization. Another important criterion is to analyze the economic relationships in land utilization, so as to determine the spatial structure of land utilization in urban areas.

1. **Thunen's formula of location differential rent**

Western agricultural economists have begun to study the spatial form and structure of land utilization since the beginning of the eighteenth century. D. Ricardo put forward the concept of general land rent in 1817. Thunen, an agricultural economist and agricultural geographer, put forward the famous "isolated state" theory in 1826, which systematically discussed the types and intensity of agricultural land management around the agricultural product consumption center (city) and how to determine the spatial structure of land utilization in urban areas, and proposed the basic content of location differential rent, that is, due to the differences in distance and freight, if the value of land with the same size but with different distances from the agricultural product consumption market is different, the utilization mode and management intensity shall also be different. As a result, the spatial difference in economic effects is generated. According to Thunen's discussion and the concentric ring model, we can make the following generalizations: Such different net returns generated from land utilization due to different distances from the product consumer market (city and consumer market) are called location differential rent. It can be simplified to the following formula:

$$R = E(p - a) - E\,fk$$

where:

R—Rent of land per unit area;
E—Yield of crops of land per unit area;
p—Market price per unit agricultural product;
a—Production cost per unit agricultural product;
f—Cost of transporting for unit distance per unit agricultural product;
k—Distance between the production place and the consumption place of agricultural products.

In the formula, k is a variable. Since it is assumed that the prices of production factors do not change with distance, E, p and a are all known quantities.

2. **Alonso's land rent model**

On the basis of Thunen's location differential rent, T. Brinkmann in the 1920s, Hoover in the 1930s and Isard, Alonso and Bofentel after World War II had carried out further research and development. They made new derivations to the theory of land rent, and went beyond the framework of agricultural utilization of land, expanding to the demand and utilization of land in all social and economic aspects.

William Alonso, an American regional scientist, assumed the same premise as that of Thunen for the model research: Homogeneous surface, that is, the quality of land at different locations is indistinguishable from each other, and the land can be bought and sold freely. There is a single-core city on the plain, the traffic in all

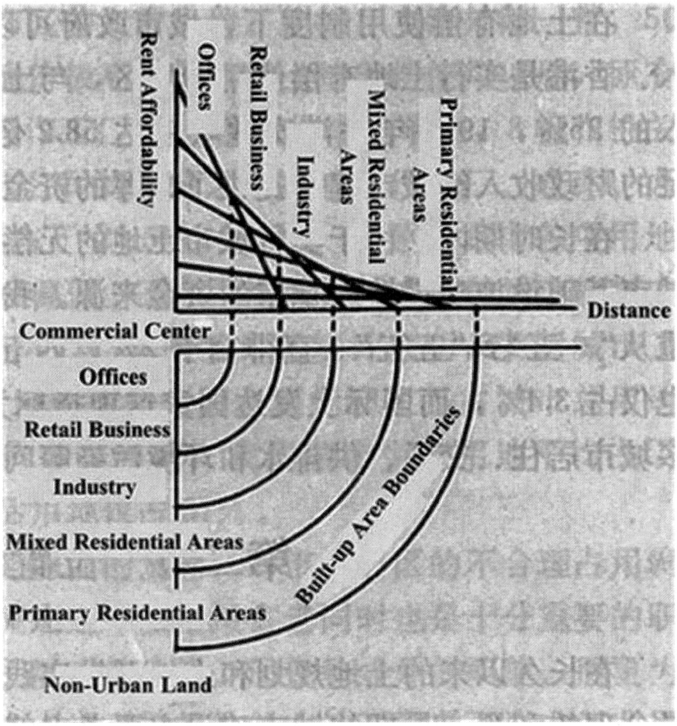

Fig. 1 Alonso's model of spatial structure of urban land utilization

directions is unimpeded, and the spatial distance and time distance are identical. Alonso's model of urban value theory can be represented in Fig. 1.

Alonso believed that urban utilization (type and utilization intensity) is affected by the accessibility of the land (relative to the center) and the resulting transportation cost, and different land utilization modes have different requirements for accessibility. As a result, the land rent curves with different slopes are formed [1].

Alonso further derived the slope of the land rent curve for different land utilization modes. In the process of deriving the following mathematical formula, his goal is to calculate the bid change curve of land users (manufacturing enterprises or financial and commercial manufacturers), but we can understand it as the derivation of the land rent curve [2].

Alonso pointed out that the manufacturers' motivation to purchase land is dominated by the operating profit (that is, net returns from land utilization in Thunen's theory), while the operating profit is related to the business volume, operating cost and land cost, that is:

$$\text{Profit} = \text{business volume} - \text{operating } \cos t - \text{land } \cos t$$

$$\text{Expressed in the formula as : } G = V - C - R \tag{1}$$

The business volume of the manufacturers mainly depends on the land location and the amount of land occupied, which can be written as

$$V = V(t, q) \tag{2}$$

The operating cost depends on the land location (t, which can be understood as the distance to the urban center), the amount of land occupied (Q), and the business volume (V), that is,

$$\mathrm{C = C(v, t, q)} \tag{3}$$

The land cost is proportional to the land rent price and area, that is,

$$\mathrm{R = P(t)q} \tag{4}$$

Substitute Formulas (2), (3) and (4) into Formula (1) respectively, and the following is obtained

$$\mathrm{G = V(t, q) - C(v, t, q) - P(t)q} \tag{5}$$

Find the partial derivatives of t and q respectively, and set them to zero:

$$\frac{\partial v}{\partial t} - \frac{\partial c}{\partial v}\frac{\partial v}{\partial t} - \frac{\partial c}{\partial t} - \frac{\partial p}{\partial t}q = 0 \tag{6}$$

$$\frac{\partial v}{\partial q} - \frac{\partial c}{\partial v}\frac{\partial v}{\partial q} - \frac{\partial c}{\partial q} - p(t) = 0 \tag{7}$$

In Formula (6), $\frac{\partial v}{\partial t}$ represents the change of business volume caused by the change of land location, and the latter items are the changes in operating cost caused by the change of land location.

In Formula (7), $\frac{\partial v}{\partial q}$ represents the change of business volume caused by the change in land use area, and the latter items are the changes in operating cost caused by the change in land use area.

When the corresponding marginal revenue of q is equal to the marginal cost, q is optimal.

$$\text{Set } G = G_0, \text{ and then } p(t) \prod G_0 \tag{8}$$

The equation of the manufacturer's bid curve is expressed as:

$$G_0 = V - C - R \tag{9}$$

$$R = p_i(t)q \tag{10}$$

$$\frac{\partial v}{\partial q} - \frac{\partial c}{\partial v}\frac{\partial v}{\partial q} - \frac{\partial c}{\partial q} - p_i(t) = 0 \tag{11}$$

According to this bid curve equation, the user with the largest bid curve slope (large business volume and low operating cost) obtains the downtown location due to its strong competitiveness. The user with the second largest slope will be located in its periphery. The rest will be pushed out until the urban fringe is reached. That is to say, for a manufacturer i at a location t_i where location balance is achieved, the land price paid is equal to the value represented by the bid curve of the location, while the price curve is defined by the value of the most satisfactory location adjacent to it, that is:

$$P_i = P_i(t)\prod t_j p_j \tag{12}$$

3. **E. V. Böventer's theory of optimal factor input and land rent**

E. V. Böventer is a German regional economist and regional scientist. In the book, Location Decision-making and Spatial Structure, published in 1972, he systematically proposed the theory of spatial structure and carried out rigorous mathematical derivation for a large number of related patterns. Among them, the theoretical derivation of location differential rent reached the highest level at the current stage and was highly praised by many scholars in the world [3].

He demonstrated this theory by centering on the following questions: In the land with a certain location and a certain area, what are the circumstances where the land rent will be generated, and when will the maximum land rent be reached? How does it affect the input of other production factors? The reason why these problems were demonstrated is that the maximum economic benefits can only be obtained when the land rent reaches the maximum value. Under the premise of "complete competition" in the market economy, theoretically speaking, the purpose and the user of any piece of land must satisfy the desire of the land owner (or land user) to obtain the maximum profit. Whoever can pay the highest land rent (or what combination of factors invested can obtain maximum benefits) can use the land.[2]

E. V. Böventer summarized the actual data in Table 1. On the land with $V_1 = V_0$, the input factor was a (a specific factor or a group of factors). When the input V_a of factor a increased from one unit to six units, the total harvest yield increased from 5 to 15 (Column 2). When the unit price of the product was assumed to be $P = 1$, the corresponding gross output value also increased from 5 to 15. The unit price of the factor was $q = 2.5$, and the corresponding total cost increased from 2.5 to

[2] It shall be noted that "whoever" here can be understood as the land user, a certain utilization mode, or the utilization intensity represented by a certain combination of elements.

15 (Column 5). The land rent that can be generated by inputting a certain amount of factors was shown in Column 6. The land rent increased at the beginning, but gradually decreased afterward, which was caused by the law of diminishing returns. Column 7 in Table 1 represents the output value of the product that can be obtained from the unit factor amount input.

It can be seen from Table 1 that when $V_a = 3$, the land rent $R = 4.5$, reaching the maximum, which was the optimal factor input scheme. It indicated that the best effect can be achieved with the utilization mode and the corresponding input intensity. If the factor input amount was continuously increased, the product (output value) that can be obtained from the unit factor amount input would decrease although the total output increased. Therefore, it was not cost-effective. Column 8 showed that the output value added by unit factor input decreased by increasing a unit of factor amount, and the quantity was 0.5 units. Column 9 represented the land rent amount apportioned by the unit factor, and its meaning was: Different factor input amounts represented different utilization modes and intensities. The large input amount indicated the high utilization intensity. Regardless of the input amount, the user should pay the land rent of 4.5 units according to the principle of land use to obtain maximum economic benefits, because it existed objectively or was potential. If the factor input is 1, the output value is 5, the production cost is 2.5, and the maximum land rent to be paid is 4.5. The sum of the expenditure is 7, which is larger than the output value, causing a loss of 2 units. If the factor input amount is 4 or 5, there is a same loss (0.5 or 2 units respectively). Only when the factor input amount is 3, the output value is equal to the cost plus the maximum land rent to be paid, which is the optimal utilization scheme (Table 1).

According to the data in Table 1, E. V. Böventer illustrated the factor input and land rent (Fig. 2).

The thick black line in Fig. 2 represented the change curve of the output value. The growth slope increased from O to A. The growth gradually slowed down with the increase of input from A point, and there was no land rent when it was parallel to the horizontal ordinate. A cost line K was drawn through the corresponding output

Table 1 Factor input and land rent

1	2	3	4	5	6	7	8	9
V_a	X	ΔX	PX	$q_a V_a$	R	PX/V_a	EE	R^*/V_a
1	5	5	5	2.5	2.5	5	–	4.5
2	9	4	9	5	4	4.5	– 0.5	2.25
3	12	3	12	7.5	4.5	4	– 1.0	1.5
4	14	2	14	10	4	3.5	– 1.5	1.125
5	15	1	15	12.5	2.5	3	– 2.0	0.9
6	15	0	15	15	0	2.5	– 2.5	0.75

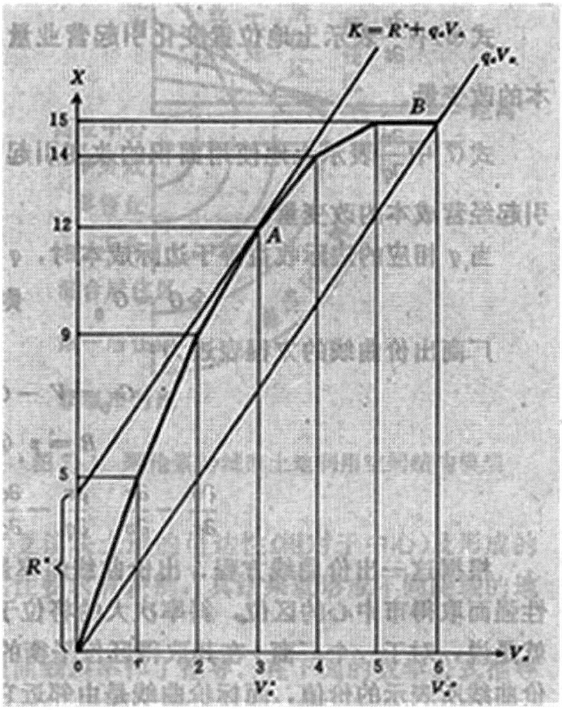

Fig. 2 Illustration of factor input and land rent

value A point of the optimal input 3, K = R* + qaVa. There was only one intersection point (A point) between this line and the output value line, and the intersection point with the longitudinal coordinate was the output value (12) at the maximum land rent.

E. V. Böventer also used the concepts of maximum harvest, marginal output value, land rent, cost, etc. for mathematical derivations of the formation of differential rent, the land rent function, etc. [4].

3 Significance of the Theory of Location Differential Rent to the Scientificization of Urban Land Utilization

For more than 100 years, many scholars have carried out a lot of research on the theory of differential rent, and also extended the concept of land rent to the field of social economy. However, the basic model and basic theory of differential location rent have not changed substantially. This theory reveals the relationship between land rent and the type and intensity of land utilization at different locations within the attraction range of the city (market) and clarifies the theoretical basis for the spatial difference of net returns from land utilization. Its scientific significance lies in the understanding of the formation mechanism of differences in the spatial structure of city-centered land utilization and how to scientifically arrange the plan for urban land utilization according to this principle.

1. **Theoretical criteria for determining land prices in urban areas and spatial structure of urban land utilization**

The location differential rent clarifies that the land rent generated at the land with different distances from the center of the urban area is different, and each utilization mode and intensity has a land rent curve with a different slope. This reveals that land at different locations has different values, and different prices shall be determined accordantly, so as to achieve the rational use of land. Those utilization modes with high input intensity per unit area (large factor input amount) and large output and able to pay high land rent, such as municipal halls, squares, monuments, churches, financial buildings, supermarkets, urban transportation hubs, etc., can afford high land prices. Therefore, they can be placed in the center of the city. On the contrary, field crops, fruit trees, etc. shall be placed in the peripheral zone of urban areas. Those in the middle, of course, are residential, industrial and recreational facilities.

The theory of location differential rent plays a role in planning practice and is also combined with the factor of accessibility. That is to say, in the urban center and its surroundings and on both sides of the urban trunk road, the higher the accessibility, the higher the differential rent, and the land price is also the highest. Accordingly, the land price on both sides of the traffic branch line shall be the second highest. For the inner city areas far from traffic lines, although the spatial distance (straight-line distance) from the urban center may not be far, the land price may be lower. Of course, the land price here increases with the improvement of urban transportation, which is also caused by the improvement of accessibility (Fig. 3).

When evaluating the urban land value from the perspective of the theory of location differential rent, the urban area is regarded as a homogeneous region, and the premise is that the traffic lines are widely distributed. In this case, the spatial distance is the same as the time distance. Therefore, the distance from the urban center represents the level of land rent. If the urban area is not homogeneous, that is, in the real city, there is often a large difference between the spatial distance and the time distance

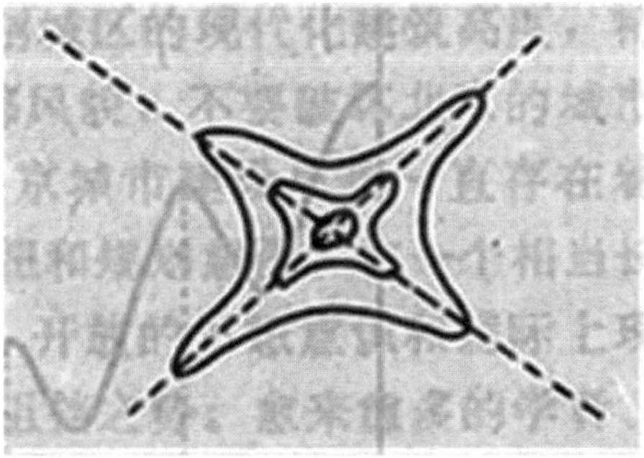

Fig. 3 Deformation of land value curve caused by urban arterial roads

between two places. In this case, the spatial distance and the time distance shall be synthetically considered when determining the land value. The result is that when the land value in the urban center is high, the land value is at the ridge when along the street and at the valley when away from the street (Fig. 4) [1].

When evaluating the land value of cities with multiple centers, the level (city-wide, district-level, etc.) and nature (including single-function, such as administrative; or comprehensive, such as commercial, financial, etc.) of the urban center shall be distinguished. Under normal circumstances, the land rent distribution curve is shown in Fig. 5a, b (from "Urban Land Location and Urban Land Value", Ning Xiaoming et al., 1991).

1. **Agglomeration factors and the countervailing power of excessive agglomeration**

The law of "location differential rent" has a strong regulatory effect on urban land utilization through land prices. According to the effect of agglomeration factors, many socio-economic objects are required to be concentrated in cities, especially in the central area of the metropolis. However, the so-called "market mechanism" generated by the law of location differential rent makes the land price in the urban center rise automatically. As a result, many utilization modes have "no right" to use the land in the central area. Therefore, they have to choose the land on the fringes and outer suburban areas of large cities and even small and medium-sized cities.

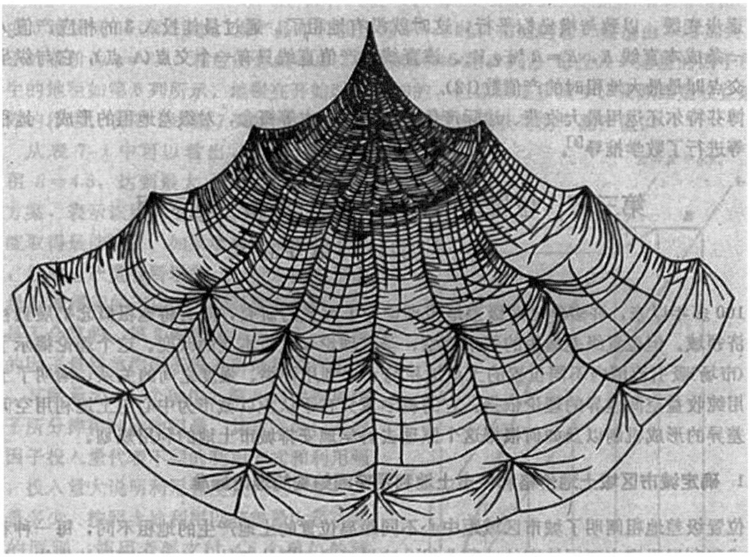

Fig. 4 Distribution pattern of urban land value

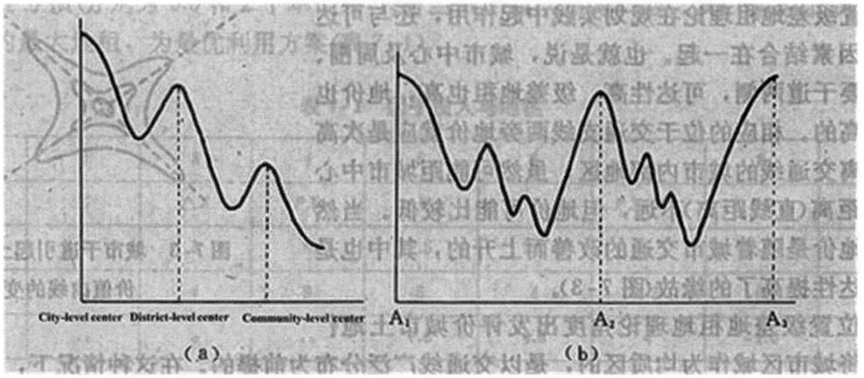

Fig. 5 Land rent curves of cities with multiple commercial centers

Therefore, the implementation of the law of "location differential rent" will promote the development of small and medium-sized cities, and promote the industrialization and urbanization of vast rural areas. This is of special significance for many developing countries.

2. **Promoting the economical use and protection of land resources**

Land resources, especially arable land, forest land, and water surfaces in densely populated areas, are precious resources for human survival and development. They are non-renewable in terms of area and some other properties. The very limited arable land in China is particularly important to the survival and development of the Chinese nation. With the progress of industrialization, many high-quality land resources have entered the category of urban land utilization. Land users can choose their own locations and floor areas in strict accordance with the land rent curve of their utilization modes by scientifically arranging the urban land utilization based on the principle of location differential rent. As a result, the saving effect can be achieved for urban land utilization as a whole, so as to prevent tangible and intangible land waste, that is, to prevent improper selection of land use location and excessive land use.

3. **Urban renewal of large cities**

China's urban construction has developed rapidly since the 1980s. In terms of the construction of large cities and metropolises, the question of how to carry out renewal without affecting the protection of styles and features of ancient cities has been raised. There have been strong responses to this question in the urban planning of Beijing, Xi'an, Kaifeng and Nanjing.

Similar questions have also been raised in the development of developed countries since this century. However, this problem has been well solved in their practice. For example, for the imperial palace is located in the center of Tokyo, Japan, the original buildings have been well preserved, and the architectural style and height have not changed. However, beyond the moat of the imperial palace, there are modern squares and high-rise buildings of large financial, commercial and industrial companies, and modern three-dimensional transportation hubs penetrating deep into the urban center.

During the modernization in China, especially in the capital, Beijing, the most popular propositions to protect the style and features of the ancient capital are rough as follows: A large number of civilian houses (courtyard houses, etc.) in urban areas of Beijing must be preserved; the height of modern buildings in urban areas shall be restricted, especially areas around the Forbidden City, Jingshan, Zhongnanhai and Tian'anmen; in order to protect the style and features of the ancient capital, the urban landscape of Beijing shall not be damaged, and the development of urban elevated traffic shall be restricted. There has always been a controversy over the idea of urban construction in Beijing since the late 1970s. Many academic thoughts and planning intentions with a pioneering spirit and the ability to understand the characteristics of modern metropolises have been suppressed by some traditional thoughts in academic circles for a long period of time. Since the mid-1980s, the influence of open ideology and the practical experience of planning and construction of modern large cities in the world has been larger and larger, becoming an irresistible trend. More and more scholars, planners and government officials have realized that Beijing's urban development must be modernized, and the overall layout of urban construction must also follow universal laws. Among them, the most prominent performance is that the transformation of the urban center and the main residential areas in the city has been strengthened, and the spatial structure of urban construction has been adjusted. These key areas (Wangfujing, Dongdan, Qianmen and Xidan) are under modernization, and high-rise buildings are emerging. These buildings have high input and high output and can obtain high net returns from land utilization. Therefore, they can afford extremely high (differential) rents. However, due to the long-term impact of the traditional thoughts on urban planning and construction that violates the theory of location differential rent before the 1980s, there are still many unreasonable points in urban land utilization in the capital, Beijing;

(1) A large number of office buildings of domestic and foreign companies as well as financial and commercial institutions are distributed on the periphery of the main residential areas in Beijing, which are too scattered. It is inconvenient to communicate with each other, with poor accessibility to citizens;
(2) There are still basically single-story houses within the Second Ring Road, which are used as residential areas. This is the most uneconomical from the perspective of the theory of location differential rent;
(3) Urban traffic is very congested, and an important reason is the lack of elevated traffic. If elevated rapid transit is established on the main trunk road (the subway is the main way to solve urban traffic congestion. However, the investment is huge, and the speed cannot be very fast), the traffic tension can be greatly eased.

The theory of location differential rent reflects the strict law. The factors of economic benefits drive the rapid changes in the spatial structure of urban land utilization in the capital, Beijing, and other large cities towards modernization and internationalization.

References

1. N. Xiaoming, L. Fayi, Urban land location and urban land value. Econ. Geogr. **11**(4), 128 (1991)
2. G. Chaolin, Several theories and methods of land utilization summary and planning. Prod. Distr. Territ. Plann. **1**, 55–59 (1992)
3. E. V. Böventer, Standortentscheidung und Raumstruktur, S.53, Hannover (1979)
4. L. Dadao, *Location Theory and Regional Research Methods* (Science Press, 1988), pp. 108–113

Chapter 8
Point-Axis Progressive Diffusion and Point-Axis Spatial Structure System

1 Socio-economic Spatial Distribution and Spatial Organization Process Model

In the book, the *Location Theory and Regional Research Methods*, the author elaborates on the general characteristics and influencing factors of socio-economic territorial organizations, proposes that reasonable agglomeration and optimal scale are the core issues of the spatial structure of productivity, and discusses the optimal scale of various socio-economic objects theoretically. This book intends to further analyze the basic types of socio-economic territorial organizations and the morphological characteristics of spatial structures, to provide theoretical models that can be used in the practical operation of land planning and regional development, and answer how to achieve the optimal organization and how to deal with the relationship between economic growth and balanced development as well as the development from imbalance to relative balance in practice.

1. **Spatial process model of socio-economic territorial organization**

We can imagine that productivity is evenly distributed at the stage where productivity level is low and socio-economic development is extremely slow (Fig. 1a); at the early stage of industrialization, industrial and mining residential areas or towns first appear at Points A and B with the development of mineral resources and the development of commodity economy, and traffic lines are constructed between A and B to meet the needs of social and economic connections (Fig. 1b); resources and economic facilities continue to be concentrated at Points A and B due to the agglomeration effect factors. Several large enterprises are established at these two points, and the traffic lines become linear infrastructure bundles of communication lines, energy supply lines, and telecommunication lines. Some economic facilities are established along the line. At the same time, new agglomerations begin to appear at Points C, D, E, F, G, etc., and the traffic lines are extended accordingly; this model is further developed, and the A-H-B-C line will become a development axis with good

D. Lu, *Regional Development and Its Spatial Structure*,
https://doi.org/10.1007/978-981-99-7683-6_8

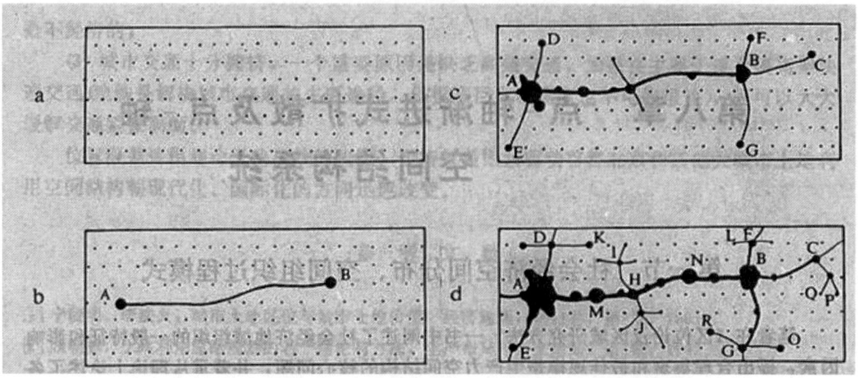

Fig. 1 Formation process model of "point-axis" spatial structure system

development conditions, high benefits as well as concentrated population, economy and technology: Points A and B form a greater degree of agglomeration, C, D, E, F, G, M, and N become new agglomeration centers, and a large number of population and economic units are concentrated along the line, forming a large dense industrial belt. Not only that, there is a second-level development axis in the other direction through Points A, B, and H respectively, and a third-level development axis formed through Points D, I, F, etc. If it develops in this way, the territorial organization of productivity will be further improved, forming a spatial structure system marked by the "point-axis" (Fig. 1).

The above-mentioned spatial process model of the territorial organization of productivity is summarized based on rich experience in regional development and is common law. People consciously or unconsciously allocate residential areas and various production facilities along the development axis by using this law.

The "point" in the "point-axis" system refers to the central places at all levels, that is, the central city (town) at all levels, the gathering point of the regions at all levels, and the central town that drives the development of regions at all levels. The "axis" in the "point-axis" system refers to a relatively dense population and industrial belt formed by connecting several central towns at different levels in a certain direction; since the axis and its surrounding areas already have great economic strength and potential, it can also be called "exploitation axis" or "development axis". That is to say, the axis is not simply a connecting line between several central towns, but a socio-economic dense belt. The reason why such an intensive belt is called an "axis" is based on the following two points:

(1) The axis is the concentrated zone of regional economic facilities and social facilities. According to the principle of spatial interaction (the source of the theory shall be the principle of Newtonian mechanics), the belt has a "force field", that is, an attractive effect, on the surrounding regions. The socio-economic objects near the axis generate a centripetal force, which does not point to a point (town)

on the axis, but several points or a line. This indicates that the axis has an agglomeration or agglutination effect on the social economy of the nearby area, and drives regional development through the objects within the sphere of influence.
(2) The socio-economic facilities concentrated on the axis have a diffusion effect on nearby areas through products, information, technology, personnel, finance, etc. The diffused material and non-material elements act on the nearby area and are combined with the elements of the area, so as to form new productivity and promote socio-economic development.

According to the above analysis of spatial distribution and spatial organization process model, it can be seen that there are four main stages in the formation of the socio-economic "point-axis" spatial structure system. The division of this stage coincides with the four stages of the development of balanced relations in industrial structure and spatial structure:

(1) At the equilibrium stage before the point-axis is formed, the subsurface is a homogeneous space. Although being distributed in an "orderly" state, the socioeconomic objects are in an unorganized state. This spatial unorganized state is extremely inefficient.
(2) Points and axes begin to be formed at the same time, the local regions begin to be organized, and the resource development and economy of the region enter a period of dynamic growth.
(3) The main point-axis system framework is formed, the social economy evolves rapidly, and the spatial structure changes greatly.
(4) The "point-axis" spatial structure system is formed, and the region enters a comprehensive and organized state. Its formation is the result of the long-term self-organization process of socio-economic factors, as well as the result of scientific regional development policies, plans and planning. From the macroscopic perspective, the spatial structure is back in the "balanced" stage. At this stage, although social and economic organizations are highly efficient, population growth and economic growth, as the signs of social development, are not at a high speed.

2. **Structure of the development axis**

The key exploitation (or development) axis generally refers to the zone where important linear infrastructure (traffic arteries, energy transmission lines, water sources, communication trunk lines, etc.) passes through and there are great socio-economic strength and development potential nearby. The width of this zone varies depending on the level, length and effect of the axis on the region. To determine the width of the development axis, it is also required to investigate the distance to and relationship with the linear infrastructure.

The development axis on the plane can be divided into the following three parts.

(1) Linear infrastructure. It is the nerve and vein of the axis and is the basis for the formation of population and industry-intensive belt. Under normal circumstances, it is a "bundle" composed of two or more kinds of linear infrastructure,

with the traffic arteries as the main body. In a few cases, another linear infrastructure (such as large-scale long-distance water transfer projects) is the mainstay. In most cases, the linear infrastructure is separated and basically parallel, and the distance between them can be measured as 30–50 km, or even more.

(2) Main part of the development axis. Generally speaking, cities, industrial and mining areas, ports, suburban agriculture and other mechanized agricultural facilities directly located on the linear infrastructure or its intersections are the constituent parts of the "point-axis" system. Here, the width of the axis is generally the diameter of the urban area (equivalent to urban and inner suburban areas) and industrial and mining areas. In megacities with a population of more than one million in China, the diameter is up to 30–50 km. For the national first-level development axis and second-level development axis (equivalent to the first-level axis of large regions and provinces), cities, industrial and mining areas, traffic ports and stations, etc. within a range of 30–50 km can also be deemed as within the scope and constituent part of the axis. It is convenient to utilize the powerful linear infrastructure for large-scale development within this distance. This scope is the key zone of industry, agriculture and urban construction, and the potential of various resources shall be exploited first. For the axis of the area below the province level, the width shall only be specified at about 10 km, or even narrower. However, since these axes exist, their role in regional exploitation and development is also vital.

(3) Direct attraction range of the development axis. The direct attraction range of the development axis is the direct attraction range of all towns, ports and stations on the development axis. There are many mathematical methods for accurately dividing the direct attraction range.[1] Regional exploitation is carried out according to the "point-axis" system, and there is no need for such a strict division. The direct attraction range of the development axis refers to the area (belt) that has frequent and close social and economic connections with the linear infrastructure, towns, industrial and mining areas and other facilities of the development axis from the perspective of reasonable transportation. Moreover, the region has little or almost no direct social and economic connections with other developmental axes at the same level. That is to say, in addition to their respective direct attraction ranges, there is also a large area of mixed attraction ranges between the two development axes at the same level. Two development axes are connected simultaneously within this range. Among them, the impact (promotion and attraction) of either development axis on the range is not great. Regional exploitation is carried out according to the "point-axis" spatial structure system, and these regions shall be placed in a secondary position. The above three constituent parts of the development axis can be represented in Fig. 2.

[1] Yang Wuyang, Liang Jinshe, Basic Theory and Division of Attraction Scope, 1984.

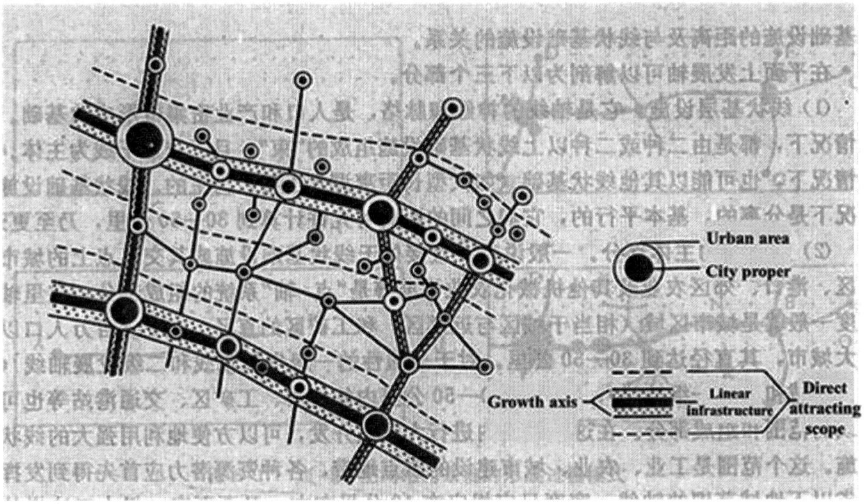

Fig. 2 Plane structure of the development axis. *Note* Only the structure of two first-level axes is marked in the figure

3. Significance of "point-axis" system model in regional exploitation and development

According to the "point-axis" system theory, requirements in the exploitation and development planning of a country or region are as follows:

(1) Within the range of a country (or region), it is necessary to determine several linear infrastructure axes between large regions, provinces, and prefecture-level cities with favorable development conditions, focusing on the development of the axis zone, especially several points (city and urban area-development centers), and give priority to the development of the resources located on the axis and within the direct attraction range of the axis.

(2) With the continuous enhancement of economic strength, more and more attention to economic development shall be placed on the development axis and development center at the lower level. Meanwhile, the development axis is gradually extended to the less developed areas (including areas within the developed areas but far from the development axis and the development center). Points that were not considered as development centers previously are determined as development centers at lower levels, and new development centers are specified.

(3) The point with better development conditions shall be selected as the secondary development center at a suitable distance from the central city. The primary center spreads technology and information to and conducts productive coordination with the secondary center. In the same way, there are tertiary and quaternary centers around the secondary center. The functions of central cities at different levels are manifested in different production scales, technical levels and product sales ranges in terms of industrial production distribution. It can

also be concluded that the economic network and spatial form of the central city shall be radial, rather than grid-like, which is the repeated coverage of the network of central cities at different levels on the subsurface.

The spatial structure that configures and improves productivity and the spatial organization that implements the entire social economy according to the "point-axis" spatial structure system can produce the practical effects in the following aspects:

(1) The "point-axis" development can meet the objective requirements that socio-economic development must be agglomerated into points in space and exert the agglomeration effect. In the three different territorial scopes of industrial area, industrial city and industrial zone, when a reasonable industrial structure and an organic combination between industry and regional infrastructure are determined, a huge spatial agglomeration effect will be produced. Such effect can only be fully utilized by developing in the form of the point-axis.
(2) The "point-axis" development can give full play to the role of central cities at all levels. A city, as the core and commander of the region, is generated from regional economic development and relies on raw materials, energy, food, labor, etc. provided by the affected areas to develop. A city supplies and commands its areas with industrial products, information, technology, policies, etc. To develop a vast area, central cities at all levels must be taken as the emphasis of development at different levels.
(3) The "point-axis" development can realize the optimal spatial combination between the production distribution and the linear infrastructure so as to achieve the necessary consistency among key traffic arteries, energy and water source construction lines at all levels within the range and towns, industrial and mining areas under key construction, thus avoiding the mutual separation in space and time that appeared in the previous practice.
(4) The "point-axis" development is favorable for the convenient connection between cities, between regions, and between urban and rural areas. Therefore, it is objectively favorable for realizing the specialization and cooperation among regions and cities, thus forming an organic regional economic network.
(5) The determination of key development axes at all levels in China can better combine the national strategy with the regional strategy, so that all regions and departments in the area have a clear and unified regional development direction, which is favorable for improving the investment effect of capital construction and the level of management organization.

The development axis with the intensive transportation industry belt as the connotation drives the development of the regional economy just like the door axis rotates with the whole door.

Since the founding of the People's Republic of China, China has carried out large-scale territorial development and regional development. In terms of the allocation of productivity and the spatial organization of social development, it basically meets the requirements of the "point-axis" spatial structure system.

In the 1950s, the Harbin-Dalian Railway was first recommended as the national key development axes, followed by the Beijing-Wuhan section of the Beijing-Guangzhou Railway and the Zhengzhou-Lanzhou section of the Longhai Railway. Among the 156 key projects (152 under actual construction), nearly 65 projects were located on the three key development axes, accounting for more than 40% of the total. The central cities subject to key development were Shenyang, Anshan, Harbin, Xi'an, Beijing, Lanzhou and Fushun, followed by Changchun, Benxi, Shijiazhuang, Zhengzhou, Luoyang, etc. In order to meet the development needs of these three key axes, China carried out a large-scale reconstruction of the corresponding trunk railways.

In the 1960s, China carried out the "third-front" construction on a large scale. Most of the key development axes were located in the central and western regions of China, such as Baoji-Chengdu, Chengdu-Kunming, Xiangyang-Chongqing, Jiaozuo-Zhicheng and Guiyang-Kunming Railways. Of course, many projects on the key development axes in the 1950s were successively completed until the 1960s. Since the layout guideline implemented at that time deviates from the requirements of the law of agglomeration economies, and the original economic foundation of the development axis was very weak, the construction projects were not far from the linear infrastructure (most were newly constructed), but they were scattered. The central towns subject to key development were not prominent. The "point" was separated from the axis development, bringing huge economic losses [1].

In the first half of the 1970s, the "third-front" construction continued, and the above-mentioned development axes were still the focus of national industrial and transportation construction. Small and medium-sized towns with industrial points as the main body began to form, and the benefits of the "point axis" began to appear partially. Since the publication of the Sino-US Joint Communique in 1972, China has begun to introduce large foreign projects. Most of these large projects were located in coastal zones and along the Yangtze River, close to seawater and freshwater sources. Among them, some enterprises used water transportation to improve their investment and operation effects. These large enterprises were mainly from the new construction and expansion of thermal power plants, oil refineries, iron and steel works and petrochemical works. Among the 10 large oil refineries and four expanded oil refineries constructed and put into operation from 1972 to 1980, eight were located along the Yangtze River. Large chemical fertilizer plants were also mainly constructed on two national first-level development axes. This was major progress in China's territorial development and industrial distribution on the basis of summarizing past experience and lessons.

Since the 1980s, China has entered a new development period of "opening to the outside world and invigorating the domestic economy", and the coastal areas have been in a favorable strategic position. The main body of the coastal open area was within the range of dozens of kilometers along the coastal zone (260 cities and counties, about 320,000 km in total), and there were 14 open cities and Shenzhen, Zhuhai, Xiamen, Shantou and Hainan Island special economic zones. About half of China's major projects were concentrated in the coastal zone, which was one of the two largest development axes in China. In the Seventh Five-year Plan, the investment in the coastal zone axis accounted for about 30% of that in China, and the net increase in the total industrial and agricultural output value accounted for about 40% of that in China. From this, we could see the status of the coastal zone axis in China. The construction scale of the axis along the Yangtze River is second only to that of the coastal zone axis. Especially since the 1980s, Jiangsu, Anhui, Jiangxi, Hubei, Hunan, Sichuan and other provinces along the Yangtze River have all proposed important points to develop the industrial belt along the Yangtze River, and the emphasis of the construction in the province was placed on the area along the Yangtze River. A series of mega-projects such as the Yangtze 300,000 t ethylene project, the Wuhan 1.7 m rolling mill and the Gezhouba Dam hydro-junction were constructed in succession. Correspondingly, the Yangtze River shipping and other water transportation facilities were constructed on a large scale. In terms of the second-level development axis, the industrial belts such as the area along the Longhai, Jiaoji, Changda and Jiaozhi Railways and the coastal area of the upper and middle reaches of the Yellow River were mainly constructed. In the 1990s, China decided to develop Pudong, Shanghai on a large scale. The large-scale construction of Pudong will bring a series of new development opportunities to the development of the Yangtze River Industrial Belt. The governments of Jiangsu and Anhui planned to concentrate their efforts on building the Yangtze River Industrial Belt in their own provinces (called "Wanjiang Economic Belt" by Anhui).

2 Structures and Types of Development Axes

According to the different types of linear infrastructure of the development axes, the development axes can be divided into the axis along the seacoast (including the axis along the coast of large lakes), the axis along the bank of large rivers, the axis along the traffic arteries on the land and mixed type. At different stages of socio-economic development and industrial structure, these development axes play different roles. Different types of development axes have obvious differences in industrial characteristics, industrial structure relevance and significance to regional development.

1. **Coastal development axis**

The coastal development axis refers to a belt-like range of about 50 km deep from the intertidal zone to the land. The temperate, subtropical and tropical coastal zones of most countries in the world have experienced long-term development and are the economic centers. Among them, the coasts and coastal areas of many countries are rich in land, mineral and marine resources, and have become the most important zones for industrialization and socio-economic modernization. With the strengthening of international economic and technical cooperation, the coastal development axis has become the main zone for the economic takeoff or revitalization of many countries. It has become an important approach and spatial development model for the development and revitalization of developing countries in the world today to carry out large-scale construction of various types of economic and technical development zones and special economic zones in the coastal development axis, which further enhances the status of the coastal development axis.

1) Historical origin of the development of the coastal development axis

It appears to be a long-term trend that human production activities and the world's population are increasingly attracted by the ocean, and this tendency exists in many coastal countries. However, the degree of attraction varies in different periods due to the different roles played by the ocean. After marine transportation and marine resources had a commercial nature, commercial activity zones were formed along the coasts of the world; with the development of commerce, industrial sectors directly related to the ocean, such as shipbuilding, ship repairing, aquatic product processing and raw material processing at colonies, appeared on the coast; with the development of the world economy, the territorial division of production activities made the coastal zones occupy the main position of domestic and international economic relations; as new resources continue to enter the above process, maritime trade and industrial activities were concentrated in the corresponding areas of the coast. Then non-ferrous metal industry, petroleum industry and agricultural product processing industry appeared in these regions, and various manufacturing and export sectors were developed. These shore sections were both the center of sea transportation and the hub of coastal transshipment. With the expansion of the connection area between land and sea, the construction of various infrastructures was attracted. Railways and waterborne traffic were developed, and then highways and all pipelines were constructed, which created environmental conditions particularly favorable for economic activities, thus attracting large populations.

In recent years, the development of science and technology has further improved the importance of coastal areas in economic and social development and stimulated various economic functions of the coastal areas, which can be seen from four aspects.

First, the world's economic relations have been generally strengthened, and the rapid development of marine transportation and the continuous advancement of marine transportation technology have increased the importance of industries and all economies of the coastal zone. Modern marine transportation can greatly shorten the "economic distance", and advanced transportation technology has greatly increased the volume of transportation, bringing new momentum to commodity transportation. As a result, many new industrial sectors have adapted to the "port situation", thus promoting the gradual development of large-scale new industries and maritime undertakings, and industries have generally "shifted" to the ocean. The shift of the fuel-energy industry to the ocean is the most obvious, especially in the petroleum industry. More than 60% of the petroleum extracted by capitalist countries is transported by sea, which has led to the great development of the petroleum industries (extraction, processing and chemical industry) along the coast of the oceans of Western Europe, the United States, Central America and the Middle East. The agglomeration of oil refining industries and the use of petroleum in the port area have attracted the power industry, energy-consuming industries (such as the aluminum industry) and various chemical industries, as petroleum and natural gas are the raw materials for the production of these industries. This has led to the development of multiple economies in the corresponding regions.

The development of the iron and steel industry in port cities is also brought out by the same factor, which is typical in Japan, as 100% of its iron ore is imported. Nowadays, Western Europe is also following the model of Japan, and some enterprises in the United States also observe this approach. This situation is caused by two factors. ① The output of iron ore in developed capitalist countries accounts for 92.9% of that in the capitalist world, while the proportion was only 60% in the 1970s; the metallurgical production in Western Europe has also changed its dependence on the import of high-quality ores. ② The tonnage of ore carriers is getting bigger and bigger, and the economic benefits are getting higher and higher. It is much more convenient to transport ore from the ore exporting countries, such as Australia, India and Venezuela, than from the coast of the old mining area through the inland railway. In addition, the iron and steel industry can complement the relatively developed coastal industries; the steel industry supplies steel to the shipbuilding industry and consumes imported oil and coal fuel, and steel can also be exported. Large metallurgical enterprises in port cities such as Dunkirk, Marseille-Fos and Rotterdam are causing changes to the distribution pattern of metal production in their countries. Large iron and steel enterprises also provide conditions for the development of the coal-coke chemical industry, construction industry and heavy machinery industry. At present, large iron and steel enterprises have become the "magnetic poles" for the development of neighboring regions.

Second, it is related to the extraction of marine biological resources, mineral resources and energy as well as seawater utilization. The continental shelf is increasingly important in industrial activities. The corresponding coastal base shall be constructed for the development of the continental shelf. The continental shelf is also the most suitable place to develop aquaculture, and control and regulate the reproduction of marine living resources. The oceanic continental shelf not only provides abundant mineral resources, especially petroleum and natural gas, but also has inexhaustible seawater for the energy industry, specifically for nuclear power stations.

Therefore, the importance of coastal areas in the development and application of nuclear energy is increasing by the day. The United States, Sweden and Germany have constructed many nuclear power plants along the coast.

According to the nuclear power development plan, France intends to build 200 nuclear power stations by 2000, with an average capacity of 1 million kilowatts. Among them, only 50 nuclear power stations rely on inland water sources for cooling, and the remaining 150 have been built and are planned to be built along the coast and nearby. The St. Petersburg nuclear power station, one of the largest nuclear power stations in Russia, is located along the coast of the Gulf of Finland.

Third, it is related to the continuous improvement of the infrastructure. The infrastructure, in a broad sense, also includes the development of deepwater ports and the construction of bridges and tunnels.

Fourth, it is related to the development of international tourism. The world's coasts are becoming the center of seawater recreation. Tourism and the flow of vacationers have formed a "smoke-free industry", which has promoted the development of coastal service facilities and formed a huge tourism economic complex.

2) Basic conditions for coastal development axis

The coastal zone can become the focus of development and construction in many countries and regions, which is determined by a series of superior conditions and resources. These can also be seen from the above-mentioned development origins. To sum up, there are the following material conditions:

(1) Excellent harbors and estuaries are the preconditions for the construction of a port, a hub that connects the hinterland on land and coastal cities. Port development is not only an integral part of coastal axis development but also the focus of the entire coastal axis development. The city formed on the basis of the port construction is the main body of the industrial development of the coastal axis and the entire coastal zone. Large and medium-sized ports are usually located in excellent harbors with deep water, current as well as good geological and geomorphological conditions for the shoreline, or near the estuaries of large rivers that connect with the vast hinterland.
(2) If located at the intersection of land and sea, it is easy to participate in social-economic activities in a large area, and get the influence of advanced technology and culture of other regions and countries. In ancient times when the transportation technology was very underdeveloped, the role of coastal location was not very obvious. Marine transportation has risen on a large scale since the "Age of Discovery" in the fifteenth century, and the coastal zones of some continents and islands have prospered rapidly, such as the northeast of the United States, South America, India, Japan and part of the coastal zones of China. Since the 1950s, the economies of some countries have taken off, and most of them have built coastal zones with extremely convenient connections with overseas. South Korea focused on the construction of the southeastern coast on a large scale since the 1970s. In recent years, it has begun to turn to the west coast that faces China. About 14 billion US dollars will be invested in the grand plan for the development of the west coast, and the construction period will extend for 20–30 years. The core project is the expressway along the west coast, and the coastal base industrial complex and ports are constructed along the expressway [2]. Although modern air transportation has been very developed, it mainly transports cutting-edge technical products, ultra-expensive goods, etc. with small size, lightweight and high value due to the impact of economic factors. Bulk cargoes such as energy (coal, petroleum and its products), raw materials (steel, chemical raw materials, synthetic materials, building materials, wood, etc.), machinery and equipment (industrial production equipment, engineering

equipment, transportation machinery), food, salt, etc. are still mainly transported by water between countries and regions. Most of the places of departure and destinations of these materials are located in the coastal zone. This is the main factor why the economy of the coastal zone still has great development momentum under modern conditions.

(3) It has land and space resources with great potential, so the environmental problems can be solved easily and reasonably. As the social economy becomes more and more developed, the living space suitable for human beings to carry out social-economic activities is getting smaller and smaller. However, there are also large areas of tidal flats and saline-alkali wasteland in the coastal areas in most developing countries and some developed countries, which are suitable for tract development. In the process of large-scale industrialization, the cost of pollution control is generally lower than that of the location far from the sea due to the vast coastal space and water bodies, and the large environmental capacity of the coastal construction industry.

(4) Seawater can be fully utilized to replace the increasingly scarce fresh water. Large-scale industrialization and urbanization have led to the increasing depletion of onshore freshwater resources. It is an important way to get rid of the shortage of fresh water supply by placing the focus of regional exploitation and regional development on the coastal zone and using seawater locally (mainly for industrial cooling).

3) General characteristics of the coastal development axis industry

The economic development of the coastal development axis mainly relies on the above-mentioned material conditions, and its industries generally have characteristics that are obviously different from those of inland regions.

(1) Birthplace and diffusion source of new technologies and new industries. Due to their special geographical location, coastal zones are most likely to accept new ideas and new technologies from other regions; at the same time, many innovations are often produced in coastal zones due to long-term economic and technological accumulation. In the process of transforming science and technology into productivity, new technologies and new industries in the coastal zone develop rapidly and gradually spread to its hinterland. This spatial process is the most obvious in those developed and developing countries that develop later. There are some examples of the former, such as the spatial shift of the territorial exploitation and development level in the United States and Australia. Examples of the latter are more common. Since the 1970s, the technological innovation and development dominated by the electronic industry has grown into large-scale new industrial clusters in the so-called "sun belts" such as the United States and Germany due to the ability of air transport to adapt to fast and flexible requirements coupled with the effects of climate and landscape factors. However, these new industries still mainly emerge first in the coastal zones in most developing countries and then are promoted.

(2) The correlation between industries is not strong. On the one hand, the industrial structure of the coastal zone reflects the structure of imported technology. At the same time, it reflects the characteristics of the resource structure and industrial structure of the hinterland. Therefore, the main collaborative relationship is "external" and "internal" (each shore section and the hinterland of the port area). However, the complementarity of industries on the development axis is not obvious. This is an important sign of the coastal zone development axis being different from other types of axes. This is particularly prominent in China: The main industry along the coast of the Bohai Sea is the heavy chemical industry, the heavy chemical industry and light textile industry along the coast of Jiangsu, Shanghai and Zhejiang are well developed. The main industries along the southern coast are light industry, electronics, etc. Although the industrial directions are obviously different, economic cooperation is basically limited to the transportation of energy to the central and southern coasts via the ports of the Yellow Sea and the Bohai Sea. However, the industrial correlation is not strong in the inner coast of the Bohai Sea. The exchange of goods between the ports of the Bohai Sea accounted for only 1–2% of the total throughput of these ports for a long time.

2. **Big river bank development axis**

The middle and lower reaches of the big rivers used to be the birthplace of ancient human civilization. Under modern economic conditions, although river banks are generally less important than coastal zones, they are still the focus of regional exploitation and economic development in many countries. This situation is particularly obvious in countries with short or no coastlines or with particularly harsh development conditions for the coastal zones. Such as the Federal Republic of Germany (Rhine River) before reunification, the former Soviet Union (Volga River and Dnieper River), Egypt (Nile River), India (Ganges River), etc. In countries with a large land area, there are generally long rivers, which flow through a large area. The development of industries along the banks of these large rivers undoubtedly occupies an important position, such as the United States and China.

The development of the banks of the Great River, as a type of development axis, has a long history. As early as the seventeenth and eighteenth centuries, that is, before industrialization, European countries built canals in succession. A series of commerce, cities and later industries converged on the banks of the Rhine, Seine, Thames and canals. The development history of the United States is more telling. The merchants from Europe first established themselves on the northeastern coast, entered the Saint Lawrence River along the canal chiseled and then the Great Lakes Region, and developed along the lakeshore. China's ancient civilization originated in the Yellow River Basin. In the heyday of the feudal society, cities, commerce, handicrafts and developed agriculture were mainly concentrated in the coastal zone of the Weihe River and the Yellow River from Tongguan to Kaifeng.

1) Basic conditions for big river bank development axis

(1) Huge transportation channel. In a society with a low level of productivity, water transportation is the most important means of transport; in many regions and countries, its significance far exceeds that of land transportation at that time, such as carts. Under the highly developed social and economic conditions in modern times, water transportation is the cheapest means of transport, which is especially suitable for the transportation of bulk goods with low value and large consumption in production. The transportation cost of these goods generally occupies a large proportion of the production cost of the product. For example, the freight of main fuels and raw materials in iron-steel production in China accounts for 12–20%, 5–30%, 30–50%, 20–30%, 5–10%, 3–5% and 45–60% of the total cost of iron and steel, thermal power, salt (to the place where industrial salt is consumed), cement, flour, cotton as well as sand and gravel respectively. According to foreign data, freight accounts for 6.0%, 0.9%, 2.5% and 6.5% of the total cost of automobiles, cigarettes, butter and sugar (beet sugar) respectively. That is to say, the production sectors and enterprises that produce and consume large amounts of energy and raw materials shall be located on both sides of rivers with large-scale navigation capability as much as possible. These production sectors include metallurgy, coal (development priority shall be given to coal mines that can utilize water transportation), building materials, thermal power plants, oil refining, petrochemicals, fertilizers, inorganic chemical raw materials, shipbuilding, processing of bulk agricultural products, automobiles, heavy machinery manufacturing, etc. The navigation capability of big rivers depends not only on the hydrological and topographic characteristics of the river, such as hydrology and seasonal changes, the longitudinal slope of the watercourse, the natural topography of the riverbed, etc. but also on the regulation scale of navigation channel, ports and their facilities. The length and water volume of the Rhine River in Western Europe are only equivalent to those of a large tributary of the Yangtze River, but its transportation volume exceeds that of the trunk stream of the Yangtze River. The largest dense industrial belt in Western Europe is located along the coast.

(2) The development and utilization of water resources and water energy resources is a powerful promoting factor that drives the economy of the development axis and the whole river basin. Water, as one of the most important natural resources on the earth, is also an important natural condition for socio-economic development and, of course, industrial and urban development. The modern economy can't develop without water. At present, the scale of water consumption is continuously expanding all over the world. According to the data from the United Nations Water Conference, the annual consumption of fresh water by the world's industry, agriculture, cities and reservoirs increased from 400 to 3000 km^3 from 1900 to 1975, with an average annual increase of 2.5%. It is estimated that the water consumption will double as of 2000, up to 6000 km^3 annually. Therefore, the issue of the water supply has become as urgent and

important as energy, economic development, environment, etc. in many countries and regions. In China, the situation of water resources has affected the structure and speed of economic development to a great extent. Especially in the vast northern region, water resources are insufficient on the whole with regard to the degree of assurance to the national economic development. In many areas, it has become a limiting factor for the further development and reasonable layout of industrial and agricultural production. In some areas of Southern China, the incompatibility of water resources and development and utilization conditions with national economic development is increasingly manifested. The characteristics of water resources distribution in China have affected not only the selection and location of industrial enterprises, but also the distribution of enterprises as well as regional industrial structure and scale on a meso scale, and the overall layout and benefit level of the national industry on a macro scale.

Most of the raw materials, fuels, auxiliary materials and all products required in the process of industrial production can be transported by means of vehicles for hundreds or even tens of thousands of kilometers. However, water, as a coolant and raw material for the industrial production process, is generally not suitable for transportation by vehicles. A complex with an annual output of three million tons of iron and steel produces about 16–18 million tons of off-site transport volume each year, but the annual water consumption reaches 148 million tons. No vehicle can afford such a huge transport volume economically. Therefore, the production sectors with high and moderate water consumption shall be located along the banks of big rivers as much as possible. The annual average water resources in China are 2.71 trillion cubic meters. Among them, the groundwater is only 700 billion cubic meters. That is to say, from the perspective of approaching the productivity of water resource allocation, the most important location trend is to approach long rivers and utilize surface water nearby. The total annual runoff of the Yangtze River is more than 900 billion cubic meters, 600 billion cubic meters even in the driest years. The total annual runoff of the Pearl River is 470 billion cubic meters. The water volume of these two rivers can be said to be "inexhaustible" in terms of industrial water consumption along the coast (axis range).

Modern cities, especially their industries, consume a lot of water. Rice and irrigation agriculture also requires a lot of water resources as a guarantee. The big rivers are taken as the water source. Industrial production sectors that consume large amounts of water shall be emphasized especially (Table 1). This table generally reflects the level of industrial water use in Northern China in the 1970s. The sectors with large water consumption are located along the rivers, becoming the main components of the dense industrial belt on the development axis, which is almost the basic feature of many big river bank development axes in the world.

The upper and middle reaches of the long rivers generally have abundant hydropower resources, which is an important material basis for the formation and growth of the dense industrial belt on the big river bank development axis. Due to the impact of this factor, such an industrial belt is most suitable for the development of those productions with high energy consumption per unit product or high energy consumption caused by a large production scale (Fig. 3). The former include non-ferrous metal smelting (standard fuel consumption per ton of product: 19–22 tons for titanium, 9–12 tons for magnesium and 8.0 tons for aluminum. Among them, the power consumption is more than 20,000 kWh for titanium, 21,000–26,000 kWh for magnesium, and 16,000–17,000 kWh for aluminum), and iron alloy (each ton of product consumes three to five tons of standard fuel. Among them, the power consumption is about 7000 kWh), synthetic materials (standard fuel consumption per ton of product: 11.5 tons for vinylon, 8.5 tons for neoprene and 3.6–4.3 tons for polyvinyl chloride. Among them, the power consumption is 7300 kWh for vinylon, 12,000 kWh for neoprene, and 6000–7500 kWh for polyvinyl chloride). The latter is analyzed based on the sector volume in China. At the end of the "Sixth Five-year Plan" and the beginning of the "Seventh Five-year Plan", the energy consumption of material production sectors in China accounted for 76.5% of the total energy consumption. Among the industries, the energy consumption of the processing and manufacturing industries accounted for 88.4%. Among the 25 industries in processing and manufacturing industries, the three industries that consume the most energy are the chemical industry, building material and non-metallic mineral product industry, and ferrous metal smelting and calendering, accounting for 16.1%, 15.9% and 15.0% of the total energy consumption respectively, a total of 47.0% of the industrial energy consumption. Other industries with large energy consumption and the proportion in the industrial energy consumption are as follows: 7.2% for mechanical, electrical and electronic equipment manufacturing, 5.1% for electricity, steam and hot water production, 4.7% for the textile industry, 2.6% for paper making and paper product industry, 2.2% for the crude oil processing industry, and 2.4% non-ferrous metal smelting and calendering industry. The energy consumption of these nine industries accounted for 71.2% of the total industrial energy consumption. From the perspective of the energy consumption structure of the industrial sector of the national economy, they were large energy consumers.[2] When other conditions are the same, these industries shall first be deployed on the big river bank development axis with abundant hydro energy.

(3) There are generally vast and fertile plains in the middle and lower reaches of the long rivers, with developed agriculture, which provides the foundation for the construction of a strong and dense industrial belt. The experience of territorial exploitation and regional economic development has shown that: In each region and zone, the level of agricultural development (including the level of agriculture in adjacent areas) and the scale of raw materials, grain and non-staple food provided to the industry often have an important impact on the

[2] National Bureau of Statistics, China Statistical Yearbook, 1988, Page 424, China Statistics Press, August 1988.

entire economy. After the big river originates from the mountain area, most of the basin area is hills and plains in the middle and lower reaches. There are favorable conditions, such as abundant water resources, abundant land resources and convenient transportation, for agricultural development.

In addition to the above three basic conditions, long rivers generally originate from mountain areas, flow through hills and plains, and have a variety of stratigraphic and geological conditions. Therefore, there are abundant mineral resources. The strong industrial belt formed by combining mineral resources with the above conditions is extremely favorable.

2) Industries of the big river bank development axis and its layout characteristics

Big rivers generally cross larger areas. Within the range attracted by the mainstream and its tributaries, there are often a variety of mineral, agriculture, forestry, animal husbandry, fishery and hydropower resources. After long-term development, a variety of industrial groups are generally formed within the river basin, such as the energy industry, development and smelting of metal ore, chemical industry, building material industry as well as planting and fishery in agriculture. The location of these sectors and industries generally depends on the distribution of resources and the geographical location of transportation in the river basin. However, the machinery, electronics, food, textile and other industries built on the basis of these resource-based industries or serving these industries are mainly distributed on the mainstream development axis. The industrial structure on the development axis has been diversified and complicated with the increasing scale of resource development and utilization of the river basin.

The industrial development of the big river bank development axis is obviously different from that of the coastal axis in terms of the formation process. Under normal circumstances, the industries developed along the river banks generally start from the basic industries, that is, the energy and raw material industries and large-scale planting industry established by utilizing the advantageous resources in the river basin and the water transportation conditions. However, the industrial chain developed from the basic industry is relatively long in general.

With a long development history and a relatively high level of river basin economy, the economic relations that are quite close are established in the mainstream and its tributaries as well as the upper, middle and lower reaches in the big river basin, becoming an economic complex of the river basin. This complex often shows two trends at different locations in the river basin. Under normal circumstances, the resource-based industries in the upper reaches account for a larger proportion compared with the lower reaches. The reason is that the upper reaches are mostly hilly and mountainous areas, with abundant minerals and energy. Relatively speaking, the economic and technical levels are generally lower than those of the lower reaches; it is easier to accept the impact of advanced technologies of other countries and regions in the place closer to the lower reaches, and the production of deep-processed products account for a larger proportion. The industrial level within the range of tributaries is generally lower than that of the mainstream. Under the combined effect of these

Table 1 Industrial sector enterprises divided by water consumption and their scales

Project	Water consumption	Industrial sector enterprises and their scales
1. Extra-large water-consuming enterprises or joint ventures	2–3 m^3/s	① Iron and steel complex with an annual output of more than one million tons ② Large regional thermal power plants and thermal power bases ③ Large petrochemical base or petrochemical fiber base
2. Enterprises with large water consumption	10,000–170,000 m^3/day	① Steel plants with an annual output of more than 200,000 tons of steel; ② Thermal power plants with an annual output of more than 200,000 kW; ③ Large oil refineries and petrochemical plants; ④ Large fertilizer plants and coking plants; ⑤ Textile dyeing complexes with more than 100,000 yarns; ⑥ Chemical fiber plants (spinning) with an annual output of 5000–10,000 tons/year; ⑦ Synthetic rubber plants with an annual output of 15,000 tons/year; ⑧ Oil fields [note (water injection)]; ⑨ Large-scale basic inorganic chemical plants and pharmaceutical plants; ⑩ Large and medium-sized paper mill and pulp plants with an annual output of more than 10,000 tons
3. Enterprises with medium water consumption	1000–10,000 m^3/day	① Small iron and steel works; ② Professional steel rolling mills; ③ Medium thermal power plants; ④ Specialized chemical plants; ⑤ Medium textile mills; ⑥ Large and medium-sized cement plants; ⑦ Coal mines and coal washing plants; ⑧ Small paper mills; ⑨ Medium and large machinery plants; ⑩ Oil chemical plants; medium-sized fertilizer plants
4. Enterprises with small water consumption	< 1000 m^3/day	① Small and medium-sized machinery plants; ② Daily chemical factories; ③ Machine tool factories; ④ Food processing factories; ⑤ Ceramic factories; ⑥ Glass factories; ⑦ Pharmaceutical factories; ⑧ Agricultural machinery factories; ⑨ Brick and tile factories; ⑩ Lime factories and knitting factories

two trends, the mainstream in the lower reaches of the big river is often the most prosperous section of the development axis.

Among the industrial groups of the big river bank development axis, the energy industry is often the main body. Especially in the industrial belts in the upper and middle reaches of some big rivers, the hydropower and metal industry with large power consumption have a prominent position.

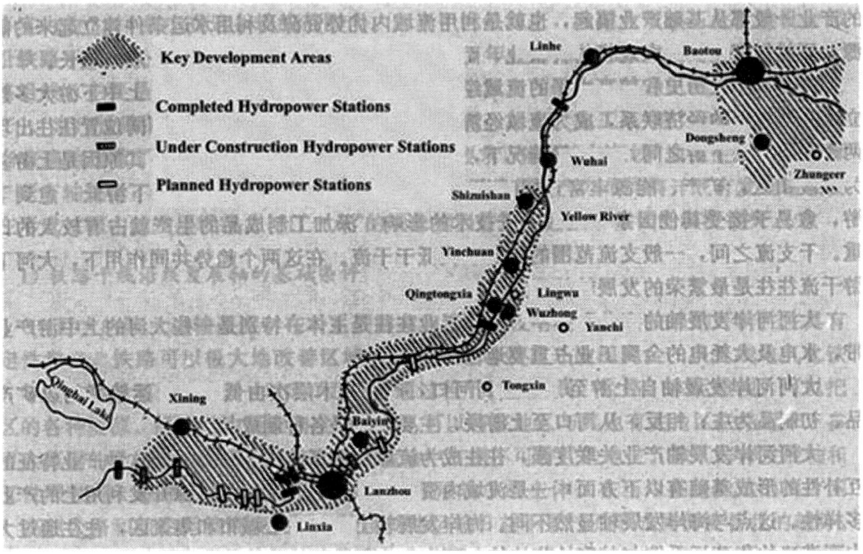

Fig. 3 Allocation of development axis and hydropower station in the upper reach of the Yellow River

The industrial technology level changes from low to high for the big river bank development axis from the upper reaches to the lower reaches until the estuary, and the transported goods are mainly mineral products and primary products; on the contrary, the main goods are various manufactured products from the estuary to the upper reaches.

The industrial linkage of the big river bank development axis is strong, which often becomes the trunk and backbone of the economic zone of the river basin. The basis for forming the complementarity of such industrial characteristics is as follows: First, the diversity of resources in the river basin and the diversity of industries based on resource development and utilization. This is apparently different from the coastal development axis. The main central cities and agglomeration areas on the coastal development axis are often connected to their hinterland through the concentrated and dredging transportation system of large and medium-sized ports, and there are generally not many economic exchanges between them. Second, as the materials shipped out of various regions along the river have to go through the same navigation channel, the navigation channel is the natural tie of economic relations between various regions in the river basin. The partners they are looking for are most likely various regions and enterprises located along the coast of big rivers. The exchange volume of goods among main ports along the Yangtze River in China reflects, to a great extent, the industrial complementarity and the correlation among regions in the river basin, among central cities in the upper, middle and lower reaches, and among sections. There is material exchange among almost all ports from the upper reaches to the lower reaches; the volume of material exchange among main central cities (ports) accounts for most of the total exchange volume; it can be seen from the data analysis of the actual survey that the materials exchanged along the river include mineral products, energy products, as well as a large number of raw material products, textile products and cash crop products. It can be seen that the economic cooperation along the river is multifaceted. However, it can be seen from the analysis of the volume of goods exchange among main ports along the coast of China that most of the goods flow occurs among a few large ports between the north and the south, and that the goods flow is relatively small among neighboring regions or small and medium-sized ports in the south, central and north; another difference from the coast of the Yangtze River is that there are fewer types of goods exchanged among coastal ports. A few types of goods such as petroleum, coal and grain are in a dominant position.

3. **Development axis along the trunk railway**

Railways play an important role in various means of transport in modern times. In many countries with vast territories, railways have been the backbone of transport carriers for a long time. At the same time, the framework of the spatial organization of the regional development is along the trunk railway, that is, the axis. It seems that the two most important functions of railways and railway transportation are as follows: complete the task of transporting goods and passengers, and be the axis of territorial exploitation. In the past, people clearly saw the former function, but had a vague understanding of the latter function, or thought that the two functions were the same.

Railways are the result of commercial and industrial development in modern times. In Western developed countries, the climax of railway construction was basically from the 1950s to the time before the First World War, while the climax of industrialization was slightly later than this stage in general. Many trunk railways opened had become the development axis of the industrialization of these countries. The "Hellweg" line connecting Cologne, Dusseldorf, Duisburg (Ruhr region), Essen, Bochum, Dortmund, Hanover and Berlin became the main axis of Germany before World War II, which is a prominent example. In the 1920s and 1930s, 3/4 of Germany's economic strength, especially the industrial and military strength, was concentrated within the range of this axis. Over the past half century, most of China's large-scale industrialization has been carried out on the development axis along the trunk railway. From the 1950s to the early 1960s, the relatively prominent ones were along the Harbin-Dalian Railway in the northeast (mainly steel and machinery industries), the middle and western sections of the Longhai Railway (mainly machinery, electric power and military production), and the northern section of the Beijing-Guangzhou Railway (Beijing to Wuhan, mainly machinery, metallurgy chemical industry). During the period of the "third-front" construction, the objective requirement for allocating the productivity along the development axis was quite commonly violated. Due to the weak foundation of productivity in the "strategic hinterland areas of China", the lack of existing axes and the implementation of the policy of excessive dispersion, the industrial productivity distribution deviated from the existing development axis or the development axis being formed. The phenomenon of "one factory, one point" was quite common, causing great economic losses.

1) Basic conditions for the development axis along the trunk railway

(1) Huge onshore transportation channel. The opening of railways is often of decisive significance in the exploitation and development of a series of countries and regions. Railways can greatly improve the accessibility of the regions and enhance openness. For a double-track electrified railway, the annual freight volume can reach 40–55 million tons. Through several railway branch lines, highways and shipping lines, various resources and economic potentials of the region can be concentrated on the development axis, and the region can be linked with other relevant regions, so as to participate in a wider range of social-economic activities. Railways occupy an irreplaceable position in China's transportation. In addition to the coastal development axis and the Yangtze River development axis, the areas along the railway are the main body of the secondary development axis and below. Nowadays, expressways and air transportation are highly developed in some developed Western countries, which makes railway transportation increasingly sluggish, thus forming a trend of partial replacement. However, a similar trend is unlikely to occur in China. The reason is that: The macro-regional differences in natural, economic and population distribution in China are very large, and the volume of long-distance exchange of goods will continue to increase. For example, coal, timber, product oil and smelting materials, which are related to the lifeline of the national economy, will inevitably rely on railway transportation; most of China's main ore fields do not rely on "water" (rivers, seas and lakes). The mineral resources in the west are obviously more than those in the east, while the distribution of productivity is just the opposite. In addition to the Yangtze River and the Pearl River, the long-distance material exchange between the eastern and western regions mainly relies on railways; developed countries have built a large number of medium and long-distance expressways, and highways have replaced a considerable part of railways for freight transportation. However, the large-scale construction of medium and long-distance expressways is not allowed by China's financial resources (the unit cost is higher than that of the railway by more than three times. More importantly, the expressways cover an astonishingly large area. The majority will occupy the arable land and the urban and rural construction land), which is also unacceptable to China's national conditions. In the long period of time in future, the construction of expressways in China shall be limited to between large cities in the large industrial and urban agglomeration areas, which belong to short and medium distances of less than 300 km, such as Beijing-Tianjin-Tanggu, Shanghai-Nanjing, Shanghai-Hangzhou, Dalian-Shenyang, Guangzhou-Shenzhen, Guangzhou-Zhuhai, etc. The traffic flow is high in these directions, and door-to-door transportation is required for many goods. Even within these large agglomeration areas, railways still play a leading role in industrial transportation. For decades, a large number of industrial enterprises have built their own industrial stations and factory-specific lines, and a series of factory formation stations extended into the urban industrial area. These systems combining industry and transportation will undoubtedly continue to be

improved and properly developed in the future; energy production and consumption in China are mainly based on coal, and the average transportation distance of coal is long. Therefore, a large part of coal transportation has to rely on railways. In recent years, railway transportation of coal has accounted for more than 35% of the total coal output. If the coal for local power generation and local sales is deducted, this proportion is as high as about 75% (the total transport volume of coal includes water transportation). In the coal bases dominated by Shanxi, the volume of coal shipped out has doubled approximately every ten years. From a national perspective, the desired results cannot be achieved for the construction of a series of coal bases without enhancing railway transportation. From the 1950s to the 1970s, the energy consumption structure of many developed countries completed the transition from coal-based to petroleum-based; although coal still accounts for a large proportion in some countries, the source of coal has substantially shifted to import. In both cases, the status of railway transportation has dropped significantly. In China, neither of these situations is possible.

(2) Inherent advantages of resources and economic foundation. Different from the development axis along the river, the line trend of the development axis along the trunk railway is the result of human choice. That is to say, through the economic line selection of railway, the trunk line shall naturally pass through important resource origins, existing and planned administrative centers, economic centers and industrial and mining areas, mountain feet, plains, etc. In order to play a greater role in regional exploitation and development, some trunk railways are especially extended along the existing avenues and highways during the line selection. The zones that it passes through have been channels for driving the surrounding development since ancient times. The Beijing-Guangzhou Railway built in the early days went south from Beijing, followed the ancient north–south traffic arteries along the eastern foothills of Taihang Mountain, and connected political and economic centers such as Baoding, Zhengding, Handan and Anyang. At that time, nearby coal and iron resources had been partially discovered. The Baotou-Lanzhou Railway paved after the founding of the People's Republic of China crosses the Yellow River three times from Baotou and via the south of the Hetao Plain and the Yinchuan Plain at the foot of the Helan Mountains, connects major small and medium-sized cities and major agricultural and animal husbandry areas in western Inner Mongolia and Ningxia Autonomous Region, and basically extends along the Yellow River. This is mainly determined by economic line selection. After decades of development, the linear infrastructure with the trunk railway as the mainstay, supplemented by the mainstream of the Yellow River and coupled with external high-voltage lines and trunk highways of Liujiaxia and Qingtongxia Hydropower Stations has become an important development axis in the northwest and even the whole country. If the line is only selected based on the natural conditions, it would cross the Yellow River and Ordos terrace from the south of Baotou, connecting Lanzhou. The Yellow River would only be crossed once, and the mileage and project scale would obviously be smaller than those of the above scheme. Today,

the Lanzhou-Xinjiang Railway has become one of the important passages that connect eastern, central and western China and connect China to the Commonwealth of Independent States and Europe, and is another main development axis in the northwest region. Its line selection not only conforms to the economic principle, but also makes good use of terrain. This axis is the economic essence of Gansu and Xinjiang. Compared with other vast areas of the two provinces, it has a very prominent inherent advantage.

2) General characteristics of industrial development of the development axis along the trunk railway

(1) The industrial structure is based on heavy industry. Since the railway construction in many countries and regions serves for industrialization, one of their characteristics is to highlight the development of heavy industry, and one of the purposes of the construction of many trunk railways is to carry out large-scale mining development. Therefore, among the energy industries established based on resource development and in the zone along the trunk railway (including the central city), the production of raw materials based on metallurgy, chemical industry and building materials is generally dominant. Of course, as the development time continues, the industry of the trunk railway development axis with a long history also shows "processing industrialization" and "high processing degree". This is a clear point of difference in the industrial structure from the development axes along the big river and the coast. Most of the world's coastal development axes are dominated by processing industries and port trade. Although some countries have configured large heavy industrial complexes along the coast after World War II, the light industrial structure is still dominating on the whole. Industries along big rivers are generally established based on resources in the hinterland of the river basin, but there is a certain distance from most of the resource sites and mines to the axis. Most of the cities located on the axes of big rivers involve the export of manufactured goods based on the development and utilization of minerals and mining. The processing industry is more dominant, and commercial trade has a prominent position. In the last three decades, some countries have utilized the ores and oil shipped from all over the world, and made full use of cheap water transportation. While building large industrial zones for raw materials near the coast and estuaries, they have also developed large industrial zones on the banks of large rivers (mainly the downstream banks), which has brought momentum to the development of heavy industries on the big river bank development axis. However, it is still a general tendency to concentrate heavy industries on the development axis along the trunk railway on the whole. The feature can be clearly illustrated by the proportion of light and heavy industries in the industrial structure of cities on the main development axes in China (Table 2).
(2) Rapid development of regional industrial scale and industrial structure is promoted. Since the construction of the trunk railway is realized at a certain stage of regional exploitation and according to objective requirements, once

the trunk line is opened, the nearby resources and economic potential will be quickly developed and exerted, and the regional economy will soon gain new impetus and "grow". The industrial development along the big river and coastal development axis has a relatively obvious progressive process. It often develops gradually with the progress of water transportation technology and the economic development of the hinterland (river basin).

According to the characteristics of this process, the industries along the exhibition axis of the trunk railway can reach a considerable scale within a short time. At the initial stage and even the takeoff stage of regional industrialization and large-scale economic development, much attention is generally paid to the opening of trunk railways just because of this point. Attention is also paid to the concentrated distribution of industries near the trunk railway, becoming a dense industrial belt, thus exerting a huge agglomeration effect.

In addition to the above characteristics, the development axis along the trunk railway with a long distance is often the transportation line of energy, metal and non-metallic ores, and bulk agricultural products. Therefore, the agglomeration zone for the development of energy-consuming industries, raw material industries and agricultural product processing industries is also in the vicinity of the axis.

The development axis along the trunk railway is generally located inland in countries and regions. Regardless of the rapid development of modern transportation and communication technologies, it is more difficult for the "inland" development axis to participate in economic activities in the region and even the world in a large range compared to the coastal development axis. In other words, the conversion and upgrading of industrial technology and industrial structure along the railway line are generally slower than those of the coastal development axis. It can also be inferred from this that the opening of the trunk railway can quickly revitalize the economy of the regions along the line (the areas driven by the development axis), and make the regions under development prosper faster. However, the development axis along the trunk railway with a long history may also lead to the relative stagnation of the regional economy.

3) Two types of development axes along the trunk railway

In general, the development axis with the trunk railway as the main body can also include remote communication channels, highway trunks, etc. Such a development axis drives the economic development of nearby areas, and the direct promotion range is within 50–100 km. The maximum cargo negotiating capability of a trunk railway (calculated by double-track electrification) is no more than 50–60 million tons per year. Such a range and transport capacity can no longer meet the needs of transportation and regional economic development in some countries and regions. Especially for countries and regions with a large area and a narrow area facing the ocean. In order to improve this situation, the second, third, or even more passages are built on the basis of one passage, in the same direction, parallel to the original passage and at a distance of 100–200 km or more. From a macro perspective, it is like a development axis. A more prominent example is the trunk railway group that

Table 2 Proportion of output values of light and heavy industries of cities on some development axes in China (1987)

Type of development axis	Name of development axis	Section	Number of cities	Light industry; heavy industry
Coastal axis	Eastern coastal development axis	Dandong-Dalian-Tianjin-Shanghai-Guangzhou-Beihai	41	58:42
Axle along the trunk railway	Harbin-Dalian development axis	Harbin-Shenyang-Dalian	15	38:62
	Lianyungang-Lanzhou-Xinjiang development axis	Lianyungang-Xi'an-Lanzhou-Urumqi	21	42:58
	Baotou-Lanzhou development axis	Baotou-Yinchuan-Lanzhou	9	25:75
	Datong-Puzhou development axis	Datong-Taiyuan-Yuncheng	7	29:71
Big river bank axle	Development axis along the Yangtze River	Shanghai-Nanjing-Wuhan-Chongqing-Dukou	34	53:47

Source China Urban Statistical Yearbook, 1988, China Statistics Press. The output values in the table are calculated at constant prices in 1980 traverses the east and west of the former Soviet Union: starting from the Moscow region, the core of the Central District, going eastward through Gorky, Kirov, Kazan, Kuibyshev and Saratov respectively, and extending to the east side of Ural Mountains, merging into three trunk lines, extending eastwards to the vicinity of Novosibirsk, and then merging into two lines. Since the 1960s, the former Soviet Union had built the Beam Railway (BAM) on the north side of the existing Trans-Siberian Railway, extending eastward from Bratsk to the Soviet Port in the Strait of Tartary (Fig. 4). This development axis along the railway line, consisting of several trunk railways, concentrated more than half of the productivity of the former Soviet Union, and stretched for 5000–6000 km from east to west. It was the largest industrial belt in the former Soviet Union. There were many short-distance trunk lines and branch lines among the above-mentioned trunk railways, and several railways and inland navigation lines were radiated in the north–south direction, concentrating the resources and socio-economic potential of the vast territory east of the Moscow region on this axis.

The development axis formed by trunk railway groups can be called the multi-channel development axis. Compared with the single-channel development axis, it not only drives more regions but also promotes the balanced development of territorial resources. It often takes a long time to form the multi-channel development axis. However, the precise arrangement shall be made in the plan as long as the zone has a huge potential to form an industrial belt.

4. **Composite development axis**

The above are all development axes with a certain means of transport as the backbone. In the practice of territorial development and regional economic development, there are often two or more types of transportation channels combined with each other in parallel, which jointly play the role of transportation channels and to drive regional development. This is the composite development axis.

The composite development axis is the product of the high socio-economic development of the country and region. At the same time, such a development axis, once preliminarily formed, can play a huge role in promoting the further development of

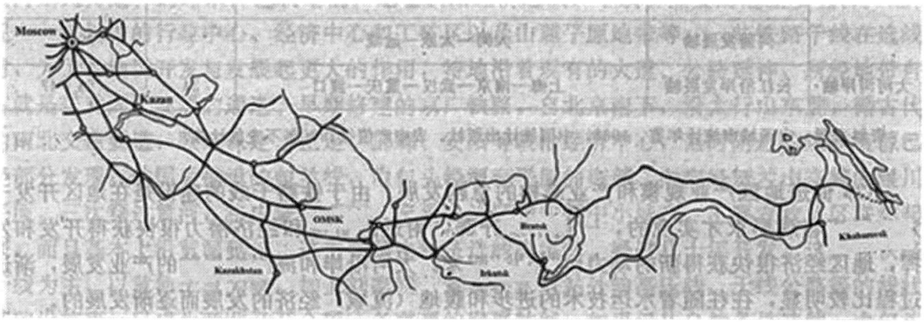

Fig. 4 Development axis formed by trunk railway groups in the east–west direction of the former Soviet Union

the country and region. As the means of transport such as railways, highways, waterways, and pipelines have their own suitable objects, the combination of several means of transport reflects the multi-level development of regional transportation needs, that is to say, the provision of convenient services to various types of socio-economic objects in the region can drive the diversified development of social economy, and release the potential of social economy at a deeper level.

The combination of railway and highway is a common thing in the composite development axis. The combination of waterways and railways shall be emphasized. For a long time in the past, people always believed that the parallel construction of railway lines along the inland waterway and the coastline (coastal route) would compete with water transportation, and affect the functioning of water transportation facilities. However, the socio-economic demands for transportation are multi-faceted and multi-layered in regions, especially in large regions and river basins. Moreover, after being opened, the railway may connect areas in a larger range through the developed land transportation network, becoming the hinterland of the big river port. That is to say, although the construction of railways may form a situation of competing with water transportation, that is, competing for the capacity of water transportation. However, it will develop the capacity of water transportation to a greater extent. Moreover, the construction of railways has brought about the prosperity of the social economy along big rivers (including urban expansion, industrial growth and comprehensive construction of other infrastructure) and increase in socio-economic exchanges, and will also obviously increase the volume of passengers and freight.

In China, the planning of the transportation network has long been affected by the above concept of "competition", and parallel construction of railways along big river banks and coasts is not common. The Baotou-Lanzhou railway line in the 1950s extended along the mainstream of the Yellow River, which had little role in transportation. However, the social economy of the zones through which the Yellow River passes has long been developed due to the benefits of the Yellow River. The railway and the Yellow River form a development axis, and the combination of the functions of the two has realized the large-scale development of the social economy of the axis and nearby zones. Parallel railways along the Yangtze River have long been neglected. Except for Shanghai-Nanjing and Ningjing-Wuhu (Nanjing-Tongling), the Dasha Line (Daye-Shahe) was completed after a long period of deliberation. With the development of the economy along the river and the need for further development of Yangtze River shipping, there will be new parallel railways along the river. The emphasis shall be placed on the shore section at the middle and lower reaches from Shanghai to Wuhan. From Nanjing (Pukou on the opposite bank) to Nantong on the north bank, the line reaches Jiujiang from Nanjing via Hefei and Anqing, and the direct railway connection from Wuhan to Chongqing shall be opened in the future.

The banks of the Rhine River are the most powerful development axes in Western Europe. The Rhine River originates in Switzerland, with a total length of 1320 km and a drainage area of 252,000 km^2. The upper reaches are the river reaches with the most intensive hydropower development in Switzerland, and the middle and lower reaches are the highly developed zones with dense populations and developed industries in

Germany and the Netherlands. The construction process of its transportation channel is: to develop the water transportation of the Rhine River first. As early as the middle of the last century, the Rhine River has become the backbone of the transportation network in Germany and the outgoing channel of Switzerland and eastern France. It is connected to several canals in the lower reaches, forming a navigation channel network with connected trunks and branches, connected river and sea ports as well as crisscrossed transportation network. Even so, a trunk railway was built close to the east and west banks of the Rhine River respectively within the border of Germany from the end of the last century to the beginning of this century. During the period of the Weimar Republic in the 1920s (1927–1929), the construction of the expressway from the vicinity of Frankfurt to the vicinity of the boundary between Germany and the Netherlands along the bank of the Rhine River began. The line continued to be built after 1933. From the 1950s to the 1960s, former West Germany entered the peak period of expressway construction. While expanding the expressway, a new expressway was built along the other bank. Apart from this, there were also trunk lines extending in the south-north direction and along the Rhine River for the long-distance oil pipelines and gas pipelines in western Germany as well as the electric power system of Europe. With the help of this powerful composite development axis, the estuary-intensive industrial area within the border of the Netherlands in the north was connected to the Ruhr industrial area, "the heart of Germany and Europe", the "Rhein-Main industrial area" in the center and the "Ludwig-Mannheim-Heidelberg" industrial area in the south. Since the 1950s, almost all of the 15 billion dollars in the "Marshall Plan" that aided the former Federal Republic of Germany has been invested in the Rhine axis. Most of the funds of the former Federal Republic of Germany have also been concentrated on this north–south axis, that's why the largest dense industrial belt in Germany and even Europe is along the coast of the Rhine axis today (Fig. 5).

3 Progressive Diffusion of Socio-economic Objects in Space

Starting from the "point-axis system" of the spatial process model of social economy, the formation and types of axes are discussed in the previous section. This section is to discuss the formation and change law of "point", an important part of the point-axis system, in space. The "point" in the spatial structure refers to various urban and rural residential areas, various industrial enterprises, comprehensive or individual economic infrastructure and social infrastructure, etc. They vary in nature, scale and function; in terms of scale, the big point includes a large city with millions of people, and the small point is like a village; in terms of nature and function, there are places of residence, places of origin, transportation hubs, commercial trade places, energy production and supply hubs, etc. Although agricultural production is distributed in a plane, the management agencies of agricultural production and the collection and

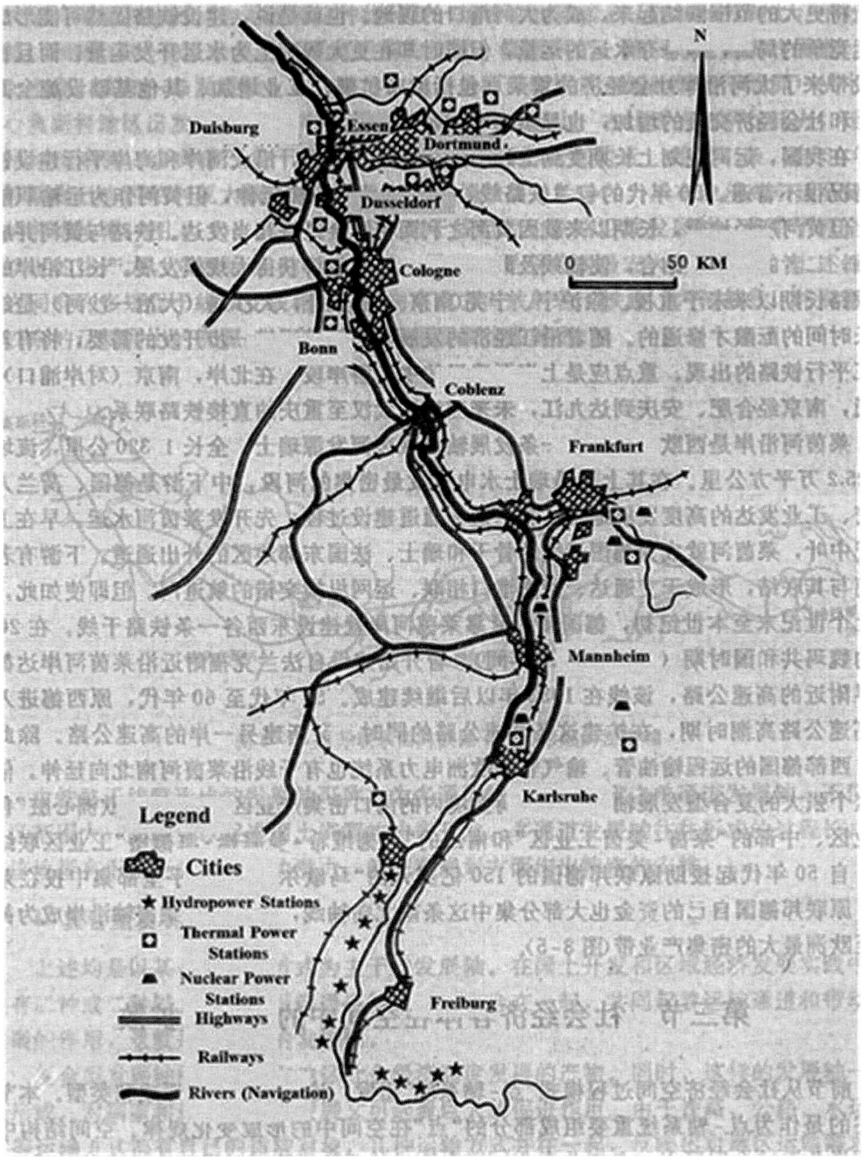

Fig. 5 Rhine development axis

distribution facilities of agricultural products are still in the form of dots. However, all of the above socio-economic facilities do not exist in isolation in terms of space, but in interaction. Various forms of spatial structure are formed just because of this interaction in space.

1. Central place theory and growth pole theory

1) Central place theory

At the beginning of this century, there was a rapid development of urbanization based on the fact that Western European countries had experienced a period of industrialization and large-scale development of the commodity economy. Under this historical context, the German geographer, Walter Christaller, published the famous "Central Place of Southern Germany" (Die Zentraler Ortean Siidteutschland, Jena) in 1933 through the investigation and analysis of hundreds of urban residential areas in southern Germany. The central content of this theory is the doctrine about the spatial structure of cities and urban functions, sizes and compositions within a certain region (country), that is, the city's "hierarchy-scale" doctrine or urban location theory, and a regular hexagon is used to vividly summarize the relationship between regional urban hierarchy and urban scale. This theory provides an important methodological basis for urban planning and regional scale.

W. Christaller used the model to make an abstract theoretical generalization of the layout principle of the research object. The presumption is that the space is regarded as homogeneous with the same density regardless of the differences in certain geographical conditions, and the principle of profit plays a completely dominant role. First, he raised the question: "We explore this reason why there are big cities and small cities. We believe that there must be some principles governing the distribution of cities. It's just that we still don't know it!" He believed that "the spatial structure of a city is the projection of human social-economic activities in space", and "the accumulation of matters to a core is the basic phenomenon of objects", that is, objects in space originate from the center and spread outwards; the center of a region, that is, the core of a region, is the commander of a specific region, which is the town. There is hardly a region in the world that is not covered by a city network consisting of cities on various scales. In most cases: If cities in a region or country are classified from large to small, there will be various levels. The empirical law shows that the towns with the minimum scale have the largest number. The higher the level, the smaller the number. W. Christaller analyzed the conditions for the formation of urban hierarchy, and believed that the emergence of urban hierarchy is governed by the following three principles: ① In order to facilitate administration management, the country is divided into several regions, which are further divided. Regions at all levels are governed by the city at its center; ② The social division of labor and the formation of a market economy cause the center point of regions at all levels to become the concentrating place of the trading market as well as institutions and facilities that provide goods and services to the surrounding areas; ③ The emergence of transportation network leads to the formation of towns at the cross points of traffic lines at different levels.

Under the combined interaction of these three basic principles, urban hierarchy and urban system are formed.

Through analysis and calculation, W. Christaller drew a conclusion as follows: A region or a country should form the following urban hierarchy: one Level A city, two Level B cities, six to twelve Level C cities, 42–54 Level D cities, and 118 Level E cities.

W. Christaller's theoretical model has provided established principles for planners, and has become an important basis for the theory of "point-axis" spatial structure. This is mainly manifested in the following two aspects:

(1) Urban and rural residential areas interact in space, forming a regular hierarchy and orderly spatial arrangement. The higher-level residential areas (towns) command the lower-level residential areas in terms of functions and supply, and meanwhile, absorb the required resources, materials and services from the lower-level residential areas;
(2) Once urban and rural residential areas form a hierarchical system, the state of spatial structure (mainly manifested as the urban system) will reach equilibrium. However, it refers to the static equilibrium theory here.

2) Growth pole theory

The idea of the growth pole was first proposed by the French economist, F. Perroux, in the 1950s, which is considered to be the cornerstone of the concept of economic regions in Western regional economics[3] and one of the bases of the unbalanced development theory. This theory holds that economic growth in a highly industrialized society is not uniformly distributed in space, but is first realized at the location where industrial production agglomeration has formed, and the impulse for economic development is generated therefrom. As the internal motion of the magnetic field is the strongest at the magnetic pole, such regional "polarization" of economic development is called the "Growth pole".

The growth pole only involves the linkage and multiplier effects between industrial sectors but does not refer to stronghold-style spatial exploitation of industrial and economic development. However, one or a few large enterprises often drive the industrial and economic development of the entire region in the process of application and development. That is to say, the regional exploitation in a certain range often starts from the point occupied by one or a few enterprises, which is the origin of the stronghold-style development model.

The industry that serves as the basis for the development and role of the growth pole is called a key industry. It is characterized by a large production scale, a very strong driving force for growth and extensive linkages with other industries. When the key industry starts to grow, other industries in the region where the enterprise (or sector) is located also start to grow. With the increase in employment, the improvement of regional taxation and the rise in purchasing power, new industries (especially

[3] Ding Sibao, "Growth Pole" Model and Economic Development in Underdeveloped Regions, Economic Geography, No. 4, 1989.

those with economic and process collaborations with key industries) are attracted to the region to invest. With the deepening and generalization of this process, the momentum for economic growth will gradually penetrate and eventually spread to the entire region [3].

The basic content of the growth pole theory can be summarized as follows.

(1) There is one or several dominant economic sectors, which have economic and technical relations with many other economic productions. Its development will lead to the corresponding development of other sectors, and the effect generated is often greater than that of its own development. This is like an industrial production area complex in an industrial zone or an industrial city. The core of such a complex is the enterprises with great development potential, which can be called enterprises forming a complex. In general, the enterprises that form a complex include oil refineries in the petrochemical complex, iron works in the iron and steel complex, assembly plants in the mechanical and electronic industries, spinning mills in the textile industry, etc. Such growth pole generally begins with the production of raw materials. With the sophistication of the structure and the development of large-scale equipment, large-scale industrial zones and industrial towns will often be formed, which will drive regional exploitation through the diffusion effect. In China's exploitation practice, one is to build an initial plant (an "upstream" manufacturing plant in general), followed by supporting development; the other is to build the supporting facilities at one time. In the first case, the initial plants, such as iron works, aluminum oxide plants, textile mills, oil refineries, coking plants, hydropower stations, etc., that is, enterprises that form a development pole or an industrial zone or an industrial complex, may be different from the industrial zones and industrial towns formed later by several to dozens of times in terms of scale.[4, 5]
(2) The result of the above development is that the industrial production in the region exceeds its own demand, with increasing output of products. Market monopoly may be formed under the conditions of the market economy mechanism. With the continuous expansion of the production scale, the accumulation of funds constantly increases, and the economic needs are also diversified, so the requirements for further development are constantly generated. Various related sectors and service industries have been developed, so as to achieve larger-scale employment and growth in national income.
(3) The growth pole acts on surrounding areas in two directions. The first is the "polarization process", that is, the agglomeration process. That is, the growth pole attracts the natural and socio-economic potential of the surrounding area, such as mineral resources, raw materials, labor force, investment, local industries or enterprises, with its strong economic and technical strength and superior conditions. The second is the diffusion process, also known as the

[4] Lu Dadao, Group Layout Types of Industrial Enterprises in Industrial Zones and Their Technical and Economic Effects, Journal of Geographical Sciences, Vol. 34, No. 3, 1979.

[5] Lu Dadao, Theory and Practice of China's Industrial Distribution, Pages 158 to 166, Science Press, May 1990.

"leakage process", that is, the growth pole invests in the surrounding areas and provides other economic and technical assistance to form affiliated enterprises or subsidiaries, so as to provide markets for primary products in the surrounding areas and absorb the agricultural surplus labor force. In the practice of regional exploitation and development, the effect on the surrounding areas is mainly "polarization" (aggregation) at the early stage of the formation of the growth pole (generally at the initial stage of regional exploitation). At this stage, absorption (deprivation) is greater than giving (promotion), and the degree of unbalanced economic development in the region increases; at the middle and late stages of the development and evolution of the growth pole, the "leakage" (diffusion) effect is dominant. At this stage, giving is more than absorption, and the regional development level tends to be balanced.

(4) The process in which the diffusion effect tends to be dominant is the key part of the growth pole theory, indicating the multiplication of the growth pole. On the one hand, it is developed with material, energy output and spatial expansion, so that the scale and strength of the growth pole are getting bigger and bigger; on the other hand, the formation of new (next-level) growth poles is facilitated by progressive diffusion.[6]

2. **"Point-axis" progressive diffusion of socio-economic objects in space**

1) Proposal of the theoretical model

The above central place theory and growth pole theory reveal the law of agglomeration and distribution of socio-economic objects ("point" like) in space. Among them, the growth pole theory also describes the diffusion of the center point to the peripheral regions and the formation of new next-level center points from a dynamic point of view. However, the diffusion phenomenon here is still based on a diffusion source. In the actual regional exploitation and regional development, there may be two or even more "points" that appear "simultaneously" within a certain range, each forming its own closed diffusion region. Such diffusion can come down to "local diffusion". When several points in space interact, the spatial diffusion will have an overall nature. "Overall diffusion" is not in the form of external diffusion of one source, and cannot be simply taken as the external diffusion of multiple sources from their own centers. However, it is the "point-axis" progressive diffusion model first proposed by the author in 1984.[7] Its theoretical foundation is the above-mentioned central place theory, the growth pole theory and the concept of the development axis (Entwichlungsachse) used in the regional planning of the former Federal Republic of

---

[6] Ding Sibao, "Growth Pole" Model and Economic Development in Underdeveloped Regions, Economic Geography, Volume 9, Issue 4, 1989.

[7] Author's report themed "The Scientific Basis of the General Map of China's Industrial Productivity Distribution in 2000" on the national symposium on economic geography convened in Urumqi in September 1984. The article was published in *Geographical Sciences,* Volume 6 (1986), Issue 2.

Germany since the mid-1970s. Before and after proposing the "point-axis" progressive diffusion model, the author studied some of the research findings on spatial diffusion by scholars in Sweden, the United States and the former Federal Republic of Germany with the help of German literature.

As is known to all, the earliest contributions to spatial diffusion were made by the Swedish scholar T. Hagerstrand and his colleagues, such as S. Godlund, E. Kant, et al. They had done a lot of research on the spatial diffusion of objects as early as the 1950s, and paid special attention to the spatial diffusion of innovation. These studies had given a great impetus to the development of geography at that time. In the 1960s, American scholars' research on diffusion reached a considerable scale. The main scholars were: L. A. Bronwn, R. Morrill; B. J. L. Berry, et al. Their research objects involved the laws of space transportation of many human and economic phenomena.[8] In German-speaking countries, the main scholars were C. Borcherdt, E. Wirth, D. Bartels, et al. Their research involved the diffusion of the central place system, the diffusion of technological innovation and information, the diffusion of foreign workers, etc.[9] According to the research objects of the above scholars and the characteristics of the proposed diffusion model, we can divide them into two categories: The first is the so-called abstract spatial diffusion or "pure" diffusion, which abstracts specific spatial objects, such as T. Hagerstrand's "pure" diffusion model (Fig. 6) [4]. The second is the specific diffusion of specific objects in the region, such as the research on the diffusion of foreign workers in the former Federal Republic of Germany by Professor E. Giese (Fig. 7). T. Hagerstrand made the following assumptions when studying the diffusion of a new object in space:

(1) The population is evenly distributed within the range of a plain region. During the actual observation and statistics, the plain can be divided into grids.
(2) The new object is located in the center of the plain.

---

[8] Morrill, R., & F. Pitts, Marriage, Migration and Mean Information Field. A Study in Geography. Ser. B., No. 24, 1967.

Morrill, R., a: Migration and the Spread and Growth of Urban Settlement. Lund. Lund Studies in Geography. Ser. B., No. 26, 1965.

Morrill, R., b: The Negro chetto: Problems and Alternatives. Geogr. Rev. 55 (1965), pp. 339–361, 1965.

Morrill, R., Waves of Spatial Diffusion, Journal of Regional Science. 8 (1968), pp. l–18, 1968.

Morrill, R., Shape of Diffusion in Space and Time. Economic Geography, 46 (1970). Supplement, pp. 259–268, 1970.

Morrill, R., & F. Pitts Marriage, Migration and Mean Information Field. A Study in Uniqueness and Generality, Ann. Ass. Amer. Geogr. 57 (1967), pp. 401–422, 1967.

[9] Whith, E.1954: Grabstock, Hacke und Pflug. Ludwigsburg.

Wirth, E., 1965: Zur Sozialgeographie der Religionsgemeinschaften im Orient. Erdkunde. 19 (1965), S. 265–284.

Whith, E.1969: Zum Problem einer allgemeinen Kulturgeographie. Raummodelle-kultur geographische Kraftelehre- raumrelevante Prozesse- Kategorien. Die Erde. 100 (1969), S. 155–193.

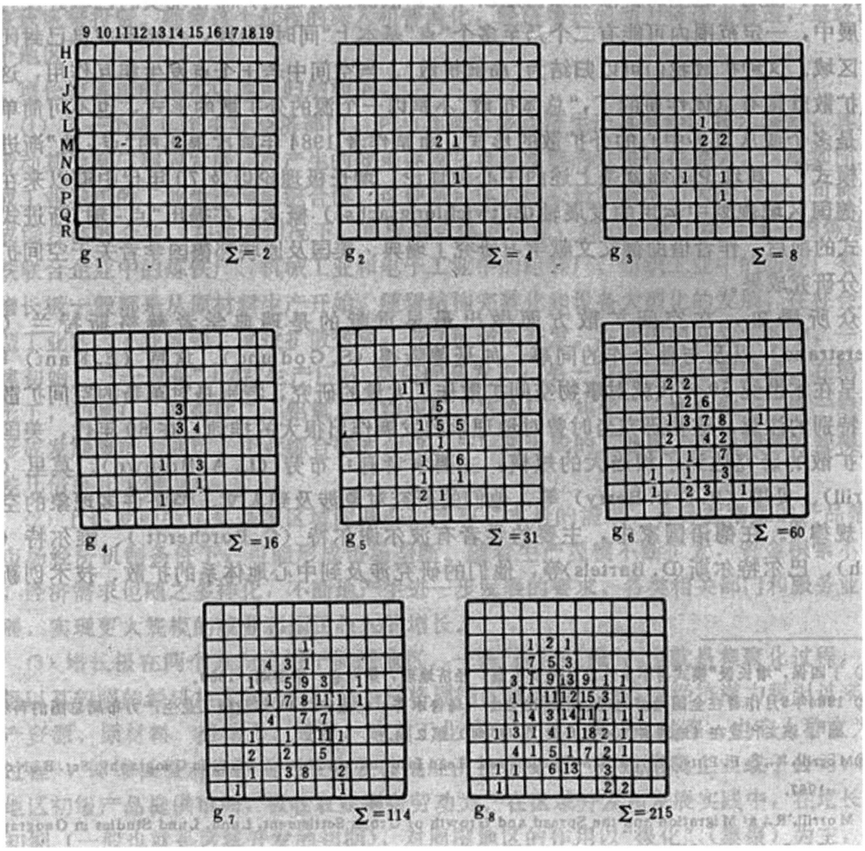

Fig. 6 Spatial diffusion (according to T. Hagerstrand)

He also stipulated that: As soon as a new object is known to everyone, it will be accepted immediately; the information can only be transmitted through private conversation.

Let P^1, P^2, P^3… represent the probability that information is transmitted from the center to the surrounding grids:

P^1	P^2	P^3	P^4	P^5
P^6	P^7	P^8	P^9	P^{10}
P^{11}	P^{12}	P^{13}	P^{14}	P^{15}
P^{16}	P^{17}	P^{18}	P^{19}	P^{20}
P^{21}	P^{22}	P^{23}	P^{24}	P^{25}

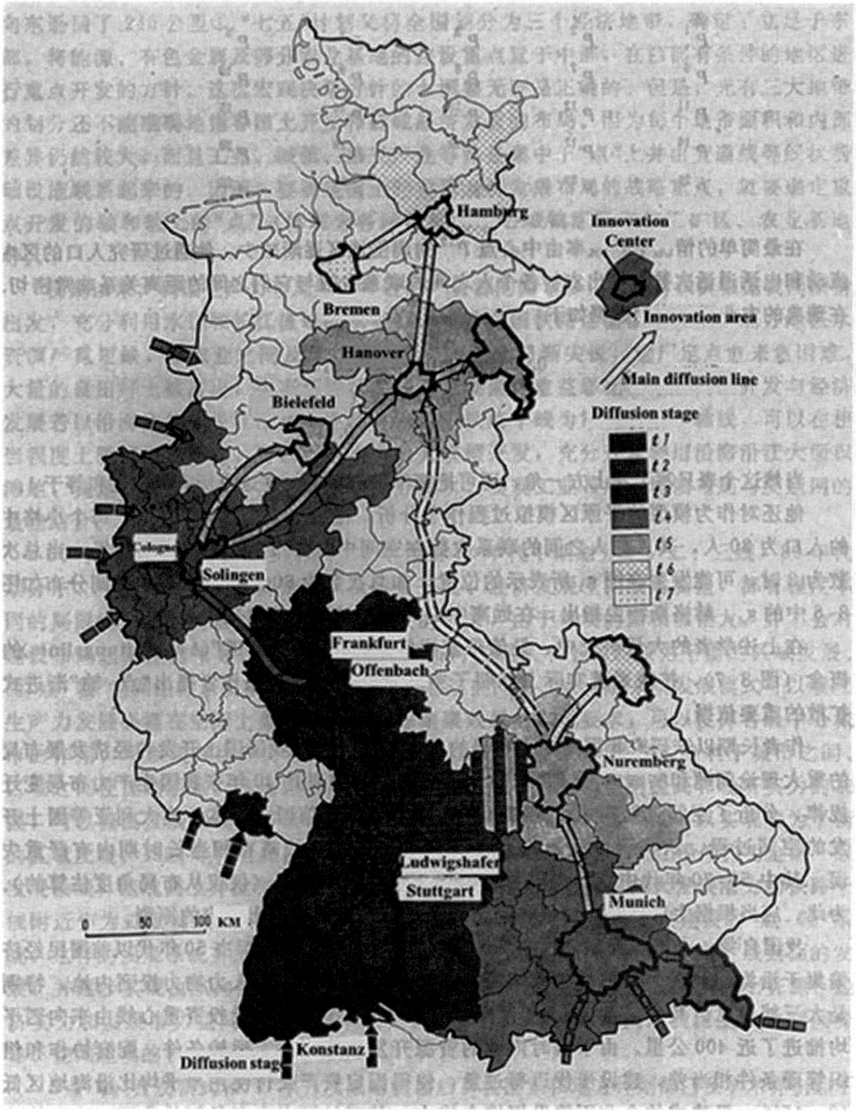

Fig. 7 Spatial diffusion of foreign workers in the Federal Republic of Germany (1960 to 1972, excluding West Berlin)

In the simplest case, the probability gradually decreases from the center point P^{12} to the surrounding areas. By studying the intra-regional mobility of the population and the number of telephone calls, he concluded that the number of contacts between individuals in rural areas is closely related to the distance between them. The corresponding data of the agricultural area in Sweden is as follows:

0.0096	0.0140	0.0168
	0.0301	0.0547
		0.4431

Of course, only the upper left corner of this table is filled in, and the probability numbers can also be filled in. Here, the sum of all probabilities is equal to 1.

He also analyzed the simulation process of the plain area which was taken as a model. The range is 5 km × 5 km, and the population in each grid is 30 persons. The distribution of the numbers of contacts between people in space is shown in Fig. 6. Here, when the total number of times is 2, it may occur at the position represented in Figure g1. When the total number of times is 60, the spatial distribution of the number of times is g_8 in Fig. 6. T. Hagerstrand pointed out that a secondary center can appear in such diffusion when away from the center.

Among the numerous studies of the above scholars, the concept of "diffusion channel" (Ausbreitungslinie) is the worthiest of attention (Fig. 7). The diffusion channel actually indicates the direction of spatial diffusion, which is an important basis for the author to propose the "point-axis" progressive diffusion.

The author has studied the law of productivity distribution in China for a long time. In order to solve the major theoretical and practical problems of China's territorial exploitation and economic development layout, the author has systematically summarized the change law of productivity distribution in China, and analyzed the spatial process model in history, particularly in the past 40 years. At the same time, the author has also analyzed the spatial process of territorial exploitation in the United States, Germany, Japan, Australia and other countries. It is recognized that: There have been serious mistakes in the territorial exploitation and productivity distribution in China for the long term. Among them, mistakes from the 1950s to the mid-1970s caused a loss of tens of billions of yuan (only estimated from the perspective of layout). In this regard, certain principles shall be proposed according to the spatial process of productivity and the objective laws of territorial exploitation.

Since the first five-year plan, especially since the mid-1960s, China has started to solve the situation that the national economy was concentrated in the coastal areas before the 1950s. In accordance with the requirements for combat readiness and balanced layout, a large amount of manpower and material resources will be invested in the mainland, especially the strategic hinterland areas of China. Among them, the gravity line of the national capital construction investment advanced by an average of nearly 400 km from east to west during the period from 1952 to 1975. Since the conditions for resource development and utilization, transportation, supporting cooperation, and organizational management in the mainland were relatively poor at that time, and the construction pace was too fast, the average delivery and utilization rate of fixed assets was 10–15% lower than that in coastal areas. The established enterprises could not exert their comprehensive capabilities, and even had a lot of extra costs for a long time. This strategic shift violated objective economic laws and natural laws, which was a profound lesson. This indicated that it will be a long-term and gradual process for China, a country with a vast field and great regional differences in natural conditions and economic foundations, to realize the strategic shift of industrial production distribution from developed regions to underdeveloped regions. Since the 1980s, China has strengthened the construction of the coastal zone and the mainland region with better conditions based on the policies of opening up and exerting regional advantages, and the gravity line of the capital construction investment has returned to the east by an average of 240 km.[10] According to the "Seventh Five-Year Plan", the whole country was divided into three economic zones, and the guideline of taking root in the east, placing the emphasis on the construction of energy, nonferrous metal and some agricultural bases in the middle and developing the western regions with conditions was determined. This great adjustment of the macro decision policy was undoubtedly correct. However, the layout of territorial exploitation and regional economic development still cannot be accurately guided by the division of the Three Regions. The area and internal differences of each zone are still large, and the industries, towns, tertiary industries, etc. are concentrated on the "point" and linked by the traffic line and other linear infrastructure. Therefore, it is necessary to determine the axis subject to key development and the "points" on the axle (reflected as central urban residential areas, industrial and mining areas, agricultural bases, etc. with various scales) before determining the strategic emphasis of territorial exploitation and economic development layout.

For a long time, more consideration has been given to the factors of getting close to mineral resources for industrial distribution in China. However, the water transportation and the water resources in the basin of the Yangtze River have not been fully utilized according to China's national conditions, causing the national railway transportation is extremely intense, the water resources in the northern region are in serious shortage, the fight for water between industry and agriculture and between cities are increasingly sharp, it is more and more difficult to fix the site for the plant, a large number of good farmland is occupied, and the ecological environment increasingly deteriorates in some agglomeration areas. If the areas along the coast and along the Yangtze River are taken as the primary axis, and several trunk railways are taken as the secondary development axis for territorial exploitation and economic development in the future, the above contradictions can be alleviated to a great extent, the large-scale development of water transportation can be promoted, the large-area beaches and wasteland along the sea and river can be fully developed and utilized, and the rich freshwater and part of the seawater in the Yangtze River can be used to achieve the best combination of industrial distribution, urban layout and transportation network.

China was divided into the coast and the mainland in the 1950s, the first, second and third fronts from the 1960s to the mid-1970s, and the eastern, middle and western economic zones in the 1980s. There are limitations in different degrees in terms of the space strategy for territorial exploitation and economic development. Their basic

---

[10] Lu Dadao, China's Macro Strategy of Regional Development, Journal of Geographical Sciences, Volume 42, Issue 2, 1987.

characteristic is that: China is divided into large blocks only. Since the area of each block is very large, it will not play a guiding role in the construction layout, and cannot clarify how to achieve regional development from unbalanced to relatively balanced. However, the "point-axis" spatial structure system model can better answer these questions. The "point-axis" system development model can meet the needs of productivity development to concentrate on a point in space, exert the objective requirements of the agglomeration effect, give full play to the role of central cities at all levels, and achieve spatial combination between productivity distribution and linear infrastructure, which is favorable for the convenient connection among cities and regions. The theory points out that with the improvement of regional exploitation and development and the strengthening of economic strength, the level of the development axis can be improved, and the axis can also be extended in space, so that a more balanced regional development can be achieved. In 1984, author proposed that the areas along the coast and along the Yangtze River shall be taken as the first-level development axis of China in the next 20–30 years, thus forming a T-shaped structure for territorial exploitation and regional development. The vicinity of the trunk railway with great development potential shall be taken as the second-level development axis, and several central cities shall be determined to form a "point-axis" system subject to key construction at different levels. Such a development system can help strengthen the horizontal connection of the three economic zones, realize the "point-axis" development from line to plane, and gradually realize the shift from east to west. Such a macro pattern accurately reflects the basic framework of the spatial distribution of China's territorial resources, economic strength and development potential, and profoundly reveals the objective laws of the spatial distribution and spatial movement of China's productivity from one aspect.

After the "point-axis" development model and the T-shaped development strategy were proposed, positive responses were aroused in national and local planning departments and academic circles. First of all, this theoretical thought was written into the Outline of Overall Planning of National Land (Draft).[11]

[11] The *Outline of Overall Planning of National Land (Draft)* compiled by the State Planning Commission on February 10, 1990 contains the following paragraphs about the development axis:

"3.06 Around 2000, the main axle of China's productivity distribution will be along the coast, the Yangtze River and the Yellow River. Combined with the second-level axes such as the areas along Lianyungang-Lanzhou, Lanzhou-Xinjiang, Beijing-Guangzhou, Zhejiang-Jiangxi-Hunan-Guizhou, Taijiao-Jiaozhi, Harbin-Dalian, Nanning-Kunming railways, it will constitute the basic framework for the overall layout of China's territorial exploitation and construction. It is of great significance to China's territorial exploitation and regulation to accelerate the construction of the development layout axis."

"3.07 The coastal axis and most regions of the eastern coastal zone will have superior geographical location, developed economy, advanced technology and better development conditions. The coastline is more than 18,000 km long, there are many ports, and the throughput capacity has reached more than 400 million tons. It is preliminarily predicted that the throughput capacity will reach 600 million to 700 million tons by 2000; there are still more than 1.3 million hectares (more than 20 million mu) of tidal flats available for development and utilization along the coast; iron ore, petroleum and some mineral resources are also abundant or have better development prospects. Over the past 10 years, the economy of coastal areas has developed rapidly, and a pattern of opening up has been formed with special economic zones—coastal open cities—coastal economic open areas.

In the territorial planning for more than 20 provinces and districts such as Shandong, Shaanxi, Hebei, Shanxi, Jiangxi, Jiangsu, Liaoning and Guizhou and a large number of regional planning that has been compiled, the overall development layout of the region has been stipulated according to the model above. For example, Shanxi (March 1989) took "the Tongpu Line running through the central part of the province and the Taijiao Line and Shitai Line running through the central-southeast part as the first-level axes for the key development and layout of the territory", and took Houxi, Houyue, Shenhe, Xiaoliu, Jingyuan, Hanchang, Shuoshi and other railway lines as second-level axes. Shaanxi (September 1989) took "the Longhai Railway as the main axis of productivity distribution in the province", and took the three zones, that is, Weibei, the coast of Hanjiang and Shenfuyu as the second-level axes of the productivity distribution.

Gui Shiyong, deputy directors of the State Planning Commission, and Wei Liqun wrote an article ["Several Ideas on Economic and Social Development Strategies by the End of this Century" in the "Research on Planned Economy" (Issue 7, 1987)], and proposed the general idea of location selection, that is, strengthen the development of coastal areas, strengthen the main pillar of national economic revitalization, start to develop the golden waterway of the Yangtze River, build the economic corridor along the river, and gradually form a T-shaped first-level development axis along the coast and along the river. At the same time, corresponding industrial and mining enterprises will be allocated along the original and new railway lines as the second-level axes for key development.

In order to theoretically reveal the scientific basis of the "point-axis" spatial structure model, it is necessary to further explore the essence, direction and type of spatial diffusion.

It is necessary to make full use of these favorable conditions, develop an export-oriented economy, and actively participate in international exchanges and competitions; speed up the infrastructure construction, develop energy and raw material industries in port areas with better conditions, gradually form several new port cities, and form the coastal economic development axis together with the original cities, which is of great significance for the national economy."

"3.08 The axis along the river and the mainstream of the Yangtze River are important channels connecting the eastern coastal areas and the inland regions of the Central and Western China, which have superior conditions for further development. In the zones along the river from Panzhihua, Sichuan to Shanghai, with a length of more than 3590 km, the water resources are abundant, and the annual runoff of the Yangtze River is about one trillion cubic meters; the hydropower resources are rich, and the installed capacity of hydropower that can be developed is about 90 million kilowatts; the navigation conditions are good, and the navigation mileage of the mainstream is more than 3000 km; the mineral resources along the river are abundant with large reserves, and resources such as iron, copper and natural gas occupy an important position in China; the Taihu Lake Basin, Chaohu Area, Poyang Lake Area, Dongting Lake Area and Jianghan Plain are important agricultural and commodity grain bases in China. In this century, it is required to focus on developing hydropower and mineral resources along the river, make full use of the shipping advantages of the golden waterway of the Yangtze River and both domestic and foreign resources, deploy a batch of industries that consume large amounts of water, energy and transport volume, and form an economic corridor along the river and centered on cities such as Shanghai-Nanjing-Wuhan-Chongqing, so as to strengthen the horizontal connection among the economic zones in the upper, middle and lower reaches, realize the complementation of resource, technology and economic advantages in the east, middle and west regions, and improve the productivity distribution in China."

2) Essence of diffusion

As W. Christaller pointed out when discussing the central place theory: The aggregation of materials to a core is the basic phenomenon of objects; at the same time, the objects in space originate from the center and are diffused outwards; the town is the core of the region and the commander of a specific region. One of the conditions for the occurrence of the diffusion phenomenon is that there must be a regional center or a certain form of gathering point for social and economic objects.

The socio-economic center of the region is generally manifested as various types of cities. Airports, financial centers, mining areas, ports, administrative centers, etc. also have the function of diffusion.

According to the principles of physics, all kinds of objects have their own potential energy in space, and they are always transporting and diffusing their own potential energy to the surrounding environment. In the process of regional development, the diffusion of such potential energy is manifested as product flow, fund flow, personnel flow, technology flow, information flow, policy flow, etc. These "flows" flow from the center point (area) to the surroundings, and are re-aggregated in the directions and at distances different from the center and combined with the local original natural and socio-economic elements, thus forming a new gathering point.

The root cause of the phenomenon of spatial diffusion is the imbalance of socio-economic spatial structure. Here, the spatial structure refers to the agglomeration degree, density relationship and intensity difference of the social economy at the spatial level. In terms of economy, cities are often the economic centers within a certain range, and various economic facilities and production units are relatively concentrated in a small range; in terms of society, cities are places where the population, administrative units, cultural, educational and scientific units within a certain range are concentrated, with a dense population. Regardless of economic and social factors, various "flows" shall be radiated to the surrounding regions in cities or various types of agglomeration areas. There are "economic gradients" and "social gradients" between the agglomeration areas within a relatively small range and the surrounding regions with relatively few socio-economic facilities. Of course, "pressure" and "diffusion force" are formed with the gradients. However, various "flows" will flow from the high-pressure place to the low-pressure place, just as molecules in the air will diffuse from places with high molecular density to the surrounding.

Spatial diffusion will be formed as long as there is a "gradient" and "pressure difference"; the results of spatial diffusion will gradually lead to the equalization of the spatial structure so that resources and space can be fully utilized step by step.

The population density changes regularly between a city and its regions, that is, a function of the distance from the urban center.[12] However, the diffusion from cities to peripheral regions is usually not the diffusion of the population from urban areas to suburban areas (of course, this process has already begun in the countries highly

[12] The formula proposed by Clark that the population density in the urban area changes with the distance to the urban center is. Among them, D_x is the population density at a distance x from the downtown, D_0 is the population density in the downtown, and b is the parameter derived empirically, that is, the density gradient.

urbanized), but mainly refers to that products, technologies, culture, finance and other "flows" flow from the city to the suburbs. The flow of population in developing countries is mostly from vast agricultural areas and suburbs to urban areas. Developing countries have weak economic strength and financial difficulties. Therefore, the limited funds can only be invested in cities. The production and working conditions in these cities are much better than those in the vast rural areas and naturally attract population migration.

3) Main way of diffusion

The "diffusion mode" here refers to the morphological characteristics of the socio-economic spatial diffusion process. Some scholars believe that diffusion is always "ink stain diffusion". It is investigated from a micro perspective, such as between a city and its suburbs. However, is such diffusion really like "ink stain" in the direction and plane through further analysis? The answer is also negative. It is not the "ink stain" type even on a "homogeneous" plain.

The spatial form of diffusion can be mainly divided into leaping and progressive.

(1) Leaping diffusion means that social, economic and cultural elements leap by a larger spatial distance from the gathering point within a large range. The elements are combined with the local original elements at a certain location, forming a new gathering point. In terms of regional exploitation, some countries and regions put this spatial development strategy into practice; however, it was found that this model was not perfect in the subsequent development, which is mainly manifested in two aspects: First, almost all diffusion refers to crossing a certain spatial distance, but leaping diffusion is often understood as crossing a large-scale space, that is, stronghold-style development is carried out in undeveloped regions (or regions not developed) far away from the economically developed area of the city; second, leaping diffusion does not indicate whether there is a diffusion channel between the diffusion source and the diffusion point (new gathering point), that is, whether there is a linear infrastructure bundle. The implementation of a leaping development strategy often leads to a situation in which new strongholds have been under large-scale development, but the linear infrastructure bundle is still very imperfect and unsuitable.
(2) Diffusion, that is, "point-axis progressive" diffusion, refers to the gradual diffusion of socio-economic "flows" along several diffusion channels (linear infrastructure bundles) from a diffusion source, forming new agglomerations with different intensities at different distances from the center. Due to the law that the diffusive force attenuates as the distance increases, the scale of new agglomerations also decreases as the distance increases. As a result of the diffusion of diffusion sources in adjacent areas, the diffusion channels are connected to each other, becoming the development axis. With the further development of the

social economy, the development axis is further extended, and the new gathering points and development axes with relatively small scale are continuously formed. This is the basic content of progressive diffusion (see Fig. 1).

4) "Point-axis progressive" diffusion and balanced development of regions

The author repeatedly pointed out that the exploitation and development levels and strength are always unbalanced in a country or a region. With the extension of the exploitation process and the enhancement of regional economic and technical strength, the degree of imbalance in the region is gradually reduced, so a relatively balanced state generally appears. It can be seen that the development process of this region is the result of "point-axis progressive diffusion" through microscopic investigation.

It has been clarified in the first section of this chapter that with the extension of axes and "points", the entire country or region is gradually covered with development axes and "points". Of course, there are obvious and regular differences in levels at this time. As the region develops further, the differences in levels of "points" and axes become smaller, and a relatively balanced state begins to form, that is, a state of spatial exploitation and development from "point" to "line" to "plane" is formed.

The linear infrastructure bundle consists of railways, trunk highways, waterways, power transmission lines, communication lines, water lines, etc. The number of facilities included in the specific "bundle" depends on the establishment time and level of the development axis. The development axis with a high level and long history is often the development axis with complete composition and strong economic strength.

The "point-axis" progressive diffusion leads to a relatively balanced development among regions, and the main reasons are as follows:

(1) The new gathering point, that is, the growth pole, has a diffusion effect on the surrounding less developed regions, which has been elaborated above;
(2) The linear infrastructure can improve regional accessibility and regional mobility. However, the improvement of accessibility and mobility will inevitably increase the opportunities for regional development. It needs to be pointed out here that: Once a linear infrastructure bundle extends to a certain area, the development gap between the area and the surrounding areas without linear infrastructure bundles will be widened (Fig. 8). However, this is an inevitable process to narrow the inter-regional differences.

In order to prevent excessive concentration in urban areas, especially the central area, during the development of large cities, the development axes in multiple directions are often constructed from the inside out, that is, from the built-up area of the large city, and the facilities that continue to be built in the city are attracted to these radial axes, so as to avoid pancake-like agglomeration. The satellite towns planned and constructed in some big cities in the world are all located in positions where the axis is far away from the city. Therefore, it can be said that the construction of a development axis that radiates outwards in a big city is conducive to the balanced development of the region where the city is located.

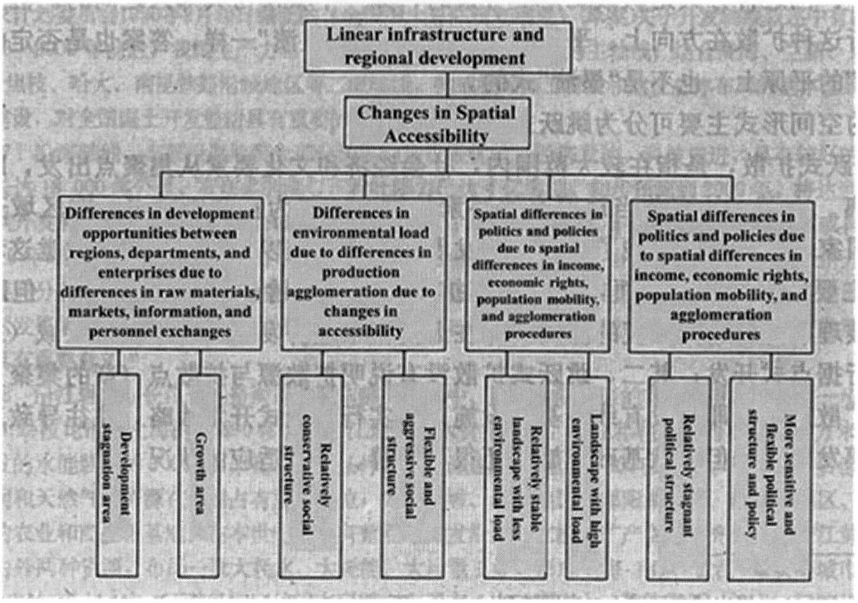

Fig. 8 Schematic diagram of the relationship between linear infrastructure and regional development

4 Point-Axis System and Point-Axis-Agglomeration Area

1. Progressive diffusion leads to the formation of a point-axis hierarchy

It has been stated in the above two sections that progressive diffusion is point-axis progressive diffusion in essence. Such diffusion is bound to form the following situation: The scale of the axis (including the types, grades, capabilities and completeness of the composition of the linear infrastructure) between the central cities that first appeared and achieved rapid growth is large, while the scale of the axis between towns with small socio-economic aggregates is small. As a result, the point-axis hierarchy is formed. This is like the central place hierarchy formed under the control of the traffic principle and proposed by W. Christaller.[13]

There are usually only a few development axes with the highest level (that is, first-level) in a region or country, only one to two in some cases. For example, China's first-level development axes are the coast and areas along the Yangtze River. The two intersect in the Yangtze River Delta, which has the most developed economy

[13] Lu Dadao, Central Place Theory and Its Application in the Regional Planning of Bayern, the Federal Republic of Germany, Urban Planning Translation Collection, pp. 75–91, China Architecture & Building Press, 1980.

and the most moderate geographical location, forming a "T" shape. It accurately reflects the basic framework of the distribution of China's territorial exploitation and social and economic development potentials. The only first-level development axis of the former Federal Republic of Germany is the Rhine axis, which passes from the northwest to the southwest. It is incomparable with other axes in terms of the scale of transportation channels, attraction range, economic strength, or urban influence scope. In some small maritime countries, the coast is often its first-level development axis. For example, the east coast of Japan is the first-level axis; its development status is even more important after the opening of the Tokaido trunk line. There are generally more second-level development axes than first-level development axes. In most cases, the second-level development axes are connected with the first-level development axes, while some are not. The development axes of the following levels are basically the same.

For a long period after the higher-level axis begins to form, its effect on the lower-level axis is mainly the attraction tendency, and the diffusion strength is weak. That is to say, the lower-level development axis and its surrounding resources and "economic potential" are constantly attracted by the higher-level axis, and resources such as social economy and manpower are gradually concentrated on the high-level axis. The agglomeration-dominated tendency will be replaced by radiation (diffusion)-dominated tendency only when the social economy on the higher-level axis is excessively agglomerated or does not coordinate with its natural foundation.

Within a certain range, the perfection degree of the development axis, including whether there is a high-level axis suitable for the scope of the region and whether the level of the axis forms a system, has a great impact on the regional socio-economic development. In some cases, a high-level development axis passes through a certain area, but there is no subordinate axis in the area for connection. Therefore, the role of the high-level axis in promoting the socio-economic development of the region is obviously limited. Moreover, the development of the industrial belt formed by the high-level axis will also be hindered. Therefore, when planning the linear infrastructure of the region, the formation of the regional system shall be considered, and the system of the lower-level region shall be connected with the system of the higher-level region. A well-developed regional development axis system shall be a semi-open system (Fig. 9). In Fig. 9a, A-P-B is the first-level development axis, P is the central city of the region, PE, PC and PF are the second-level axes formed through P, and BC, CD, DA, AE and EF are the third-level axes. If the figure is converted and abstracted from the perspective of diffusion or aggregation force, it can be changed to Fig. 9b. According to the diagram, it is a semi-open system. Such a regional development axis hierarchy connects the industrial belts at all levels in a region into a "whole", and connects this local "whole" with the industrial belt system of the higher-level region, becoming a part of the organic network in a wider range.

In the practice of regional development, many instances can be found to illustrate that the formation of the above point-axis system model is representative and has sufficient explanatory power. Figure 10 is an instance.[14]

---

[14] Yu Binyang, Research on the Development of Regional Urban Agglomerations, October 1991.

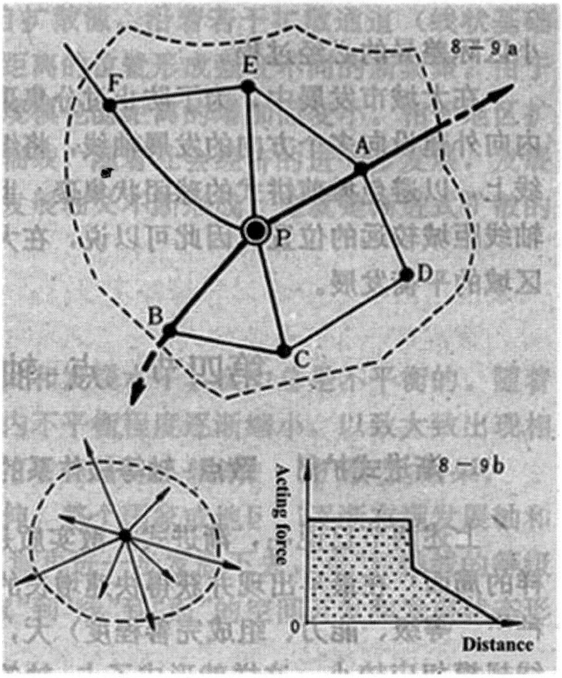

Fig. 9 Regional development axis hierarchy

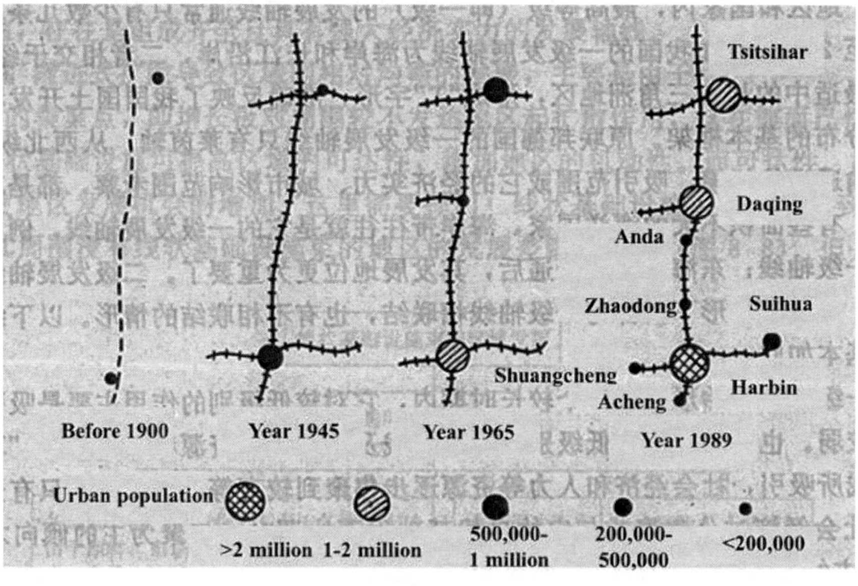

Fig. 10 Formation of the point-axis spatial structure system between Harbin and Qiqihar

2. Axis development of urban areas and formation of agglomeration areas

1) Imbalance in the development direction of urban areas

The spatial structure of the social economy is also not homogeneous within the urban and suburban areas of cities. The plane development of urban areas also follows the "point-axis" law. However, the "point" within a city is not the point in a region (refer to a town). It is generally the most attractive part of the city, such as the downtown square (the municipal government is also near the square in most cases), commercial buildings or concentrated business districts, key cultural relics and historical sites, major short-distance rest places, major industrial areas or satellite towns, large sports facilities, etc. The traffic artery and other linear infrastructure that connect them form the urban development axle. The trunk roads attract some medium and large urban facilities to be arranged on both sides in the initial phase after their opening. However, the establishment of these facilities provides conditions for the development of other urban components (such as factories, shops, residential areas, cultural and educational institutions, etc.).

Among the urban regional development axes, the axis that passes through the urban center and extends to nearby cities is the most striking. For example, if there are more than two cities within a few kilometers or tens of kilometers, the trunk roads connecting the cities will often develop into such axes. Most of them are the main axes of urban development in terms of their functions. The east–west Chang'an Street, as a part of the trunk road connecting Beijing and Tianjin in the city, is the most obvious example (which, as a development axis, also includes railway and communication parts). The trunk road between Shenyang and Fushun is also the main axis of urban development after entering Fushun. For Beijing, the ancient capital of several dynasties for hundreds of years, its main development axis was in the north–south direction, and the main trunk road and imperial city buildings were concentrated on this axis before the founding of the People's Republic of China. After the founding of the People's Republic of China, the main axis is actually changed to the east–west direction, that is, the east–west Chang'an Street with a length of more than 40 km. In addition to Tian'anmen Square, main communication facilities, railway stations, high-end hotels, business districts, government agencies, etc. are concentrated within the range of this development axis (about 1 km on both sides of the trunk road). The change in the direction of the main axis of Beijing is conducive to strengthening the connection with the economic center and port city of Tianjin.

The imbalance in the urban regional development direction reflects an objective law of urban development, that is, the spatial structure of cities is constantly expanded and improved according to the "point-axis system", which is similar to regions. A city can choose a favorable direction to arrange its main axis and deploy main facilities within the range. In this way, its functions can be better exerted, and its functional division can be scientific. The main axis of the city generally extends for a long distance, or the distance is not longer than the secondary axis. However, the linear infrastructure has a large scale and high level, and the main facilities of the city are concentrated on this axis. The secondary development axis is short sometimes, with small linear infrastructure and a low level of urban facilities.

2) The main axes of adjacent cities are connected and developed into agglomeration areas

With the development of economic industrialization and social urbanization, industrial and urban agglomeration areas are constantly emerging in the world. There is no precise definition regarding the concept of agglomeration area. In China, the urban planning departments sometimes call it a "dense area" or "urban contiguous area". In Germany, it is usually referred to as "Ballungsgebiet". The names are different, but there is one thing in common: Two or more cities not far from each other connect their suburbs (mostly referring to the inner suburbs. Sometimes the middle suburbs are also called contiguous areas) due to extended development, thus forming an interconnected urban agglomeration; they are connected geographically and functionally. A unified whole is also gradually formed in terms of infrastructure.

By analyzing the formation of urban agglomeration areas from the plane, the thing in common is that the urban public buildings, manufacturers, and residences are constantly developed along the main development axis between cities, which in turn drives a large number of tertiary industries to the development axis. As a result, two or more cities with a distance of several kilometers or tens of kilometers are connected together.

In order to achieve proper control of metropolises and big cities, the common practice in the world is to develop satellite towns. The satellite towns developed by using the existing infrastructure are closer to the central city in some cases, and the linear infrastructure connecting the central city and the satellite towns has a higher level. Various industries are gradually agglomerated, which may also combine the two parts into an agglomeration area finally.

3) Urban planning of expansion along the axis

Through analysis of the change process of the spatial organization of socio-economic objects within the city, it indicates that the development and construction and various facilities within the city also follow the principle of "axis" development. Therefore, it is an important planning method to complete the urban layout with the help of the development axis. Since the founding of the People's Republic of China, most of the new cities in China have developed from "one street". The initial "one street" is often the central axis in all directions of the city afterward. A large number of urban construction projects are carried out along axes at all levels.

The city of Beijing in the Ming and Qing Dynasties was planned and constructed according to the principle of the axis. The most prominent is that there was a north–south main axis, with a length of 8 km, running through the whole city. There was the city gate tower, the big memorial arch, the great stone bridge, the long corridor, the imperial road, the arched door as well as layers of main halls, inner courtyards and palace gates on this main axis. In addition to this main axis, there were also axes such as Dongsi-Dongdan-Chongwenmen and Xisi-Xidan-Xuanwumen in the north–south direction. Because the imperial city was in the middle, there were few trunk roads in the east–west direction, so it was inconvenient to travel [5].

Washington, the capital of the United States, was also planned and constructed according to the principle of the axis. The main central axis of the city is in the east–west direction, and the main public buildings in Washington are concentrated in and around the section from the Capitol in the east to the Lincoln Memorial in the west. The distance of this section is about 3500 m, most of the middle is the green square, and the Washington Monument built to commemorate the founding father Washington is standing in the center of the square. The top of the monument can be directly reached by taking the elevator, which is the best place to overlook the city. The White House is 900 m in the true north of the Washington Monument. There are several large museums such as the National Museum on the north and south sides of the axis [6] (Fig. 11).

In the overall urban planning, in addition to taking full consideration of the development scale of the city, the axis direction and scale of the main levels of the city mainly depend on the following factors:

(1) Consistent with the main expansion direction of urban land. In urban planning, urban land expansion is uneven in all directions. As the main direction of land expansion, it is generally the zone where the main urban construction parts are concentrated. After the founding of the People's Republic of China, the east–west development is taken as the main direction for Beijing. Due to the establishment of the industrial area in the eastern suburbs and the military authorities in the western suburbs, land has been extended rapidly to the east

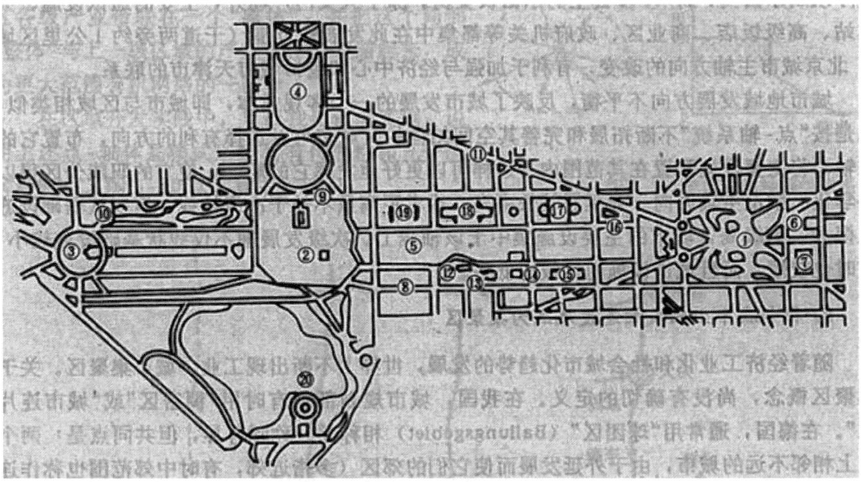

Fig. 11 Washington central axis and main surrounding buildings. 1. Capitol; 2. Washington Monument; 3. Lincoln Memorial, 4. White House; 5. Green Square; 6. Supreme Court, 7: Library of Congress; 8. Independence Avenue, 9. Constitution Avenue; 10. Vietnam War Memorial; 11. Pennsylvania Avenue; 12. National Museum, 13. Craft Museum; 14. Sculpture Museum and Gardens; 15. Space Museum; 16. Museum of Fine Arts; 17. Art Gallery; 18. Museum of Natural History and Humanity; 19. Museum of American History; 20. Jefferson Museum

and west. Chang'an Street, passing Tian'anmen and with a length of more than 40 km, has become the main axis of the city. In addition to wide streets, there are also subways to coordinate, and the traffic flow ranks first in the city.

(2) The topographic conditions are suitable for development as the main axis of the city. In most cases, the topography is not a factor that determines the direction of the main axis of the city for the plain cities that are not restricted by the topography. However, for cities located in river valleys, the main axis of development is often parallel to the river, or the river becomes one of the linear infrastructures on the main axis.

(3) Consistent with the exit direction of the main external traffic of the city. The main external traffic channel of a city usually connects the city with the main satellite towns nearby, which is the main flow direction of the external passenger and cargo flow of the city, and the main flow direction of information. Therefore, the capacity and level of the linear infrastructure must rank first, which is very attractive to the selection of various urban socio-economic locations. It is conducive to convenient connections among cities to plan and construct a development axis that is consistent with the direction of the main exit of the city, which also promotes the development of related cities.

Before determining the main axis of urban development, other factors shall also be considered, such as historical factors, important historic sites, monuments, important urban centers and public activity places that were formed in history. It is best for the main axis to pass through their vicinity, or connect them to each other.

3. Agglomeration area—an advanced form of spatial agglomeration

As a result of the "point-axis" progressive diffusion, the compound agglomeration of population, economy and infrastructure will often be formed. When the population density and economic strength reach a certain level, the agglomeration area will be formed. Here, an agglomeration area is an enlarged "point" or collection of points, which is the highest form of spatial agglomeration.

The agglomeration area is often formed after the intersection of several higher-level development axes. It can also be said that the formation of the agglomeration area is accompanied by the development of relevant axes. There are generally three cases for the formation process of agglomeration areas: First, several industrial cities (or transportation hub cities) are not far apart, and the long-term continuous (uncontrolled or weak control) development makes them get closer to each other, thus forming a whole. Or they are a "spherical aggregate" linked together when being viewed from the macroscopic area. Second, several small and medium-sized cities are formed around a big city. As a result of development, they get close to each other, thus forming agglomeration areas. Here, there may be satellite towns built to evacuate the economy and population of big cities, and some are medium-sized cities formed due to the development of the surrounding mineral resources, water resources and transportation hub facilities.

Of course, there are different levels for agglomeration areas. However, most of the so-called agglomeration areas are observed from a country or even an area composed of a group of countries. For example, central and southern Liaoning (Shenyang, Anshan, Benxi, Fushun and Liaoyang), Beijing-Tianjin-Tangshan, Yangtze River Delta, Pearl River Delta, Wuhan area, Xi'an area, etc. in China, Donbass and Kuzbass in the former Soviet Union, Philadelphia, Appalachia and Great Lakes Region in the United States, Ruhr area, Frankfurt area and Munich area in Germany (agglomeration areas characterized by high-tech industries and formed since the 1970s), etc.

1) Basic characteristics of the agglomeration area

(1) Agglomeration areas are generally formed after long-term sustainable development, and have a vast "hinterland" and economic resources, as well as a strong economic power and dense population, thus becoming the socio-economic center of large regions and even countries. For five cities in the central region of Liaoning, that is, Shenyang, Anshan, Benxi, Fushun and Liaoyang, the land area of the main region where the population and economy are distributed is less than 10,000 km^2. However, during the "Seventh Five-year Plan" period, the population accounted for 36% of the whole province, the total industrial output value accounted for 53–55% of the whole province, while the total profits and taxes accounted for more than 63% of the whole province. The Beijing-Tianjin-Tangshan region (including Qinhuangdao and Langfang regions) is an economic center of North China and the whole country. The land area is only more than 50,000 km^2, accounting for less than 4% of the land area of five provinces, cities and districts, that is, Beijing, Tianjin, Hebei, Shanxi and Inner Mongolia. However, the population and the total industrial and agricultural output value accounted for about 42% of those of North China, and the total value of foreign trade acquisition accounted for more than 50% in 1990. It was the portal to the sea in the North China Region. A large number of products and technical facilities in the Beijing-Tianjin-Tangshan region drove the development of the North China Region, and played the role as the regional economic center. The Yangtze River Delta Region with Shanghai as the center (north to Nantong, south to Hangzhou and west to Changzhou) has more than 430 large, medium and small towns, including 57 counties, with a land area of 84,000 km^2 and a population of about 55 million, accounting for 0.88% and 5% of China respectively. The financial contribution of the total industrial output value and the foreign export amount were close to 1/5 of those in China respectively. In the total industrial output value, only about 48% was used for consumption in this region, that is, most is used for export and transferred to other regions of the country. The population density of this agglomeration area is about 610 persons/km^2, and the per capita arable land is about 0.7 mu. With a land area of more than 8200 km^2, Wuhan is an important industrial production base for iron and steel, machinery, electrical appliances and light textiles. Among them, three-quarters of machinery and electrical products and two-thirds of light textile products are exported to other places.

(2) In order to meet the needs of strong economic power and dense population, the agglomeration areas have developed traffic and transportation networks composed of various means of transport, as well as developed infrastructure for communication, energy production and supply, water supply and drainage, etc. The agglomeration areas are generally the collecting and distributing center and the transit center for bulk passenger and cargo flows. Therefore, large-scale traffic and transportation hubs, such as trunk railway formation stations, large passenger stations, international airports, large ports, etc., are mostly placed in agglomeration areas; the structure of the transportation network in the agglomeration area is generally complete, and the popularization of each new means of transport always begins in the agglomeration area. Moreover, under the premise of scientific planning, various means of transport are rationally divided according to their own characteristics and functions, and gathered in the center of the core city, extending to the periphery in a circular shape, and covering the service area repeatedly in layers. In the center of the core city, an overhead crossing of trunk railways, regional railways, subways and urban arterial roads is often formed. According to the technical and economic characteristics of various means of transport, there are different service radii outwards. In developed countries, there have been expressways between most of the large and medium cities (Table 3).

In the agglomeration area, most of the cities have powerful secondary energy (electricity, hot water, gas, etc.) production and transmission facilities, which connect the relevant cities to form a powerful energy supply system.

(3) The agglomeration area generally has higher labor productivity and national economic benefits due to the developed infrastructure, high processing degree of industrial structure as well as high level of science and technology. This is very obvious in some developed countries such as Japan. It is very prominent in China. Through the analysis of the comprehensive input coefficient and production coefficient of large and medium-sized urban agglomeration areas, it can be concluded that the comprehensive input coefficient decreases with the increase of urban population size, and the comprehensive output coefficient is the opposite. The profits and taxes per 100 yuan provided by the independent accounting enterprises in cities of various sizes in 1987 are as follows: 27.5 yuan for cities with a population of more than 2 million, 24.0 yuan for cities with a population of 1 million to 2 million, 21.2 yuan for cities with a population of 500,000 to 1 million, 17.5 yuan for cities with a population of 200,000 to 500,000, and 16.9 yuan for cities with a population of less than 200,000. This indicator basically reflects the general trend of high economic benefits in agglomeration areas.

(4) The agglomeration area has a higher level of population urbanization. This feature is more prominent in developing countries than in developed countries. Take the three largest agglomeration areas in China as an example (1987): The proportions of the non-agricultural population in the total population are as follows: 45.8% for the Beijing-Tianjin region, 57.9% for the central Liaoning

region, and 35.0% for the Yangtze River Delta region, higher than the national average by 50–150% respectively.

2) Development problems faced by agglomeration areas

The agglomeration area is an advanced development form of the socio-economic spatial structure and is the place where the main economic strength and development potential are located in the world and many countries. However, a series of development problems and obstacles are caused in succession in many agglomeration areas due to the progress of science and technology and the development of social economy in various countries and regions.

(1) Lag in the evolution of industrial structure. The advanced level of industrial structure as well as timely transformation and upgrading are extremely important for the development of a country or region and the position in the fierce competition among countries. After World War II, the developed countries in the world experienced recovery and development for 15–20 years, the economic and technical levels were significantly improved, and the pace of technical innovation was greatly accelerated. The industrial heartland of these countries played an important role in recovery and development, but most quickly entered the

Table 3 Representative traffic and transportation systems in agglomeration areas of developed countries

Scope		Category of the transportation route	Transport means and organization
Internal transportation in agglomeration areas	In the central city	Subways, city streets, trunk railways, local railways and sidewalks (that is, cycle path)	Subways, trains, buses, private cars and bicycles
	Between the central city core and the suburbs	Trams, local railways, trunk railways, trunk lines, general highways and waterways	Subways, trams, buses, private cars and water transportation
	Between central cities	Local railways, trunk railways, expressways, national highways and waterways	Ordinary express trains, special express trains, public and private cars, and water transportation
External traffic in the agglomeration area		Trunk railways, local railways, waterways, airports, expressways and national highways	Special express train and suburban ordinary express train, cars and private cars, airplanes and water transportation

Note In the agglomeration area of developed countries, the underground railway will drive out of the ground when reaching the suburbs, thus becoming a tram or suburban train "structural crisis". The main reason for the formation of the "structural crisis" is as follows: The main agglomeration areas in the world began to form in the second half of the last century and the beginning of this century. Most were built on the basis of the large-scale development of mineral resources (such as the Birmingham region in Britain, the Ruhr region in Germany, the Appalachia region in the United States, and the central Liaoning region in China (Shenyang, Anshan, Benxi, Fushun and Liaoyang, etc.). These agglomeration areas are characterized by traditional heavy chemical industries, and the structure has been "too heavy" in the past 20–30 years. There are also some agglomeration areas developed with industrial cities at the ports as the core (such as the Kyoto-Osaka-Kobe area in Japan). In both cases, the industrial advantages of agglomeration areas are reflected in traditional industries, such as the coal industry, iron and steel industry, electric power industry, heavy machine manufacturing industry, etc. However, these traditional industries were gradually impacted and even replaced by some emerging industries after the "oil crisis" in the 1970s and the large-scale new technology innovation process in the 1980s. However, this replacement generally occurred in the original non-agglomeration areas. The second reason is that it is very difficult to adjust the structure of most agglomeration areas due to the impact of "geographical inertia", and it is also difficult to achieve results in the short term. The third reason is that a number of large enterprises developed in these agglomeration areas in the process of economic recovery and rapid growth at the beginning of this century, especially in the 1950s and 1960s. In those years, large enterprises showed many advantages, namely, large production scale, strong ability to occupy the market, stable position in the competition, a strong sense of employment security for employees, etc. However, there have been fierce international competitions in the upgrading of industrial structure and the development of emerging industries since the 1970s. In some developed countries, traditional industries have gradually shifted to developing countries, which has aggravated the structural crisis in the areas where traditional industries were concentrated.

(2) Short supply of infrastructure. There are great demands for infrastructure due to the excessive concentration of population and socio-economic facilities in large and medium-sized urban agglomeration areas. However, the industries in the agglomeration area are highly concentrated, with scarce land and high land price. The productive and social infrastructure requires a lot of land, and some have to be transformed to solve the problem of aging of the original facilities. Therefore, the economic cost is huge. The infrastructure of the agglomeration area generally includes: transportation routes and corresponding port stations, power production and supply, water source engineering and water supply, waste treatment and discharge, stacking; education, medical and health facilities, sports facilities, financial and information industries, rest and resort facilities, etc. Insufficient supply of infrastructure leads to the weakening of external and suburban attractiveness and a decline in the resident life quality. Information transfer is restricted due to the short supply of infrastructure, which is not conducive to foreign and external investors' setting up factories and investing here, and even affects the

raw material supply and product output of production units. The insufficiency of education, especially vocational education, directly leads to an unfavorable position in the competition for technological innovation. There is a short supply of environmental improvement facilities and vacation facilities, which is also related to the layout of most agglomeration areas away from the sea. The bulk raw materials and products must be concentrated and distributed by railways or highways because they are not located at the seaport. Compared with the scheme of being located at the coast, the connection between the international market and the domestic market is much worse.

(3) Prominent contamination and deterioration of the ecological environment. For many agglomeration areas with traditional industries as the main body, the emissions of "three wastes" are high due to high energy consumption, high water consumption and large land occupation. Water, air and soil are contaminated to different degrees. For example, the central Liaoning region (Shenyang, Anshan, Benxi, Fushun and Liaoyang). The agglomeration area is located in the middle reaches of the Hun River and the Taizi River, and is dominated by heavy industrial production with high energy consumption, high water consumption and high pollutant emissions. The so-called "three major sectors" with high pollutant emissions: chemical industry, metallurgy and light industry, and "six major enterprises": Non-ferrous smelters, thermal power plants, cement plants, petrochemical plants, paper mills, etc. are quite concentrated here. The sewage irrigation area in the Shenyang-Fushun irrigation area and Liaoyang, Anshan and Yingkou irrigation areas is more than three million mu. Among them, the seriously contaminated irrigation area is more than one million mu. The domestic and industrial water sources are generally contaminated in Shenyang, Fushun and Liaoyang, areas along the Taizi River and the Hun River and the vicinity of the irrigation area. The air pollution in the urban areas of the five cities is severe. According to the economic development planning, the emissions of "three wastes" around 2000 may double compared with that in 1988, and the part that may be controlled (reaching the emission standard but not without harm) will only be equivalent to 20–30% of the total. There will be an intolerable state in some urban areas and water pollution areas. In the 1950s and 1960s, the pollution problems in some agglomeration areas in industrialized countries such as Britain and Japan once shocked the world. They used modern technologies, paid huge economic costs, and took a series of measures to adjust the industrial structure and spatial structure, which solved the pollution problems in the agglomeration area to a certain degree. However, the poor quality of the living environment and the ecological problems are still the main problems that plague these agglomeration areas.

(4) A large amount of land is occupied by industries and cities, causing the landscape structure unbalanced and the space crowded. At the initial phase of the development of agglomeration areas, the built-up areas are generally expanded from one or two, or several points. With the economic development and the construction of various facilities, the space becomes less and less, the adjacent cities are connected together, the green land, the scenic area and the vacation

area are shrinking day by day, and the originally coordinated landscape structure becomes unbalanced. As a result, the pattern of the relationship between human and nature has changed.

Due to the existence of the above problems under the conditions of the market economy, the following non-benign circles are formed: The lag in the evolution of the industrial structure has caused the products of some industries and sectors to lose the market, so that a large number of workers are unemployed. Coupled with the environmental degradation, residents migrate from this area to other areas; the loss of employees and residents has led to the reduction in taxes of the local government and the decline in the technological innovation capability; as a result, the local investment capacity in industry and infrastructure has declined, which further affects the development of technological innovation; this further hinders the upgrading of industrial technology, thus leading to a more disadvantageous position in the pursuit of the market (illustrated in Fig. 12). Under the current planning system in China, there are still the above-mentioned problems to different degrees in the agglomeration areas with large enterprises as the main body, but the outflow of population, workers and technicians is not obvious.

Job losses, unemployment, tax outflows, decreased investment capacity, low infrastructure level, a slow or negative growth of national income, dense population, declining environmental quality, overweight structure or lagging evolution.

3) Possible ways to solve the development problems of agglomeration areas

As mentioned at the beginning of this chapter: An agglomeration area is an advanced form of the evolution of socio-economic spatial structure, a comprehensive socio-economic system formed after long-term development and spatial agglomeration, and the main body of economic strength in a country or a large region. There are

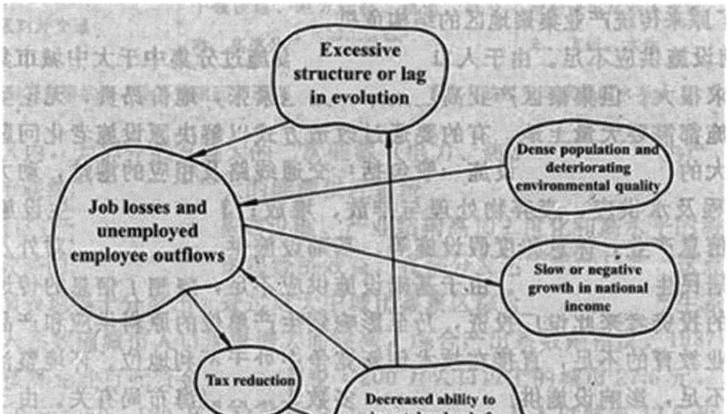

Fig. 12 Development barriers caused by "structural crisis" in agglomeration areas

no strict standards for the division of agglomeration areas. Any area with the above characteristics and having the above problems at a certain stage can be called an agglomeration area. It is a relative concept from this point of view.

Since the agglomeration area is the core area of the social economy of many countries, the governments of all countries attach great importance to the sustainable development of the agglomeration area, which is the main object of the regional policy of each country. In developed countries, although the government attaches great importance to the principle of balance and provides various supports to less developed areas (mainly remote areas and those areas with extremely unfavorable natural conditions), the focus is still on the development of agglomeration areas. In this aspect, Japan, Britain, France, Germany and the United States have all taken many measures, and obvious effects have been achieved.

To solve the development problems of agglomeration areas, the effective countermeasures shall be taken for specific problems:

(1) Adjust the industrial structure. The specific measures are as follows: control or even reduce the scale of traditional industries; gradually innovate and adjust the production technology and product structure of traditional industries as per the changes in market demands; and promote the development of small and medium-sized enterprises. It is important to introduce new technologies from foreign countries or other regions to accelerate the transformation of the industrial structure. After World War II, the long-term sustainable economic development of Japan was mainly supported by several agglomeration areas, and the industrial structure of these agglomeration areas has undergone great changes in the process of high-speed growth (Table 4).
(2) Improve infrastructure conditions. With the development of science and technology and the increasingly fierce competition, the conditions of the infrastructure (completeness and structural advancement) in a country or a region have an increasingly prominent impact on economic and social development and have a particularly huge effect on industries, technological innovation and sustainable development in agglomeration areas. Here, the most important thing is to have modern traffic and transportation facilities as well as a convenient communication network. For large agglomeration areas of international significance, there shall be first-class international airports, international communication hubs and several expressways passing through or by the agglomeration areas, as well as large seaports (if the agglomeration area is not near the sea, there shall be a transportation channel directly leading to the seaport). Modern transportation and communication facilities not only ensure the transportation of raw materials for industrial production, energy, semi-finished products and finished products but also ensure that companies, enterprises and institutions in the agglomeration area can occupy the market as soon as possible. There was little difference in economic strength and technical level between the two agglomeration areas of Keihin and Hanshin in Japan before high-speed growth. In recent years, international flights in Osaka are considerably fewer than those in Tokyo since there is no extra-large international airport in Osaka. In addition, due to its position

as the political center and other factors, the economic development of Keihin has obviously surpassed that of the Hanshin area. Some large companies in the Hanshin area moved to Tokyo, so as to obtain economic and technical information at home and abroad as soon as possible. Because of this, Osaka is carrying out land reclamation in a large area, so as to build a new international airport at sea.

(3) Implement technological innovation, and develop small and medium-sized enterprises. In both developed countries or developing countries, the agglomeration areas with problems are developing the plan for technical innovation, introducing new technical processes, transforming original facilities, and developing new products; it is necessary to improve the status of cutting-edge products and high-tech products for developing emerging industries. There can be many forms of technological innovations in terms of spatial distribution: achieve technical transformations in the original enterprise; establish economic and technological development zones as well as industrial technology centers, etc.; establish joint ventures with companies with advanced technologies with separate layouts.

In recent ten years, it is an effective measure for timely adjusting the industrial structure, increasing employment, improving financial conditions and implementing technological innovation to implement the "middle-class policy" and develop a large number of small and medium-sized enterprises.

First of all, it shall be emphasized that there are no uniform regulations on the classification of enterprise scales in various countries, and the standards for classifying scale levels also vary for different sectors and industries. The small enterprise in Sector A may be the medium-sized enterprise in Sector B by the number of employees. U. Haas (Haas) divides the scale of mining and processing enterprises into the following seven levels:

1 to 9 workers	10 to 19 workers
20 to 49 workers	50 to 99 workers
100 to 199 workers	200 to 499 workers
Over 500 workers	

The first two levels can be called "small enterprises", the last group is called "large enterprises", and the middle groups are called "medium-sized enterprises" [7]. The World Bank classifies the scale of industrial enterprises in some countries as follows (Table 5). In the table, those with less than 33 workers are called "small enterprises", those with more than 243 workers at the last level are "large enterprises", and those in the middle are all "medium-sized enterprises" [8].

It can be seen from Table 5 that the large enterprises in Yugoslavia and Hungary accounted for a large proportion, that is, dominant; the small enterprises in Japan, the United States, China and Britain accounted for a large proportion of different degrees.

After World War II, the enterprise scale continued to expand in some industrialized countries (especially in the 1950s and 1960s), and the jobs were increasingly concentrated in large enterprises. For a long time, people believed that: large enterprises have a higher sense of employment security, better social welfare, higher wage levels, more opportunities to be promoted within the enterprise, and higher competitiveness.

However, this argument has been increasingly questioned. It has been proven through numerous studies in the Western industrial countries (especially the United States and Britain) that small enterprises have created a large number of jobs, while the jobs in large production units have significantly reduced since the late 1960s.

Over the past 20 years, many scientists have concluded that small and medium-sized enterprises create jobs more than average. D. Birch analyzed the long-term statistical figures from 1979 to 1980 [9], conducted a statistical analysis on the percentage of jobs created by 5,600,000 enterprises, including enterprises of all scale levels in all industries, from 1969 to 1976, and obtained a series of the most important empirical results.

According to Birch's analysis, small enterprises with less than 20 workers created 2/3 of the net job growth, while large enterprises (mainly refer to those with more than 500 workers) only created 10% of the net job growth. Obviously, the regional economic growth mainly depends on the new enterprises that have been located. In this case, the scale structure of the enterprise is of important significance.

Many studies have proven that small and medium-sized enterprises have great spatial flexibility during the process of location and migration of new enterprises. Therefore, they can constitute the objects supported by the regional economic policies. The following conclusion can be further drawn from this point: A healthy region is not in a structure state dominated by large enterprises, but in a structure state dominated by small and medium-sized enterprises. Therefore, the small and medium-sized enterprises play an important role in the new enterprise policies.

Small and medium-sized enterprises can create more jobs and promote the rapid economic growth of the region. Of course, small and medium-sized enterprises also have some disadvantages. Many scholars believe that the "infant mortality rate" of small and medium-sized enterprises is quite high. On the other hand, some small and medium-sized enterprises have developed into large enterprises due to very high profits. In addition, the small and medium-sized enterprises are often in a weaker position in terms of product export.

(4) Improve the environmental quality, and build environmental and tourism facilities. Since the 1960s and 1970s, the main developed industries have recovered from industrial pollution, and people's environmental awareness has been greatly enhanced. In addition, the role of economic factors such as freight has declined relatively due to the high development of the social economy and

the rapid expansion of the scale of emerging industries. Environmental characteristics and conditions have become important location factors for industrial distribution.

When choosing the investment sites, investors may not pay much attention to the convenient cooperation conditions and urban life supply conditions in the agglomeration area, but tend to the green natural environment. These areas have fresh air, clean water, sufficient sunshine, as well as abundant natural and artificial vacation and tourism facilities.

The governments, entrepreneurs and planning agencies at the places where the agglomeration areas are located have noticed this trend. We shall maintain and expand the green land in the agglomeration area through strict planning and policy control, especially the main cities in the agglomeration area shall be separated by green land; we shall increase the short-haul rest and entertainment facilities, and build forest parks at appropriate distances. At the same time, we shall construct regional sewage treatment facilities and an air quality monitoring network. In order to take these measures and improve the environment quality, a series of laws on environmental protection must be developed.

4) Development problems of agglomeration areas and examples of solutions

There are various types of large-scale industrial urban agglomeration areas in the world, which have different natural foundations and different development stages. Of course, the economic and technical levels, employment status, as well as domestic and foreign markets of the region and country are also different. All these determine that the ways to solve the development problems of agglomeration areas are different. However, the solutions also have similar aspects since many of the problems in the agglomeration areas are common. The central Liaoning region (Shenyang, Anshan, Benxi, Fushun and Liaoyang) in China and the Ruhr region in Germany are taken as examples as follows. Among them, the way to solve the problems in the agglomeration area of central Liaoning is only to make planning and assumption, and the way to

Table 4 Transformation of the industrial structure of main agglomeration areas in Japan

Stage	Agglomeration area Leading industries				
	Kitakyushu	Setouchi	Hanshin	Chukyo	Keihin
Before high-speed growth	Steel	Shipbuilding and fiber	Fiber and metal	Fiber and ceramics	Machinery and general industrial products
Recent 10–20 years	Food	Steel and petroleum	Electricity and general industrial product	Automobile	Electricity and printing

Note "Fiber" in the table mainly refers to the textile industry, and the electrical products include household electrical appliances

Table 5 Proportion of enterprises of different scales in some countries (enterprises with five or more workers)

Enterprise scale (number of employees)	China (1982)	Britain (1979)	United States (1977)	South Korea (1981)	Japan (1972)	India (1976)	Yugoslavia (1981)	Hungary (1981)
5–33	59.2	65.2	56.4	70.6	80.2	51.9	6.6	2.2
33–75	19.5	15.7	20.3	14.4	10.7	35.3	15.8	4.8
75–189	12.2	10.8	12.4	9.2	6.1	7.8	32.1	18.7
189–243	8.5	1.4	3.8	1.5	0.8	0.8	12.0	9.2
≥ 243	0.6	6.9	7.1	4.3	2.2	4.4	33.5	65.1

Source World Bank, Main Report on China's Development Issues, p. 40, 1985, Beijing

solve the development problems in the Ruhr region is only to focus on narrating the situation of implementing the policy of technological innovation and the development of small and medium-sized enterprises.

A. Shenyang, Anshan, Benxi, Fushun and Liaoyang region in the Central Liaoning

The built-up areas and suburbs of the five cities in the central Liaoning are basically connected, covering an area of more than 6000 km^2. In 1988, the urban population was 7.1 million, and the total industrial output value was 42.8 billion yuan (above the township and at the constant price in 1980), accounting for about 55% of the total in Liaoning. This agglomeration area was formed on the basis of the development of coal and iron resources since the 1930s, and has become the largest heavy industry base in China now. In the heavy industry, the mining and basic raw material industries (steel and non-ferrous metallurgy, petroleum processing and chemicals) are taken as the main body. For a long time, central Liaoning has been an important base for raw material production and supply in China, as well as the largest manufacturing base for industrial facilities, power transmission and transformation equipment and transportation equipment. That is to say, the heavy industry products provided have been supplied to the domestic market for a long time. In addition, the industrial structure of the agglomeration area has two characteristics:

a. The industrial structure belongs to the "resource-processing" type

The industrial foundation of central Liaoning was initially based on resources. The characteristics of the resource structure have obviously affected the status of the industrial structure until now. Based on the development of coal, iron, petroleum and other resources, steel, coal, electric power, petroleum processing and chemical industry, building materials, etc. with strong strength have been developed. In order to serve these industries and utilize the raw materials provided by these sectors, the production of machinery, synthetic materials and chemical products has been developed. With the evolution of the industrial structure, some industries that are less dependent on resources, such as household appliances, instruments, fine chemical products, electrical equipment, electronic products, clothing and other processing and manufacturing production have also developed rapidly. However, since the production of steel, coal, electric power, building materials, chemical industry and petroleum that directly utilizes raw materials and fuel resources in Liaoning accounts for 40% of the industrial output value, it still belongs to the "resource-processing" industrial structure.

b. The scale structure of industrial enterprises is mainly large and medium-sized

Due to the fact that the industrial structure belongs to the resource-processing type, and coupled with historical reasons, another characteristic of the industry in central Liaoning is that it is dominated by large and medium-sized enterprises. Especially, large and extra-large enterprises occupy a very important position in the national economy of the province and the national industry. In 1988, there were 864 large and medium-sized industrial enterprises in Liaoning, accounting for 8.1% of the whole country, and 65–70% were concentrated in the five central cities. The total output value was 54.809 billion yuan, accounting for 9.7% of the large and medium-sized enterprises in China, which ranked second. The total industrial output value of large and medium-sized enterprises accounted for 64.8% of the total industrial output value of the province, much higher than those of all other provinces, cities and districts (the national average was 48.8%). Among them, the proportions of the output values of large enterprises in the total industrial output value are as follows: 46.8% for Liaoning and 29.8% for the whole country (all are independent accounting enterprises above the township level). In the Seventh Five-Year Plan, due to changes in the economic system, many large enterprises have encountered a series of obstacles in the source of raw materials, product sales, technological transformation funds and the development of export-oriented production, which has affected the development of these enterprises and the improvement of the technical level.

Since the reform and opening up, the development of the central Liaoning has encountered many difficulties, mainly including the following points:

(1) Industrial production with high energy consumption, high water consumption and serious pollution accounts for an excessive proportion, and the infrastructure is seriously insufficient. Over the 40 years since the founding of the People's Republic of China, Liaoning played the role as the largest heavy industry base in China, which mainly relied on metallurgy, petroleum processing, machinery and certain chemical industries. Except for the mechanical industry, the rest of the eight pillar industries in the province belong to the heavy chemical industry. These sectors account for most of the investment in industrial construction. From 1949 to 1987, heavy industry investment in the province accounted for 90.5% of all industries. Among the heavy industries, mining and raw material industries were large in scale, occupying a large amount of energy and water, and were also the most important pollution source. In 1988, excluding the energy consumption of petroleum processing and electric power industry, ferrous and non-ferrous metal mining and dressing, smelting and calendering industries accounted for 58.0% of all industrial energy consumption (32.761 million tons

of standard coal). The two major iron and steel companies in Anshan and Benxi accounted for 45.5%, with Anshan Iron and Steel alone accounting for 32.0%. The heavy industry accounted for 84.2% of all industrial electricity consumption in 1988. Among them, the electricity consumption of ferrous and non-ferrous metal industries reached 10.817 billion kWh, accounting for 29.51% of the industrial electricity consumption. According to incomplete statistics, heavy industry accounted for about 84% of all industrial water consumption (about 3.5 billion m^3). In terms of freshwater consumption by sector, the petroleum processing and petrochemical industry ranked first, with an annual consumption of about 780 million m^3, 750 million m^3 for iron and steel and 580 million m^3 for thermal power. The three sectors accounted for 66.2% of industrial water and electricity consumption. According to incomplete statistics, the above three sectors accounted for about 70% of the industrial wastewater discharged in the province in 1988 (1.68 billion tons). The three sectors, that is, industrial waste residue, dust and waste gas, accounted for a larger proportion. In general, the "three wastes" emitted by the industries in Liaoning ranked first among all regions in China.

(2) The phenomenon of equipment aging is relatively serious, and the industrial structure level is low. Liaoning is an old industrial base in China, and most of the large and medium-sized enterprises were newly built and expanded in the 1950s and 1960s. The technical transformation of the industries in the province has been large in scale since the Sixth Five-Year Plan. However, due to insufficient funds, many enterprises have undertaken the tasks of supporting other regions and exporting, causing that a large number of industrial equipment have not been updated as necessary, and the industrial structure has not been transformed and upgraded quickly.

Liaoning relies on raw material production based on resource development. The output value of mining and raw material industries accounted for 53.3% of that of the heavy industries, and 35.8% of the total output value of all industries, higher than the national average by 10% and 14.2% respectively. In the original value of industrial fixed assets in 1988 (more than 90 billion), about 90% were within the category of traditional heavy chemical industries which were capital-intensive and labor-intensive. The 157 industries (original standard) in the province were divided into three types of intensive industries. If measured by net output value: 44.1% for labor-intensive and 46.7% for capital-intensive; if measured by the original value of fixed assets: 37.1% for labor-intensive and 53.5% for capital-intensive. The industrial equipment was seriously aged, the net value rate of fixed assets of the industries in the province was only 63%, while the national level was 67.4%; many products were not upgraded, and the economic benefits were poor. In 1988, the profits and taxes per 100 yuan provided by the industrial enterprises owned by the whole people were lower than those of five provinces and cities, such as Shanghai, Tianjin, Zhejiang and Jiangsu, and the labor productivity only ranked 11th in China. The heavy industries in Liaoning mainly served the mining and raw material industries in China, accounting for 38.1% of all industries, while the national average only accounted for 23.7%. Emerging industries and high-tech industries were developing slowly. At present, only aircraft manufacturing, synthetic materials, chemical drugs, household appliances, etc. have formed industries and have a certain position in the country. The output value of the electronic industry only accounts for 5% of that of the whole country. Especially, microelectronics and computers that represent the contemporary electronic industry have not formed emerging industries yet; fine chemicals and instruments are also relatively weak. Optical fibers, bioengineering, structural ceramic materials, new alloys, etc. with a high technical level have basically not started yet.

The technical structure level of industrial products is also low, which is manifested in the few deep-processed products and few high-precision products. In 1986, the output value of alloy steel only accounted for 15.7% of the gross output value of steel, and the output value of fine chemicals only accounted for 27% of the gross output value of chemicals. Among more than 7600 types of electromechanical products manufactured in the mechanical industry, only 38.8% reached the level in the 1970s, and 61.2% reached the level in the 1950s and 1960s.

Large-scale technological transformation and equipment introduction have been implemented since the "Sixth Five-Year Plan", but a large number of production processes have not been matched to form capabilities.

(3) The supply of energy, water, transportation and other infrastructure is tight, and some resources are increasingly depleted. The trend of energy shortage in Liaoning has been obvious since the early 1970s. However, non-ferrous metals, iron and steel, calcium carbide and ferroalloys, chemicals, etc. that consume a large amount of energy have continued to grow substantially afterward. In 1983, the output of ten common non-ferrous metals was 208,100 tons, accounting for 18.1% of the national output. In 1988, the output reached 337,000 tons, accounting for 16.6% of the national output. The output of calcium carbide and ferroalloy and their proportions in the whole country also increased. Among them, the output of ferroalloy is as follows: 22,300 tons in 1965, 87,800 tons in 1980, and 128,400 tons in 1988; at the same time, the proportion of coal output in Liaoning in the country dropped from 9.8% in 1965 to 4.7% in 1988. Therefore, the net coal import amount of the province continued to increase. In 1986, the net import amount accounted for 18.7% of the total net import amount in the whole country. In recent years, the electricity shortage in the province is close to 10 billion kWh, which has obviously affected industrial and agricultural production.

In the central urban agglomeration, Dalian along the coast, Jinzhou, Fuxin and other cities, the fight for water between industry and agriculture and between industry and cities is prominent. It is difficult to meet the water source requirements to construct and expand many large and medium-sized enterprises. The industrial and urban water supply distances are getting longer and longer, and the cost is getting higher and higher.

The capacity utilization rate of important trunk railways such as Shenyang-Shanhaiguan, Changchun-Dalian, Shenyang-Jilin, Shenyang hub, Anshan station and Liaoyang station has been very high and was in a state of saturation and supersaturation.

Some main industrial sectors in Liaoning developed on the basis of large-scale development and utilization of the resources in the province. However, some resources are facing a situation of depletion successively due to the long history and high intensity of development. Most of the non-ferrous metal (copper, lead and zinc) minerals have been developed and utilized, and the mines will be closed successively; more than 80% of petroleum resources are used; coal output will gradually decline since 2005 due to insufficient resources. All of them have an impact on the transformation of the industrial structure of the urban agglomeration and the selection of the replacement industry in Central Liaoning.

Due to the impact of the above problems, the economic speed of central Liaoning has been slow since the 1980s. Some enterprises have difficulties in the source of raw materials, the production operation is insufficient, there lacks the circulating fund due to product overstock, the factory is overstaffed, the progress of technical transformation is slow, and the export-oriented production is in a difficult situation. Due to the insufficient fund for the factory, the tax revenue of the local government does not meet the planned requirements, thus weakening the ability for pollution control and improve the ecological environment, and there is a trend that the pollution will be further aggravated.

According to the analysis of the existing problems in the agglomeration area, and considering the important position of the industries in Liaoning in the country, China shall support large and medium-sized enterprises in energy supply, loans, taxation and other aspects in the future. Two measures shall also be taken in terms of structural adjustment of the agglomeration area:

(1) Gradually adjust the industrial structure. The industrial structure pattern of the agglomeration area in the central Liaoning is formed on the basis of the development and utilization of the coal, iron ore, building materials and chemical raw material resources in the province and after decades of matching, improvement and development. For a long time, the iron and steel industry has been taken as the largest leading industry in the province. From 1949 to 1988, the total investment in the capital construction of the iron and steel industry in the province was nearly 10 billion yuan, accounting for about 18% of the total investment in the capital construction of industries in the province, ranking first. In order to develop the iron and steel industry on a large scale and meet the large demands for energy, industries such as coal, electricity, transportation tool manufacturing and heavy machinery manufacturing were first driven. The development of the iron and steel industry has also played a great role in promoting building materials and some light industries. However, with the evolution of the industrial structure, these sectors and industries have also developed into relatively independent sector systems, and the scope of markets and services has been greatly expanded. The petrochemical industry, developed on

the basis of the comprehensive utilization of oil refineries in the past and the introduction of equipment in the 1970s, will play an increasingly leading role in the national economy, and shall be taken as the leading industry of Liaoning for a long period of time in the future. Liaoning has many types of mechanical industries and strong technical strength, and plays an important role in equipping all sectors of the national economy, promoting the technical transformation of old industrial bases, increasing export, etc. With the development of the industrialization process, the growth of the national economy will depend on the mechanical industry more and more. The main industries in Liaoning are basically resource-based. In the development process of the resource industry, there are generally two trends: First, with the reduction of some non-renewable resources, related industries may retreat to a secondary position; second, it is required to improve the processing depth of the resource-based industries and the technical level of products. According to the specific situation of Liaoning, non-ferrous metallurgy, iron and steel, petroleum processing, building materials, paper making, textile and other sectors must develop new products, new processes and new technologies, and make every effort to create more value under the premise of basically not increasing (or not greatly increasing) the scale of resource utilization. Among them, the scale of resource utilization in the non-ferrous industry shall be strictly controlled and even gradually reduced.

The main direction shall be transitioned from the iron and steel industry to the mechanical industry, especially the heavy-duty machinery with great relative advantages over that of other regions of the country. In this regard, the experience of the Ruhr region is available for reference. Several major structural adjustments have been made in this region in response to the "structural crisis" in the past 30 years. However, the number of employees and sales amount of the machinery industry, especially the heavy machinery industry, has never declined, and the position in the domestic and foreign markets has always been maintained. During the structural adjustment in the future, it is necessary and possible to gradually increase the proportion of light industry (only 25.5% in 1986). The current proportion is the lowest among all large industrial agglomeration areas in the whole country. The potential for further increase of this proportion is as follows: The petrochemical products are increased substantially, and the light industry established on the products of the metallurgical industry has a broad prospect; the scale and structure of industrial enterprises are adjusted. In the late 1980s, the output value of large enterprises accounted for 49.9% of the industrial output values of the five cities, higher than 78% of the national average, while that of small enterprises accounted for 33.8%, only 63% of the national average. Under the planned commodity economy, the small and medium-sized enterprises have considerable advantages in accelerating economic growth, increasing employment, adjusting product structure, promoting innovation achievements, and developing the export-oriented economy. In Liaoning, attention shall be first paid to the development of small and medium-sized enterprises in the five cities.

(2) Adjust the spatial structure of the industrial distribution, and gradually advance to the Liaohe River Delta. We shall construct new industrial sites, small industrial towns as well as economic and technological development zones in the suburbs of big cities (mainly the middle and outer suburban areas), such as Sujiatun and Xinmin in Shenyang, Songsantaizi in Anshan, Wolong and Waitoushan in Benxi, etc.

The industries in Liaoning are over-concentrated in a few cities in the middle and along the coast. If large-scale development in agglomeration is still required for a long term in the future, it is bound to pay a high price. For this reason, the mesoscale spatial structure adjustment shall be made in the industrial distribution in the province from the beginning of the next century. The general strategy is to advance from the center to the coastal zone (Fig. 13). The specific direction is: to advance to the region with Yingkou and Panjin as the center in the Liaohe Delta. In the process of implementing such adjustment and development of new districts, two types of enterprises shall be mainly established: First, major projects in the raw material and energy industries, such as the construction of a large-scale iron and steel base, thermal power station, cement plant, etc. near the Bayuquan Port; second, technology-intensive and export-oriented production and the construction of various forms of development zones.

As the target area for the transfer of industrial distribution in Liaoning at the beginning of the next century, the following conditions shall be met for the area: rich resources and broad space, which are ideal conditions for building a large-scale coastal raw material complex. The scale of crude oil processing and natural gas utilization is likely to expand as the area is located in the Liaohe Oilfield. Among them, according to the requirements for rational utilization of resources, thickened oil and high pour point oil shall be processed on-site; the whole area has more than one million mu of seaside beach land, an intertidal zone with a large area and the deep-water coastline near Bayuquan, which can provide vast land for industrial, urban and transportation construction. Reed planting, salt industry, aquaculture, etc. provide the industry with the production of raw materials, and the deep-water port is constructed. The water source supply can be achieved in several ways as the area is located in the estuary area.

The construction of railways and ports has greatly improved the geographical location for transportation. By the end of this century, the throughput capacity of the new and old ports of Yingkou will reach 10 million tons. In addition to the Shenyang-Dalian Expressway, it is suggested to build the Haicheng-Yingkou-Panjin-Jinzhou arterial road. Coupled with the reconstruction of the Gouhai Railway, this area has become a hub that connects the Liaodong Peninsula with inside Shanhaiguan and other areas in Northeast China.

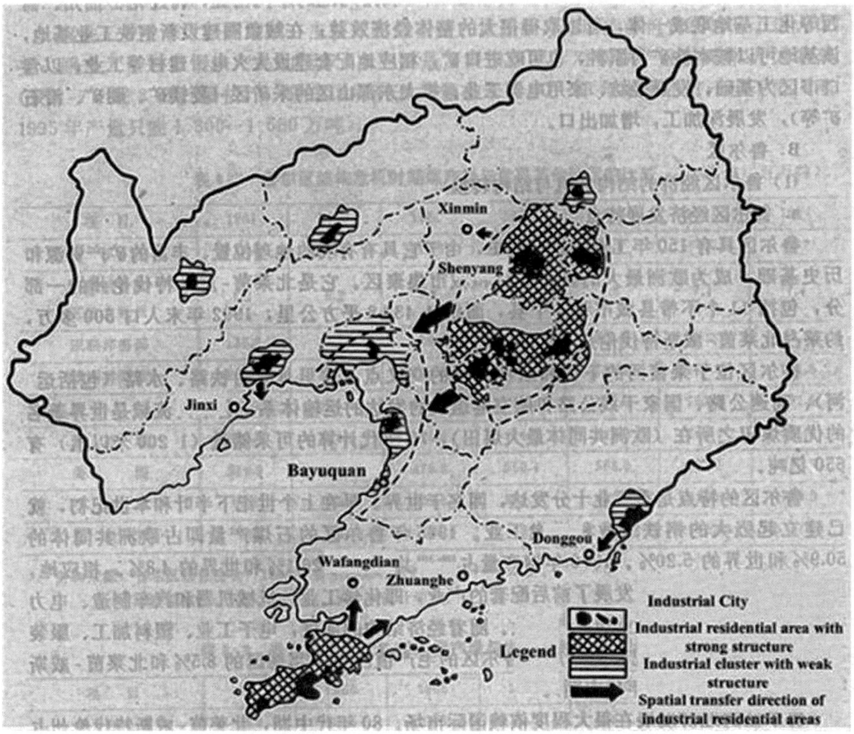

Fig. 13 Urban agglomeration area of industries in Liaoning and its spatial structure adjustment direction

This area has a large environmental capacity, and the problem of industrial pollution can be reasonably solved.

Industrial development direction and layout of the area: focus on the development of the petroleum and natural gas chemical industry. On the basis of the existing large and medium fertilizer plants and the 115,000 t ethylene plant, a set of large-scale oil refining and ethylene plants and corresponding post-processing systems (including the salt chemical industry) will be built in Panjin at the end of this century or the beginning of the next century. Taking advantage of the middle position of this area, Liaohua, Jinzhou, Jinxi and other chemical bases are integrated, which can achieve great overall economic benefits; build the new iron and steel industrial base in Bayuquan, which can use Anben's iron ore as raw material or imported ore. Accordantly, build the large thermal power, building materials and other industries; based on the urban area of Yingkou, develop light textile and household appliance industries; expand the mining area in the eastern mountainous area (magnesite, boron, talc ore, etc.), develop deep processing, and increase export.

B. Ruhr region

(1) Structural crisis and structural transformation of the economy in the Ruhr region

a. Characteristics of economic development in the Ruhr region

The Ruhr region has a 150-year history of industrial development. Due to its favorable geographical location, abundant mineral resources and historical foundation, it has become the largest heavy industrial area and urban agglomeration area in Europe. It is a part of North Rhine-Westphalia, including 11 cities without counties and four counties, with an area of 4432.8 km^2, it had a population of more than five million at the end of 1992, accounting for about 1/3 of that of North Rhine-Westphalia.

The Ruhr region is located at the intersection of the lower reaches of the Rhine River and European traffic, and there is a developed transportation system consisting of railways, waterways (including canals), expressways, national trunk highways and airports. The Ruhr River Basin is the place where the world-famous high-quality coal field (the largest coal field in the European Community) is located, and the recoverable reserves (above 1200 m) calculated in the 1970s were 65 billion tons.

The Ruhr region is characterized by very developed heavy industry, which is world-famous. The strong steel industry and coal industry had been established as early as the second half of the last century and the beginning of this century. In 1965, the output of stone coal in the Ruhr region accounted for 50.9% of that of the European Community and 5.20% of the world. In 1970, the output of steel in the Ruhr region accounted for 26.1% of that of the European Community and 4.8% of the world. Accordantly, supporting industries have been developed based on raw materials and energy, that is, the chemical industry, machinery, machine and automobile manufacturing, electric machine manufacturing, rare metal industry and electrical industry. With the evolution of the economic structure, the electronic industry, plastic processing, clothing industry and service industry have undergone large-scale development. The gross output value of the Ruhr region accounted for 8.5% of that of the former Federal Republic of Germany and 31% of North Rhine-Westphalia (the mid-1980s).

The economic development of the Ruhr region depends on the international market to a considerable degree. In the mid-1980s, North Rhine-Westphalia accounted for 3% of the world's trade volume, and was an important developed foreign trade region in the world. The leading products exported are machines, automobiles, electrical and electronic products as well as chemical products; the main materials imported are: agricultural products, food and various products for entertainment. For foreign investors, there are mainly investors from Japan, the Netherlands, the United States, etc., the Ruhr region is an attractive area due to the efficient transport conditions and labor productivity.

It shall be emphasized here that: International market and economic conditions have a strong impact on the economic development of the Ruhr region. For the Ruhr region, it is important to constantly improve the technical level and the product's quality in term of its competitiveness.

Dense cities and populations have brought heavy ecological burdens to the natural life base. The area occupied by the urban and rural residential areas accounts for 35% of the total area today. This figure was less than 7% 90 years ago and only 24% in 1960. This requires adjustments in the economic structure and spatial structure of the Ruhr region, so as to reduce the ecological burden.

b. Development of structural crisis in the Ruhr region

It has been pointed out above that the Ruhr region has developed the upstream supply and downstream processing industrial sectors and service industries based on raw materials and energy, which indicates that the Ruhr region has constantly undergone structural transformations during its economic development. The coal sales crisis occurred in 1958, followed by the steel sales crisis in the 1970s and 1980s, which led to the profound structural transformations in the Ruhr region. In the past 30 years, obvious effects have been achieved, which was manifested in high-speed economic growth and a basically stable employment market. However, this structural transformation process has not ended yet.

The following things can indicate the structural crisis of the Ruhr region and its impact on the social economy!

The production of the iron and steel industry has been greatly reduced. The main reason is that a large amount of oil is imported, and the energy consumption structure in Germany has greatly changed; for the countries that used to import a large number of steel products (mainly developing countries), the demands for steel products have been greatly reduced (Tables 6 and 7). It is estimated that the coal mining volume in the Ruhr region will be further reduced in the next 10–15 years (the output can only be 13 million to 15 million tons as of 1995).

The number of jobs was greatly reduced during the adjustment and transformation process. The structural crisis can be clearly seen in the changes in employment. From 1961 to 1973, the number of employees in the Ruhr region fell from 2.4 million to 2.2 million, while employment slightly rose in other regions of North Rhine-Westphalia. The number of employees fell further after that, especially in the industrial sector (Table 8). From 1958 to 1973, the number of jobs in the industries in the Ruhr region fell from 1,098,000 to 810,000, that is, reduced by 26%. The number of jobs in the Ruhr region was reduced by 380,000, mainly in the industrial sectors of coal and steel. It is estimated that another 85,000 jobs will be lost in the coal and steel sectors in North Rhine-Westphalia as of 2000.

Due to structural problems and poor environmental conditions, the population in the Ruhr region began to reduce since 1959, and the reduction in the population was as high as 75,000 in 1967 and 47,400 in 1984. Changes in the population are strictly related to the economic development of a region, and affect the economic growth, financial conditions and investment scale to a great extent.

Compared with other regions in Germany, there is only tiny growth in the Ruhr region in terms of the gross domestic product. Before the coal sales crisis, the Ruhr region was the region with the fastest growth in the former Federal Republic of Germany. The gross domestic product in the Ruhr region increased by 143% from

Table 6 Comparison of coal output in the Ruhr region during the period of structural crisis with that in other regions in the world

Project	1951	1960	1965	1970	1975	1981	1984
Ruhr region	110.6	115.4	110.9	91.1	75.8	70.0	61.2
North Rhine-Westphalia excluding the Ruhr region	6.1	8.2	7.8	9.6	7.6	7.1	7.4
Former Federal Republic of Germany	135.2	142.3	135.1	111.4	92.4	88.5	84.9
European Community	231.4	233.9	218.0	164.6	122.3	113.2	107.8
Poland	82.0	104.4	118.8	140.1	171.6	163.0	191.5
Former Soviet Union	202.4	374.9	423.9	474.0	540.0	544.1	555.0
United States	519.9	391.5	475.8	550.4	568.0	691.5	742.5
China	54.6	170.5	231.8	354.0	482.2	621.6	789.0
Around the world	1 498.1	1 983.2	2 128.0	2 182.7	2 430.9	2 802.5	3 018.1

Unit: million tons
Source Association of Residences in the Ruhr Region, Statistics on Cities and Counties in the Ruhr Region in 1985, Pages 157 to 159
National Bureau of Statistics of China: Statistical Yearbooks in 1980 and 1986

Table 7 Comparison of steel output in the Ruhr region during the period of structural crisis with that in other regions in the world

Project	1951	1960	1965	1970	1975	1981	1984
Ruhr region	10.4	23.6	24.6	28.5	24.3	23.9	22.2
North Rhine-Westphalia excluding the Ruhr region	0.7	1.1	1.7	2.0	1.8	1.9	1.9
Former Federal Republic of Germany	16.1	34.1	36.8	45.0	40.4	41.6	39.4
European Community	37.7	72.8	86.0	109.1	104.8	109.2	103.2
Japan	6.5	22.1	41.1	93.3	102.2	101.7	105.6
United States	95.4	91.9	122.5	122.1	108.0	111.3	85.9
Former Soviet Union	31.3	65.3	91.0	116.0	141.5	149.0	155.0
China	1.1	5.8	12.2	17.8	23.9	35.6	46.8
Around the world	210.5	330.2	458.9	517.6	646.8	710.0	711.4

Unit: million tons
Source Association of Residences in the Ruhr Region, Statistics on Cities and Counties in the Ruhr Region in 1985, Pages 157 to 159
National Bureau of Statistics of China: Statistical Yearbooks in 1980 and 1986

1957 to 1971. The figure was 193% for North Rhine-Westphalia and 237% for the former Federal Republic of Germany.

Through structural adjustment and modernization measures, the Ruhr region has achieved high-speed economic growth over the past few years, basically remaining flat with the state and the Federal Republic of Germany.

Table 8 Changes in the number of employees by sector during the period of structural crisis in the Ruhr region

Ruhr region	1970	1972	1974	1976	1978	1980	1982
Agriculture and forestry	33.8	39.0	24.9	24.0	23.2	28.0	27.3
Manufacturing enterprise	1 285.7	1 229.1	1 197.9	1 098.4	1 039.3	1 065.0	1 013.9
Commerce, transportation and communication industries	410.6	421.7	406.7	383.2	391.4	399.0	384.7
Others (service industry)	470.5	490.0	514.2	551.9	565.4	569.0	608.9
Total	2 200.6	2 179.8	2 143.7	2 057.5	2 019.3	2 062.0	2 034.8
North Rhine-Westphalia excluding the Ruhr region	4 761.9	4 788.6	4 760.1	4 671.5	4 720.1	4 886.0	4 848.8

Unit: 1000 persons
Source Association of Residences in the Ruhr Region, Statistics on Cities and Counties in 1985, Page 150

(2) Innovation and middle-class policies of the state government of North Rhine-Westphalia

The structural transformation and economic development of the Ruhr region are taken as urgent tasks in both the state and the former Federal Republic of Germany. However, it will take a long time to solve the structural problems in the Ruhr region. The state government, together with various community units, has developed the development outline for economic innovation and ecological restoration since 1970.

The goals and measures of the innovation policy are as follows: expand the economic structure, modernize the production structure, and ensure jobs and economic growth;

Promote technological innovation in the economic field (especially in the industry), improve the technological level of skilled workers, and improve the competitiveness of industrial products;

Promote the modernization of ecological construction, and improve environmental conditions;

Implement middle-class policies, and provide key support to small and medium-sized enterprises.

a. Develop small and medium-sized enterprises and improve the scale structure

The Ruhr region is dominated by large enterprises if measured by the number of employees in each enterprise. According to the census data of jobs in 1970, the Ruhr region was 16% higher than the North Rhine-Westphalia and 20% higher than the average value of the former Federal Republic of Germany in terms of the proportion of large enterprises.

Over the past ten years, small and medium-sized enterprises developed much faster than in the 1950s and 1960s by implementing middle-class policies. While the total number of enterprises is constantly increasing, the enterprise scale has become smaller. According to incomplete statistics, there were a total of 5303 enterprises in the mining and processing industries in the Ruhr region in 1978. Among them, there are 2744 enterprises with less than 20 workers and 241 large enterprises with more than 500 workers. As of 1982, the number of large enterprises had fallen to 226, while the number of small enterprises had risen to 3388 (with a net increase of 664); the number of employees in large enterprises decreased by 41,009 from 1978 to 1982, while the number of employees in small enterprises increased by 1696.

It shall be emphasized here that the number of workers in medium-sized enterprises decreased significantly (by 14,088) from 1978 to 1982.

In 1982, two-fifths of workers worked in the mining and processing industries, and 18% worked in the iron and steel industry. The number of enterprises in these sectors increased by 8% compared with that in 1978, but the number of workers decreased by about 7% during the same period. The distribution obviously shifted in terms of industry. Large enterprises are mainly distributed in mining, petroleum processing, iron and steel industry, rare metal industry, building materials, wood, paper making, paper pulp, etc. Large enterprises play a large role (over 20%), which also includes glass manufacturing and processing industries here.

Different from the above, the small and medium-sized enterprises are mainly distributed in the following industries: sand and gravel mining and processing, brick and tile factory, cold-rolling and transformation plants, metalworking, trams, electrical industry, motor repair, metal manufacturing, musical instrument, toy manufacturing, chemical industry, office machinery and digital processing machinery, wood processing, paper products, precision machinery, printing industry, plastic processing, rubber processing, leather processing, textile industry, etc.

In addition, the foundry, optical industry, and instrument industry are small enterprises traditionally.

The development of small and medium-sized enterprises in the mining and processing industry sector in the Ruhr region has been described above. The changes in employment of various economic sectors in the Ruhr region will be analyzed below. W. Eckhart investigated 832 enterprises, and obtained detailed results (Table 9).

It can be clearly seen from Table 9 that for enterprises with less than 50 workers in the processing industry, enterprises with less than 20 workers in the commercial industry, and enterprises with less than 50 workers in other service industries (health, education, social legal and economic consulting, tourism and entertainment), the jobs increased during the period from 1975 to 1983.

In the structural transformation of the Ruhr region, the main growth sectors were: the electrical and electronic industry, and automobile and machinery manufacturing. A total of 92,000 jobs were created in the following sectors during the period from 1958 to 1973. Among them, the electrical and electronic industry, the automobile industry, the machine manufacturing, the chemical industry and the plastic processing accounted for 26%, 21%, 19%, 12% and 7% respectively.

b. Develop new industries, and improve the technical structure

In the 1960s and 1970s, extremely rapid progress was made in the world's industrial technology, and it became a common trend to adjust the industrial structure and

Table 9 Changes in the number of employees in the Ruhr region by sector and by enterprise scale from 1975 to 1983

Economic sector	Enterprise status in 1975			Net change in the number of jobs in enterprises of all levels and scales (1975–1983)									
	Number of enterprises	Jobs	%	≤ 20 persons		21–50 persons		51–100 persons		101–500 persons		Total	
				Quantity	%	Quantity	%	Quantity	%	Quantity	%	Quantity	%
Unclassified enterprise	5	26.5	0.2	+ 22	+ 83.0							+ 22	+ 83.0
Agriculture and forestry	15	205.5	1.4									− 7	− 3.5
Energy and mining													
Processing industry	194	4314.5	30.2	+ 825	+ 85.3	+ 185.5	+ 42.0	− 57.5	− 6.0	− 607	− 31.2	+ 346	+ 8.0
Building industry	117	2070	14.5									− 100	− 4.8
Commerce	316	3159	22.1	+ 558	+ 39.4	− 75	− 21.4	+ 51	+ 10.3	− 180.5	− 20.0	+ 353.5	+ 11.2
Traffic and communication industry	31	2771.5	19.4									+ 52.5	+ 1.9
Credit and insurance industry	8	196.5	1.4									+ 76.5	+ 38.9
Other service industries	146	1447	10.8	+ 867.5	4–124.8	+ 39.5	+ 35.7	− 125	− 41.5	+ 12	+ 2.7	+ 794	+ 54.9
Total	832	14,287.5	100.0	+ 2665.5	+ 67.3	+ 59.5	+ 3.6	− 291.5	− 13.1	− 896	− 22.1	+ 1537.5	+ 10.7

Source Eckart, W., V. Einem, E.: Dynamik der Arbeitsplatzentwicklung, Regionalvergleich Ruhrgebiet-Frankfurt, 1986–1987 promote the boom or recovery of the regional economy through technological innovation. The Ruhr region took the development of new industries as an important way to solve the "structural crisis". Here, new industries include high-tech industries as well as industries that did not exist in the Ruhr region before the large-scale structural transformation and were of little significance at the time. Specifically, new industries include the following three categories: high-tech industrial enterprises, such as electronic industry, laser technology, etc.; those that had already existed during the structural transformation in the Ruhr region in the 1960s: precision machine manufacturing, clothing industry and precision ceramics industry; industries that significantly expand the scale and make technological innovation through location transfer (transfer from urban center to suburbs or from the core city in the Ruhr region to edge city) and the introduction of new processing equipment. These sectors belong to the traditional production field, but do not include the mining or iron and steel industry.

The development of new industries was the main result of the middle-class policy and innovation activities in the Ruhr region, and almost all new industrial enterprises were small and medium-sized ones. For example, the Bochum Tenschade Industrial Zone consisted of 68 companies, the total number of employees was only 2470, and each company had only 36 persons on average. There were only a few exceptions, such as the Bochum Ober automobile factory. The development of new industries has greatly promoted the structural transformation in the Ruhr region.

c. Establish industrial technology centers

In the past few years, eight industrial technology centers have been established in the Ruhr region, and they are distributed in Bochum, Dortmund, Essen, Duisburg, Gleesenkirchen, Hamm, Oberhausen, etc. The main goals of the industrial technology centers are:

Achieve a close combination between science and economy, that is, realize the rapid transformation of research results into production practice;

Strengthen the long-term development of technologies that facilitate structural transformation and create jobs.

The establishment of the industrial technology center in the Ruhr region is the result of the implementation of technological innovation and middle-class policies in North Rhine-Westphalia. These centers are the development and testing places for samples and parts in many key technical fields that have great potential in the cities of the Ruhr region: electronic industry, material engineering, development and application of software systems, laser technology development, measuring instruments and sensors, medical equipment and information industry.

The location features of the industrial technology centers in the Ruhr region are as follows:

Good linear basic conditions, that is, it is convenient to connect expressways, railways and substations. Some industrial technology centers are directly located near the intersection of expressways or near railway stations and urban ports;

It is an important locational premise to be located near a university or a college, so that the technical support can be obtained through various professional professors.

The production direction of the industrial technology center shall adapt to the existing economic sector in the city where the center is located. In this way, the center can become an economic, scientific and urban community facility, and a strong impetus may be obtained for the new production processing at the center.

d. Build environmental facilities, and improve environmental quality

In addition to the sewage treatment facilities in enterprise units, the sewage from the central and northern parts of the Ruhr region is also concentrated in the Emscher River. A large-scale comprehensive sewage treatment plant has been established in the lower reaches, with an annual treatment capacity of 600 million m^3; the green isolation belts are established among main cities, and the establishment of factories is prohibited in the belts; excavate artificial watercourses, deepen riverbeds, raise river banks, and build drainage pumping stations; control the uneven subsidence of the ground caused by coal mining; except for a few power plants, the fuel oil and natural gas have been taken as the fuel for most power plants, and the electrical precipitation and flue duct desulfurization device have been provided for the iron and steel works, coal mines, coking plants, cement plants, etc. as regulated; a strict atmosphere and sewage discharge monitoring network has been established. A series of close-range rest and entertainment facilities have been built on and along the Ruhr River.

e. Adjust the spatial structure through the overall planning of regional development.

The Ruhr region established the planning agency—Residential Association of Ruhr Coal Mining District (KVR) in Germany as early as 1920. The overall planning and numerous professional planning of regional development have been compiled and revised five or six times so far, such as industrial location selection planning, green land planning, rest and entertainment facility planning, etc. Each city has prepared the overall planning at the main development stage. As coal mining gradually moves northwards, a series of new cities have been established in the central and northern parts in succession, so as to avoid the further expansion of old cities along the "Sacred Way" in the south; some residential centers, "commuter towns" and commercial centers that are not connected to each other have been established around large and medium-sized cities, and jobs are created nearby to attract residents of large and medium-sized cities to move there; in addition, measures have been taken to decentralize the functions of cities, making the industrial structures of cities such as Duisburg, Essen and Dortmund have their own characteristics.

References

1. D. Lu et al., *Theory and Practice of China's Industrial Distribution* (Science Press, 1990), pp. 24–39
2. L. Gao, The situation of South Korea's participation in regional economic cooperation from the development of the west coast. Coast. Econ. (1), 16–17 (1989)
3. Y. Cao, Theoretical overview of growth core and growth center. Geogr. Res. (3) (1990)
4. T. Hagerstrand, *Aspekte der raumlichen Struktur Von Sozialen K ommunikationsnetzen und der Informationausbreitung* (Kiepenheuer & Witsch, Berlin, 1970), S. 367–379.
5. E. Giese, Raumliche Diffusion auslMndisches Arbeitnehmer in der Bundesvepuklik Deutschland 1960–1972. Die Erde **109**, 92–110 (1978)
6. X. Xu, J. Zhu, *Modern Urban Geography* (China Architecture & Building Press, Beijing, 1989), p. 211
7. R. Hou, Visit to Washington (III). Geogr. Knowl. (11) (1990)
8. P. Klemmer, H. Schrumpf, Kleine und mittlere Betriebe in Riihrgebiet unter besonderer Berlicksichtigung des Produzierenden Gewerbes (1984), S. 46
9. World Bank, Main report on China's development issues, Beijing (1985), p. 40

Chapter 9
Technological Innovation and Spatial Structure

Productivity is the most active factor in the process of continuous development of human society from a low level to a high level. It is a universal law at all historical stages of human society that production relations shall adapt to the development of productivity. However, the development of productivity ultimately depends on the interaction among the three factors of productivity: people, labor tools and labor subject, that is, depending on the direction, scale and level of the development, processing and utilization of labor subject by those who master and use labor tools and labor skills. Here, the improvement of labor skills as well as the invention and application of new labor tools often play a key role. That is to say, technological innovation is an important lever for the progress of human society.

It was elaborated in Chap. 6 of this book that the spatial structure of the social economy also undergoes regular changes in the process of development and change of human society. Then, how does technological innovation affect the spatial structure of the social economy?

1 Technological Innovation Is the Fundamental Driving Force for the Evolution of Socio-economic Spatial Structure

1. Technological innovation causes changes in industrial structure

In the socio-economic development of human beings, the industrial structure constantly evolves from low level to high level and from simple to complicated. In history, the relative prosperity of a country or region is often determined by relatively advanced industries. In today's world, success in competitions is related to the

D. Lu, *Regional Development and Its Spatial Structure*,
https://doi.org/10.1007/978-981-99-7683-6_9

reasonable industrial structure, as well as the timely adjustment and transformation of leading industries, which have become extremely important factors for national and regional economic development and social progress.

Comprehensive meaning of industrial structure: The production sector or industry that can drive the economic development of a country or region (that is, the degree of industrial relations is high), with a high rate of increase in labor productivity and high added value of the product in a certain period of time, that is, leading industry[1]; the proportional relationship between various industries; the input–output link between industries [1]. It is an objective requirement for the coordinated development of all industries and the smooth development of leading industries to correctly specify the leading industries, and accordingly arrange the composition and scale of basic industries and supporting industries according to the requirements of the correlation between industries, so as to achieve an organic connection among them.

In the long feudal society, the main industrial sectors were farming, animal husbandry, fishing, etc. The development of production technology was mainly reflected in the smelting and processing of iron, the use of gunpowder in the military, and the gradually significant improvement of science and technology. The handicraft industry emerged on the basis of agricultural development, and the labor tools and raw material processing methods were continuously improved. Compared with the backward slave society, the iron smelting technology led to the appearance of iron plows and other tools, which was the fundamental factor that made farming the main industry at that time.

With the development of production, the sprout of capitalism emerged in feudal society. Among them, the simple cooperation of labor socialization began to appear mainly due to the improvement of labor tools, the social division of labor and the formation of commodity production. The workshop handicraft industry appeared in Europe from about the middle of the sixteenth century to the end of the eighteenth century. In a handicraft workshop, handicraftsmen of different specialties or the same specialty were united. In the workshop, carriage manufacturing, woodware industry, shoe-making industry, button industry, etc. became important handicraft industries at that time. In history, the workshop handicraft industry prepared the necessary conditions for the transition to the machine industry: First, the workshop industry made the work division reach a very high level and simplified many labor operations. These operations were so simple that the machines could replace the workers' hands. Second, the specialization and great improvement of the labor tools made it possible to transition from hand tools to machines. Third, since workers had specialized in certain operations for a long time, the workshop handicraft industry had also trained skilled workers for the large-scale machine industry [2].

[1] There are great differences in the conversion cycles of leading industries in a country or region. With the evolution from the agricultural society to the industrialized society and the development from the primary stage to the advanced stage in the process of industrialization, the life cycle of the leading industry is becoming shorter and shorter. Since the 1950s, the dominant industries in some developed countries have been transformed every about 10 to 15 years.

The British Industrial Revolution from the late eighteenth century to the early nineteenth century marked a fundamental change in production technology, and machines were the material and technological basis for this change. The shuttle was invented in 1733, the spinning machine was invented in 1766–1767, and then there were machines powered by water. The steam engine was invented in Britain in 1784 [2]. Due to the invention, promotion and application of machines and power, heavy industries such as large-scale machine building industry, metallurgical industry and coal mining industry, following the textile industry, had successively become the main industries at the early and middle stages of industrialization. In 1931, German Hoffmann pointed out in "Stages and Types of Industrialization" that with the progress of the industrialization process, the ratio between the net output value of the consumer goods industry and the net output value of the capital goods industry is constantly declining, that is, the increase in the proportion of the heavy industry is a common phenomenon in all countries, which is "Hoffman's Law". It reflects the situation of European industrialization, especially that of Germany in the first half of industrialization.

Due to the progress of industrialization, especially the development of the large-scale iron and steel industry, the mechanical industry as well as coal and coking industries, social-economic activities, population and industries are concentrated in cities. Large cities in Europe (with a population of more than 100,000) increased sixfold in the nineteenth century, and more than half of the population was concentrated in cities in the middle of the nineteenth century in Britain and the early twentieth century in Germany. There were more and more developed traffic, business and financial facilities in and around the cities, and the cities played an increasingly prominent role as the core commander of regional economic and social development. In this way, the long-standing socio-economic spatial structure in the agricultural society was fundamentally changed.

After World War II, some developed countries carried out economic recovery and reconstruction, and a series of developing countries started the process of industrialization. Industry played an increasingly important role in the national economy, and cities also played an increasingly important role in organizing socio-economic life, that is, economic industrialization and social urbanization. The technological innovations that supported the large-scale industrialization of these countries mainly included: large-scale blast furnaces and steel-making furnaces, large-capacity thermal and hydraulic generators, heavy and precision machine manufacturing, chemical deep processing, etc. The leading industries were the steel industry, energy industry, petrochemical industry and mechanical industry in most cases. With the emergence of electronic technology, computer technology, high-purity metallurgy, fine chemical industry and laser technology, high-tech industries developed rapidly in some developed countries. Among them, those of the processing combination type included the following industries:

(1) Aircraft and aerospace industries. The aircraft manufacturing industry, aircraft engine manufacturing industry as well as other manufacturing industries of other aircraft components and auxiliary devices.

(2) Optical industry. The manufacturing industry of raw materials for glass product processing, manufacturing industry of glass fiber and its products, the manufacturing industry of wired communication machinery and appliances, and the manufacturing industry of analytical instruments.
(3) Machine building industry of medical electronic equipment. Medical equipment manufacturing industry, manufacturing industry of dental mechanical appliances, manufacturing industry of medical mechanical appliances for animals, medical material industry, and manufacturing industry of lenses and prism for optical machinery.
(4) Industrial robot industry. The manufacturing industry of robots for industrial use.
(5) Integrated circuit industry. Semiconductor element manufacturing industry, integrated circuit manufacturing industry, as well as manufacturing industry of other components for electronic equipment and communication equipment.
(6) Computing device industry. The manufacturing industry of mechanical appliances for offices, the manufacturing industry of wired communication equipment, the manufacturing industry of electronic computers and their auxiliary devices, as well as the manufacturing industry of other electronic application devices.
(7) Industrial machinery. The manufacturing industry of other engines, manufacturing industry of auxiliary devices for metal machine tools and metal processing machinery, manufacturing industry of other special industries, manufacturing industry of air supply fans of air compressors and gas compressors, manufacturing industry of chemical machinery and its devices, manufacturing industry of other electronic devices and manufacturing industry of electronic tubes.

2. **Industry, location factors and location complex**

The location factors here are different from the location conditions of a single station. The location factors refer to the external and surrounding environmental conditions that certain industries (sectors and industries) shall have for production, so as to cooperate with supporting production and service systems. Certain industries (sectors and industries) must be allocated in areas with certain conditions, which will lead to the formation of several socio-economic objects with their own characteristics. That is to say, there is a coupling relationship between various industries and their location factors.

In areas where farming is taken as the leading industry, there is a need for arable land and exploitable water sources, and it is necessary to form villages where the farming population lives as well as roads that connect farming areas with villages and agricultural product consumption centers. In areas where the production level is not high and commodity production is not very developed, a very balanced distribution is formed for the size of the village and the length of the road. During the modern mechanized large-scale agricultural production, due to the improvement of labor productivity and the formation of large villages (urban prototypes), large farm areas, advanced roads and large water conservancy facilities, the unbalanced state of socio-economic spatial structure is becoming more and more prominent.

The location factors of large-scale industrial development are mines, transportation lines, factories, water source projects, cities, etc., which constitute a special industrial landscape during the industrialization period. It can also be called a complex of industrial location (factor). Different industrial sectors and industries have different structures of the location complexes formed.

Iron and steel industry. The required location conditions are usually: coal, iron and other mineral resources, water sources, strong transport arteries and transportation hubs; in addition to huge smelting plants and transportation facilities, the iron and steel industry generally requires mines and the formation of medium-sized and even large cities, as well as the coordination of electrical, mechanical, chemical and other industrial sectors. With the expansion of the scale of the iron and steel industry, the comprehensive economic strength of the location (city) has increased, thus becoming a comprehensive industrial area.

Electric power industry. Among them, the location conditions required for the thermal power base are: fuel minerals (coal, oil, etc.), water sources, ash yard and transportation facilities; location conditions required for the hydropower base are: a certain amount of hydropower resources and favorable development conditions, load center.

Chemical industry and petrochemical industry. The required location conditions are: chemical raw material resources, transportation facilities and water sources, and large chemical complexes often form a "chemical city".

The enterprises in the general mechanical industry, light industry and textile industry have no special requirements for location conditions; the location factors of the industries, apart from the factories themselves, are mainly workers' residential areas and towns formed.

With the development of technology and the social economy, the industrial structure is constantly evolving. At the early and middle stages of industrialization, traditional industries, especially the heavy chemical industry and energy industry, were dominating. According to the requirements of scale economy, large-scale industrial and urban agglomeration areas were often formed. In such agglomeration areas, in addition to large and medium-sized cities and other densely distributed residential areas, there were large-scale production facilities, power engineering, transportation hubs and routes, water supply and drainage engineering, environmental treatment engineering, required green land, etc. The total area could reach several thousands of square kilometers or even tens of thousands of square kilometers. For example, the Ruhr region is an industrial and urban agglomeration area with a heavy chemical industry as the main body, covering an area of 4400 km^2. With various industrial factors (including population) very concentrated, it is one of the large-scale location complexes rare in the world at present (Fig. 1). The central Liaoning region in China, that is, Shenyang, Anshan, Benxi, Fushun and Liaoyang, is also an industrial and urban agglomeration area with a heavy chemical industry as the main body. The urban and suburban areas of the five cities have an area of 6000 km^2, and the foundation for forming the industrial structure is very similar to that of the Ruhr region. The Ruhr region has implemented many technological innovations since the 1950s, which

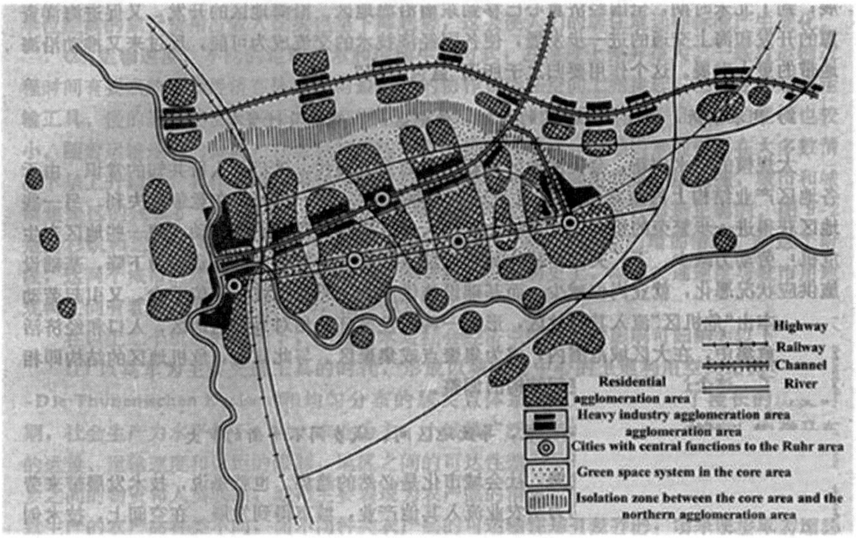

Fig. 1 Spatial structure form of the Ruhr region

has greatly improved the level of structural hierarchy of the industry. The proportion of emerging industries in the GDP has increased significantly, and the spatial structure form of the Ruhr region has also undergone great changes. Large-scale heavy chemical enterprises in Dortmund, Bochum, Essen, Mieheim and other cities have been greatly reduced. Some have merged, and some have moved to the west along the Rhine River for relocation. Therefore, the spatial structure state has been more balanced. However, the agglomeration area in central Liaoning will continue to develop with agglomeration as the main tendency for a long time.

3. **Technological innovation and balanced development among regions**

Many countries and regions in the world consider the balanced development among regions within their jurisdiction as an important socio-economic development goal. However, the practice at home and abroad has fully proved that it is a very long historical process from unbalanced development among regions to a relatively balanced one; moreover, the lower the level of economic and technological development in a country or region, the longer the process of achieving relatively balanced development among internal regions. The key to this is the impact of the technical level of social production.

1) Technological innovations lead to changes in production modes, thus affecting changes in the spatial structure.

In the history of human society, the results of major technological innovations are bound to cause changes in production modes. For example, the development and widespread application of new machines in the era of the Industrial Revolution accelerated the process of resource development, so factories appeared in large numbers, and the working class was born. These changes were accompanied by the urbanization of the population. With the formation and development of large cities, socio-economic activities are increasingly concentrated in a few large cities, and the spatial structure state becomes significantly unbalanced. For another example, the invention and popularization of computer technology have promoted the development of the information industry and brought two possible trends: First, many industries that greatly rely on computing technology and information industry, such as banking, insurance, commerce, vocational education, etc., are concentrated in large and middle-sized cities; second, due to the development of information and the weakening of the distance factor, there is a tendency to decentralize some industries.

Technological progress is one of the fundamental driving forces for social progress. With the progress of technology, people's horizons have been expanded, and the space for human activities has been expanded. Throughout the history of mankind, water transportation was developed on a large scale before land transportation. Due to the development of navigation technology and other socio-economic reasons, the centers of population and economic activities in many countries have moved from inland to coastal zones. This law can be well illustrated by the historical development of ancient China. The political and economic centers were in the Central Plains in the early stage. The maritime traffic had gradually developed since the Tang Dynasty, which directly and indirectly promoted the development of commerce, agriculture and handicrafts in the southeast coastal areas; the national economic center had shifted to the southeastern coastal areas by the Northern Song Dynasty. The development of coastal areas, in turn, promoted the development of marine resources and the further development of maritime traffic, making it possible for countries to achieve economic and technical exchange, which in turn promoted greater development of coastal zones. This role shall be attributed to the so-called "blue civilization".

2) Technological innovations lead to the transformation of industrial structure, which causes differentiated prosperity among regions.

Large-scale technological innovations will promote the upgrading and updating of the regional industrial structure, which is common sense. Due to the differences in the industrial structure of various regions (advanced or not) and the conversion time, some regions lose in the competition, while other regions gain the opportunity for further prosperity. This will cause population, capital, etc. to flow in space, leading to crises in some areas: reduction in the labor force and tax, and closing of factories. As a result, the capability for social and private investment declines, the supply of infrastructure deteriorates, and jobs are reduced. However, the deterioration of the supply of infrastructure and the reduction of jobs cause labor and capital to flow from the "regions in crisis" to other regions, forming a vicious circle. Population and economic activities are increasingly concentrated in relatively prosperous regions, which become gathering points or agglomeration areas in a large region. At the same time, the structure of the regions in crisis becomes relatively weak, and this process leads to the adjustment of the spatial structure.

3) Technological innovations promote the development of urbanization, which leads to the expansion of imbalance among regions and between urban and rural areas.

With the development of science, technology and the economy, social urbanization is an inevitable trend. That is to say, technological development brings about innovations in labor tools, and a large number of agricultural laborers flow from agriculture to other industries, thus achieving urban development. In space, the process of technological innovation and its application and promotion is unbalanced, and the birthplace of innovations and the regions applying and promoting them first often gain greater benefits; the process of urbanization is accelerated, which also leads to further imbalance.

4) The large-scale application of achievements of technological innovations may improve the social welfare standards, enhance environmental awareness, and promote the weakening of imbalance among regions.

The large-scale application of achievements of technological innovations promotes the improvement of the economic level; the level of social accumulation and capital inflow increases, which leads to the expansion of socio-economic strength and investment capacity. At the same time, people's pursuit of a clean environment and the requirements for vacation and tourism are placed in an increasingly important position, while the role of economic factors that affect the spatial layout and spatial structure of socio-economic factors declines. These two points lead to the large-scale development of undeveloped areas and underdeveloped areas.

4. **Progress of transportation technology and spatial structure**

The innovation of transportation technology is an important result of social development and technological progress. At the same time, it is also a powerful promoting factor for socio-economic development, and one of the driving forces for changes in spatial structure. It can be found that there are suitable production modes and transportation technologies in the transportation industry at any stage of socio-economic development. However, different production modes in the transportation industry have different effects on the pattern of spatial utilization and the spatial laws of social-economic activities of human beings.

The improvement of transportation technology affects the process of socio-economic spatial structure, and its internal mechanism mainly includes the following two aspects:

(1) Transportation cost. Many socio-economic objects are under spatial interaction. A certain amount of transportation cost must be paid accordingly in case of any spatial flow of people and goods. Different objects have different requirements for the means of transport and different sensitivity to cost. Due to the progress of transportation technology, the general trend is that the proportion of freight per unit weight of goods in its own value is declining, the corresponding industrial location selection will have greater flexibility, and the relationship of spatial combination will be multi-schematic. However, in the era of relatively backward transportation technology, there were fairly rigid regulations on the locations of

many industries. With the progress of transportation technology, the proportion of travel expenses per unit distance in income also tends to decline, which will change the criteria for selecting people's living places.

(2) Transportation speed. Different transportation technologies and transportation production modes vary greatly in terms of speed. Some socio-economic objects that have time requirements for the transportation process are often close to the most important collaboration goals in terms of space. With the support of backward means of transportation and slow transportation speed, decentralization is bound to be the main tendency of spatial organization, and the scale of agglomeration is also small. As transportation speed increases, the spatial mobility of socio-economic objects is enhanced, and there is a rising trend for the role of transportation factors in most cases. As a result, the degree of agglomeration of socio-economic objects in space is often strengthened, and the scale of urban agglomeration areas is expanded. Such a trend is true for both regions and urban areas. Figure 2 shows the increase in the average speed of urban traffic from the 1930s to the 1960s in four cities, such as Los Angeles and Chicago. The scale of the two cities increased substantially at this stage. In the history of the urban development of Berlin, there is an amazing consistency between the urban traffic speed and the scale of urban land use (Figs. 3 and 4).

Under the influence of the above two mechanisms, the impact of transportation technology on the socio-economic spatial structure can be summarized as the following trends:

(1) In the era when carriages are taken as the main means of transportation, a spatial structure of land use (Thunen circle-Die Thiinenischen Kreisen) centered on towns and a system of evenly distributed residential areas were formed. This era experienced a long historical period, during which the level of social productivity was low, with large-scale agriculture as the mainstay, small towns were widely distributed, but there were few large cities. The accessibility between regions was poor due to the limitations of transport volume, speed and distance of carriages. The main function of carriage transportation was to facilitate the material and personnel connections between urban and rural areas, and the city was mainly manifested as a consumer market for agricultural products. Different types of agricultural products were produced due to different modes of land utilization, and the transportability of different types of agricultural products was different. As a result, a regular circle of land utilization around the city was formed. According to Thunen's research, carriages were the main means of transportation in the Middle Ages in Europe, and the average transportation distance was only tens of kilometers, That is to say, the radius of the outermost circle of the Thunen circle was only a few tens of kilometers [3]. A city would not be formed if it was dominated by traffic alone. However, other factors, such as political and historical factors and geographic location, still had a strong impact. Large cities have undergone rapid development due to sea navigation and canal digging since the seventeenth century.

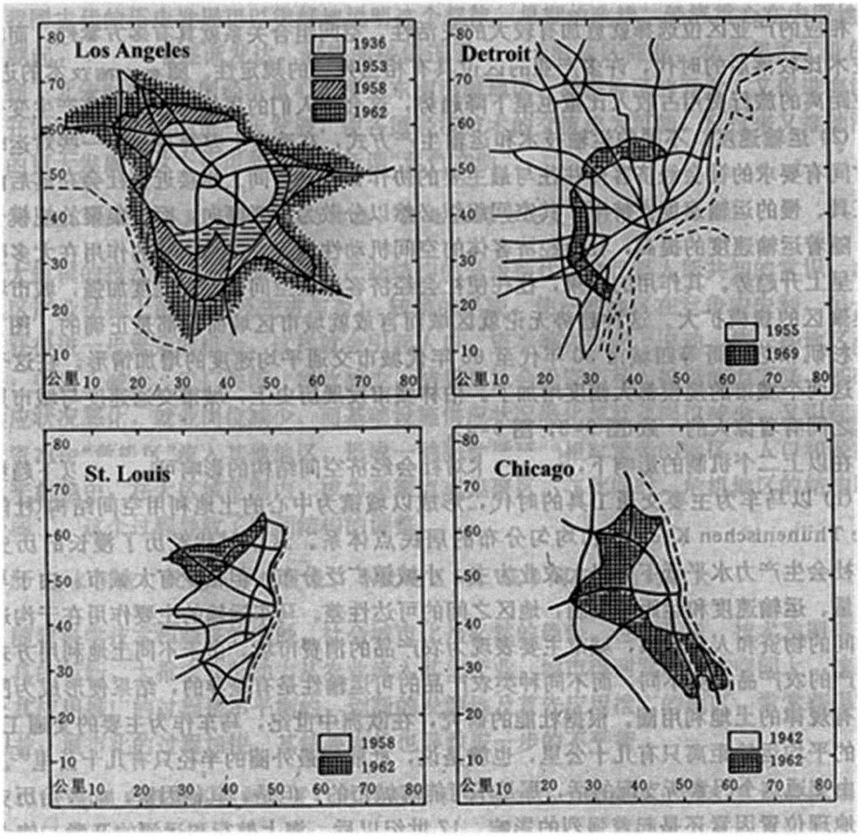

Fig. 2 Expansion of urban area caused by increased traffic speeds in the four cities, such as Los Angeles and Chicago

(2) The development of inland water transportation, especially the large-scale ocean transportation technology, accelerated the development of coastal zones and coastal areas, A large number of complexes with ports, industries and cities as the main body appeared, and the economic center of the country and region shifted to the coastal areas. In the development history of the transportation industry, water transportation lasted the longest. At the early stage of industrialization, Western countries set off a climax of canal digging, causing cities and factories to rise along the river; large water transportation hubs and large cities were formed at the places where rivers or rivers and seas met. In the historical process of the evolution of the spatial structure of regional development, the factors that caused huge adjustments and imbalance shall be attributed to the development of large-scale water transportation. Almost all industries that emerged in modern times along the southeast coast of China are concentrated in port cities, such as Shanghai, Fuzhou, Xiamen, Guangzhou, Ningbo, etc.

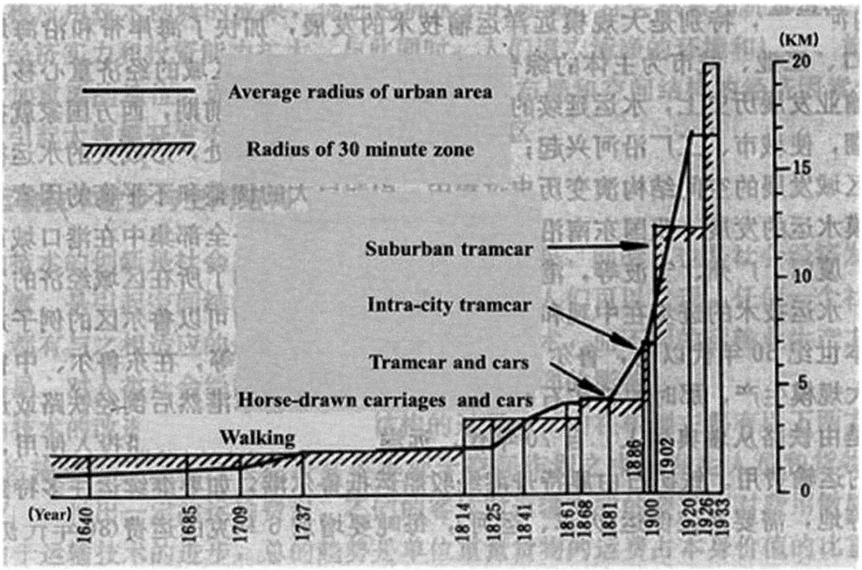

Fig. 3 Radius of the historical urban area of Berlin and the 30-min distance from the urban center

The economic development of port cities has driven the economic development and population increase of the regions where they are located. The progress of water transport technology affected spatial structure at the meso and micro scales, which can be sufficiently illustrated by the example of the Ruhr region. Before the 1960s, there was large-scale production in the East Ruhr, the Middle Ruhr and the West Ruhr for the iron and steel industry, chemical industry, etc. in the Ruhr region. At that time, raw materials such as iron ore were transported by water to the Ruhr port and then transported to the east by railway or canal, or transported in by railway from Sweden. Large ocean-going ships and large inland barges have been put into use since the 1970s, which has greatly reduced the transportation cost of ore. Iron ore was transported from the port of Rotterdam to the port of Ruhr by barge, and hauling was required (train and canal) to continue shipping to Dortmund, Bochum, Essen, etc. As a result, the freight was added by six marks per ton (the early 1980s). The layout of the steel industry in the Ruhr region was greatly changed due to this factor, and almost all iron production was gradually concentrated in the West Ruhr along the Rhine River. Due to the major adjustment of the iron and steel industry, the population, infrastructure, etc. were concentrated along the Rhine River, which became a powerful development axis for the Ruhr region and even the former Federal Republic of Germany.

(3) The large-scale development of railway transportation strengthens the integrity of the regional social economy, and the areas through which many trunk railways pass become powerful development axes. The great development of railways

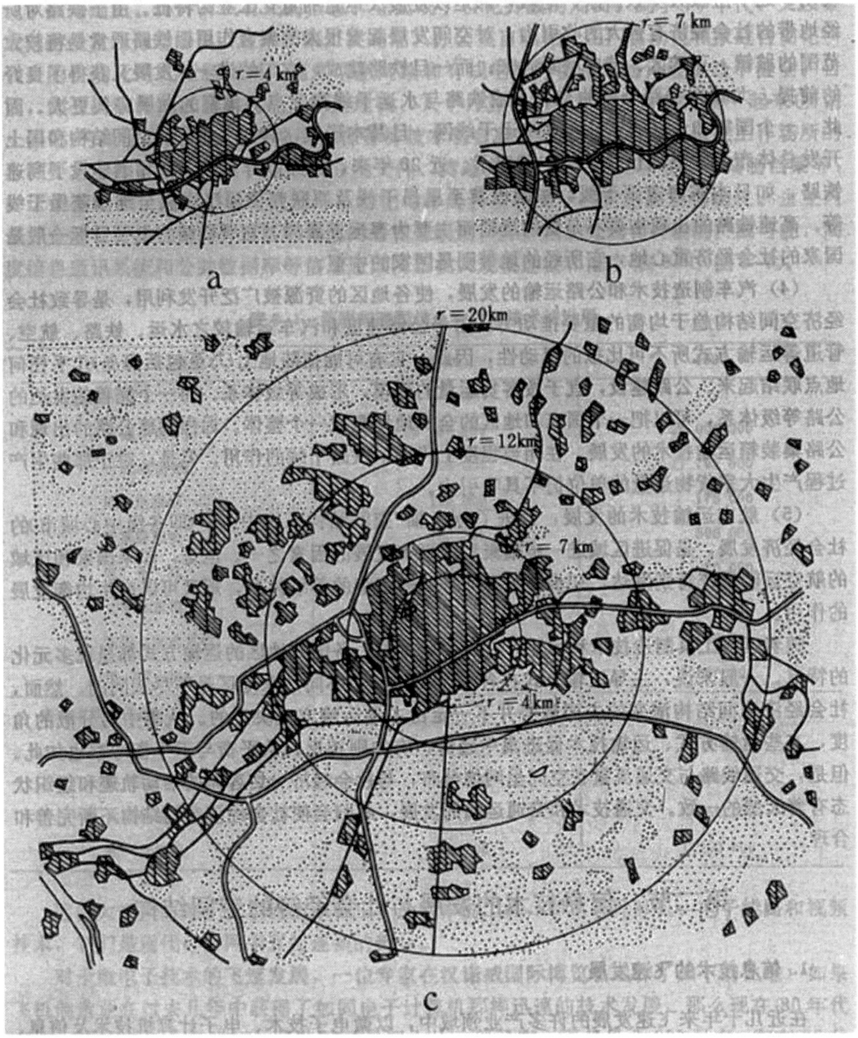

Fig. 4 Expansion of the historical city plan of Berlin

in all countries in the world mainly occurs in the early stage and the climax of industrialization. At this stage, most countries and regions are characterized by the development of heavy chemical industries with large traffic volumes. The railway has a great aggregation effect on the spatial development due to its strong attraction to the social economy in the zone where the railway passes through. Railways usually connect towns, ports as well as industrial and mining areas in a wider range; once the railways are built, there will be favorable preconditions for their further development. The development momentum of agglomeration is even greater, especially at the intersection between trunk railways or the intersection between the trunk railway and the trunk line of water transportation. Therefore, once the railway network and water transportation trunk network of a country or region are basically formed, the spatial structure of the social economy and the framework of the overall layout of territorial exploitation are basically determined. In the past 20 years, some developed countries in the world have successively built high-speed railways, such as the Tokaido Shinkansen in Japan, the Paris-Lyon line in France, the Hannover-Wurzburg line in the former Federal Republic of Germany, etc. The high-speed railway is primarily developed for passenger transportation, rather than freight transportation. The main targets that it connects are generally the social and economic centers of the country, and the zones it passes through are the hubs of the country.

(4) The development of automobile manufacturing technology and highway transportation leads to the extensive development and utilization of resources in various regions, which is an important driving force for the equilibrium of the socio-economic spatial structure. Due to the highway speed and the incomparable mobility of automobile transportation compared with water transportation, railway, aviation, pipeline and other means of transport, it may connect almost any place on land (and even islands). It is advisable to form a hierarchical system for highway construction according to the scale of passenger and freight volumes. And a perfect and developed hierarchical system for highways can link all regions of a country and region into a whole. With the emergence of long-distance expressways and the development of highway container transportation technology, the role of trunk railways has been replaced to a great extent. However, it remains unattractive for units that generate the bulk freight volume in the production process.

(5) The development of air transportation technology has promoted the socio-economic development of "points" (that is, central cities at all levels) in the "point-axis" spatial structure system, and is one of the factors that promote further agglomeration and unbalanced development of regions. However, if the air transportation system of the country and region is particularly developed, and a hierarchical system of airlines and airports has been preliminarily formed, it can also play a role in balanced development.

With the continuous improvement of transport means manufacturing technology and transportation facilities, the means of transport in various countries and regions have shown diversified characteristics. Generally speaking, the dominant means of transport are different at all historical stages and in different countries and regions. However, the main trends in the evolution of socio-economic spatial structure are not necessarily determined by the dominant means of transport. From the perspective of concentration and decentralization, some means of transport and transportation technologies promote the trend of concentration, while others are the opposite; the same is true from the perspective of balance and imbalance. However, traffic routes and traffic flow are distributed in a network-like manner in space, which is naturally consistent with the activity trajectories and organization states of all socio-economic objects. The development of traffic technology and transportation will always constantly improve and rationalize the socio-economic spatial structure.

2 Development of Information Technology and Spatial Structure of the Social Economy

1. **Rapid development of information technology**

Among the many industrial fields that have developed rapidly in recent decades, the development of microelectronics technology, electronic computer technology and information and communication technology ranks first. These technologies are used not only for massive computing in the scientific and industrial sectors but also for socio-economic management as well as the transmission of data and other information. Today, we are at a great historical turning point in the development of a highly information-based society, and the cutting-edge information technology revolution centered on electronic computers is the driving force that promotes this turning point. Humans could declare that: The information era has begun. The world's first computer was developed in 1946, and the climax of the information technology revolution was ushered in after the 1960s. This period was characterized by the formation and development of an information network that combined electronic computers and communication technologies. Information systems in many sectors such as banking, transportation, medical care and public administration began to be practically applied, and data communication was realized. The computer-related information industries were in full bloom. In the 1970s, the second climax was ushered in with the invention of the integrated circuit as the detonating agent. The era of automation represented by office automation, factory automation and home automation has arrived, with information and communication facilities such as high-level information networks and cable television tending to be sound. At present, the high-performance home composite terminals that integrate multiple functions such as word processors, personal computers, and faxes (so-called integration) have appeared in some countries and regions, which combine various technologies into a universal office system, so as to adapt to various requirements of the workplace: making phone calls, writing letters, making copies, distributing documents, transmitting documents, collecting and processing data, compiling archives, etc. The enterprise automation that unifies factory automation and office automation is further realized on this basis. The telecommunication networks within the enterprise, between enterprises and public institutions and among regions are interconnected, which opens up broad possibilities for information processing and transmission. The information communication networks such as high-level information communication systems and public databases are being constructed and promoted on a large scale (Table 1 and Fig. 5).

Table 1 Development of information and communication technologies in the former Federal Republic of Germany

Category	1980	1990
Communication technology		
Direct phone (piece)	20,850,000	28,000,000
Extension phone (piece)	7,700,000	11,700,000
Subscriber telegraph	140,000	175,000
Data terminal	100,000	500,000
Teleprinter	9000	70,000
Text fax machine	–	130,000
Graphic text fax machine	6000	3,000,000
Video machine	40,000	9,000,000
Office and data technology		
Computer	95,000	8,00,000
Terminal	215,000	900,000
Literature processing system	60,000	650,000
Private computer	45,000	1,250,000

There are three most fundamental tools in this large-scale electronic technological innovation, that is, electronic computer, electronic circuit and video technology, which are the interconnected parts of the modern telecommunication network.

For the rapid development of microelectronics technology, an expert made the following analogy at the Hannover Messe: If the aircraft manufacturing industry has gained technological development as rapid as electronic computers during the past few years, the price of an A310 small aircraft would not be 100 million marks at present (in the late 1980s), but only 2,500 marks; the aircraft can circle the earth within 20 min, consuming only 25 kg of fuel. Others thought that this was "a breathless development speed".

The rapid development of information and communication technologies will generate a huge impact on the social economy, leading to corresponding changes in commerce, manufacturing, schools, families, socio-economic organizations, spatial distribution, etc. In terms of employment and industrial structure, information will form an industrial field with an increasingly large capacity, and there will be more and more jobs. Figure 6 shows the sharp rise in the number of employees in the information industry and the changing trend of other industries in the former Federal Republic of Germany from 1950 to 1980. This figure is widely representative in developed countries.

The development of post and telecommunications in China lagged behind for a long time. With the implementation of the reform and opening-up policy, a large number of foreign merchants and foreign tourists came to China, foreign trade, as well as economic and technical exchange, developed rapidly in China, and the situation of incompatibility of the post, telecommunications and communications became more

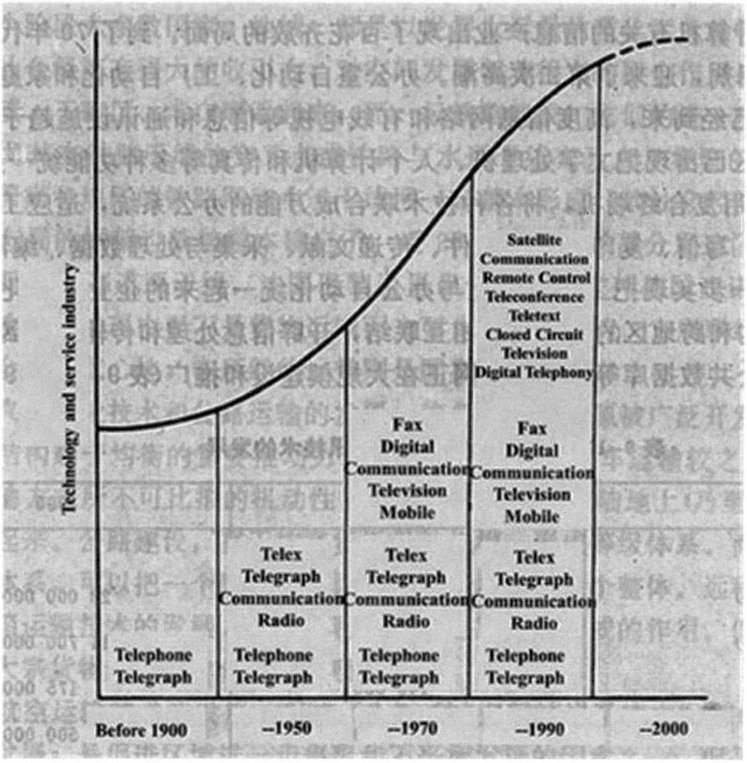

Fig. 5 Rapid development of communication technology

prominent. Since the "Sixth Five-year Plan", the state and local governments have made great efforts to improve the post, telecommunications and communication systems. The annual average growth rate of the total post and telecommunications business volume in China was 7.8% from 1952 to 1988. Among them, the average annual growth rate was 16.5% since 1980. In terms of equipment, carrier telephones, teletypewriters and fax machines have increased at a rate of 8–10% in recent years. Automated office systems are also being developed and promoted. It can be predicted that the information and communication technologies will also undergo high-speed development in China; the scale and number of employees in the information industry, as an emerging industry, will also develop in a "breathless" way in China.

2. **Impact of information technology development on the socio-economic spatial structure**

1) Information becomes the new production factor and location factor.

The rapid development of information and communication technology affects all aspects of social production and social life. The development practice of many countries and regions has indicated and will continue to indicate that: Information

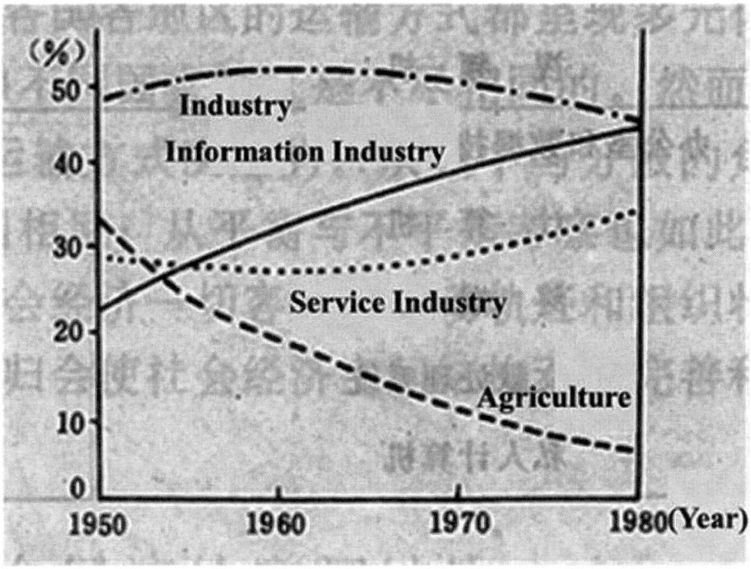

Fig. 6 Growth of number of employees in the information industry of the former Federal Republic of Germany

has become an important factor that affects industrial production development and economic growth, and its significance is constantly increasing. For decision-makers of enterprises or regions, access to accurate and comprehensive market information and technical information is an important precondition for making correct decisions. It is extremely important to improve the information channel in the market economy or the planned commodity economy. Many experts have pointed out that information technology is becoming a pillar of economic development. The speed of economic development, as well as international and regional economic and trade competitions, are manifested as the development speed of information technology and communication technology to a great extent.

At the same time, information is becoming a location factor that affects industrial distribution and socio-economic spatial structure. The role of this factor is increasingly obvious in countries and regions where the economy is more developed and information and communication technologies are more widely used. In less developed countries and regions, the scale of promotion and application of information technology and communication technology is small, and the degree of application in socio-economic practice is poor. Therefore, the significance to the evolution of socio-economic spatial structure is not obvious. Information, as a location factor, can make the spatial location and combination relation of socio-economic objects shift in various trends, forming complicated relationships with other factors. Generally speaking, its acting direction is basically the same as that of environmental factors and labor force factor, but different from that of resources, traffic and transportation, as well as agglomeration economies. Therefore, the specific function of the information factor shall be evaluated in combination with other factors in practice.

2) Concentration of industries and institutions that heavily rely on information in space.

As a location factor, information, like other location factors, has different effects on various industrial sectors and social and economic objects. The industrial sectors and institutions that heavily rely on information, such as communications, finance, commerce, tourism services, some manufacturing industries, corporate head departments, scientific research and design departments, etc., tend to be more and more concentrated in space. On the one hand, this kind of concentration seeks a location in the center of large and middle-sized cities. However, it is more about building characteristic location complexes in the suburbs of large cities (mostly financial and traffic centers), such as, new "management city" (also known as "company city"), "commercial city", etc., which is the concentration on a micro-scale. Another is the concentration on the macro- and mesoscale, such as the concentration of high-tech industries, including information and communication technology industries, in the Sun Belt of the United States and the piedmont of the Alps in Germany. A large number of production enterprises and management institutions are relatively concentrated in these areas, but are not close to each other. The southern region of Germany, with Munich and Frankfurt as the center, is located in the center of Europe and is an international communication and air transportation hub. It is convenient to connect the markets of major countries in Europe. Moreover, there are developed service industries (especially, Frankfurt is an international financial center) as well as pleasant scenery and sunny weather in the region.

3) Production, management and living are decentralized in space.

This is another aspect of the impact of the development of information and communication technologies on the spatial structure. A lot of work can be done in a decentralized way in a highly developed information society. Especially various office work of companies, such as document drafting, production, photocopying and distribution and meetings, many links in the product design and sales process, social counseling, finance, vocational education, etc., which do not need to be concentrated in the city in a traditional way. At that time, the production and living areas will be connected through wireless and wired cables (or optical fibers) and ordinary telephone wires (Fig. 7), thus forming an integrated communication network. There is no need to be present on-site to manage and dispatch in the production plants and workshops, supervising product standards and quality, and understand the implementation of plans as well as the workers' production conditions and equipment running status in the workshop, and there is even no need to listen to the report in the office.

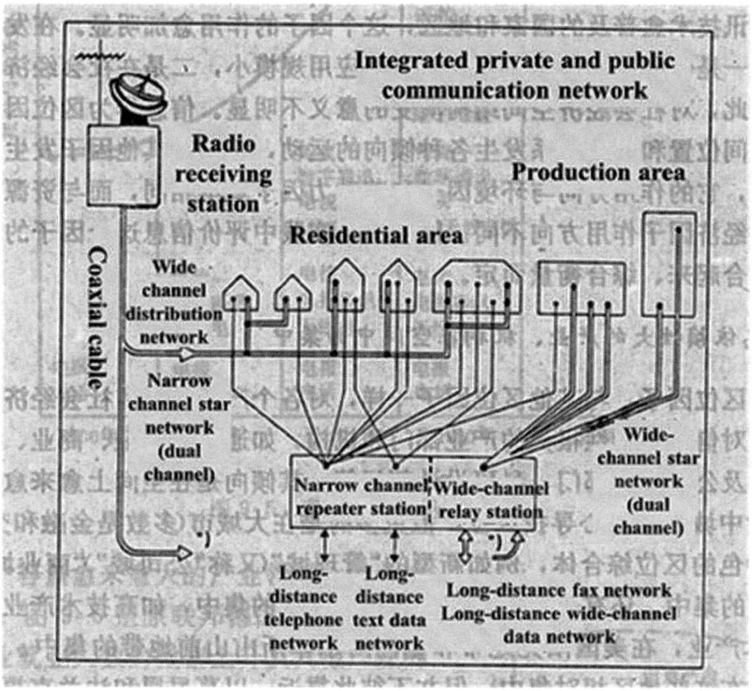

Fig. 7 Schematic diagram of integrated communication network

However, the production, management and living places cannot be too decentralized in space, because direct contact with personnel and a direct understanding of the site are still required in most cases. Therefore, the above facilities are not necessarily concentrated in the city, but more of them are generally located in the suburbs of the city, which is one of the main reasons for the trend of sub-urbanization that has emerged in recent years.

4) Satellite communications and video conferences further reduce the spatial distances in the world.

The transport capacity of the latest satellites has increased by more than 100 times since the first batch of satellites entered space in the early 1960s. People can communicate with each other via satellites from any location on the earth, and the time for information transfer is only 1/4 s. A large insurance company in the United States often takes 12 min to send the computer-processed material via satellite, while it takes 31 h to transmit the material through the ground line.

The satellites are mainly used to hold the video conference recently. Workers from companies, government agencies and educational institutions can hold meetings through satellite lines all over the world and in front of TV monitors. The relevant personnel not only do not need to travel long distances to attend meetings but also can talk with and meet each other. Shareholder meetings, design reviews and business negotiations can be held and even product samples can be seen via satellite lines without even leaving the office building.

It takes no time to construct a bridge to overcome the distance through video conference. It is a special form of meeting, through which people at different locations can communicate in terms of image, sound and data. The video conference is held through the image screens. However, it is the same as the traditional meeting form in principle, and people can see, listen, talk in an unhindered way, show documents, examine each other's goods, explain and appreciate works of art, exchange documents, and perceive the reactions of the meeting partners. The video conference is generally held in a special conference room. In recent years, many European countries, North America and Japan have established video conference communication networks, many large cities have established public video conference rooms, and some private companies and management departments are also establishing corresponding conference rooms.

The video conference has the following advantages: saving time greatly by avoiding unproductive travel time, expanding human capabilities, shortening the decision-making process, and improving the efficiency of conferences, thus making possible the meetings and conversations that cannot be held due to bad weather and high travel costs.

The economic advantages of video conferences are also quite obvious. According to the operating practice and analysis in Europe, assuming that the travel cost per person per time is 1000 dollars and the monthly maintenance cost of the video conference room with an investment of 500,000 dollars is 20,000 dollars, if there are four persons in each official contact, personnel travel is suitable in the circumstances where there is one meeting or conference each month, and the video conference will be economical in other circumstances (Table 2).

3 Development of High-Tech Industry and the Spatial Structure Formed

1. Development of high-tech industry

With the development of new technologies such as microelectronics, electronic computers, optical communications, new materials, bioengineering and aerospace since the 1970s, the equipment investment of high-tech industries in developed countries has continued to increase, and the production scale has continued to expand. After the mid-1980s, direct equipment investment in high-tech sectors such as semiconductors and mechatronics in Japan accounted for about 10% of the total equipment investment. If the indirect equipment investment in sectors that provide electronic equipment and general machinery for high-tech enterprises is included, the proportion can be as high as 20%. According to the calculation of the Agency of Industrial

Table 2 Comparison of costs of video conference and personnel travel

Number of meetings per month	Travel fee for four persons (dollar)	Monthly TV conference fee					
		1 h	2 h	3 h	4 h		
1	4000	20,800	21,600	22,400	23,200		Travel is cost-effective
2	8000	21,600	23,200	24,800	26,400	Investment of 125,000 dollars	Video conference is more economical
7	28,000	25,600	31,200	36,800	42,400		
9	36,000	27,200	34,400	41,600	48,800		
13	52,000	30,400	40,800	51,200	61,600		
26	104,000	60,800	81,600	102,400	103,200	Investment of 500,000 dollars	

Source Future communications, printed by Philips, 1987

Science and Technology in Japan, the proportion of output value of high-tech industries in the gross national product will increase to 15–20% in the mid-1990s from 2% in the early 1980s. At present, the output value and the number of employees in the electronic industry in some developed countries such as Japan have greatly surpassed those in the automobile industry and the iron and steel industry.[2]

There is no special division for specific products and sectors in high-tech industries. The industries listed in the first section of this chapter are the processing and assembly type, which has a primary position among high-tech industries. In addition, there are bioengineering technology, fine chemistry, communication equipment, aerospace engineering, etc. Although high-tech industries are varied with complicated products, there are some common characteristics in their production process and products. In addition to the characteristic of representing the latest scientific and technological achievements, there are also the following characteristics:

(1) The products are generally light, thin and small. In the field of microelectronics technology, some integrated circuits and integrated blocks are only a few square micrometers in the area and only a few milligrams in weight or even lighter.
(2) The products are varied with small batches and small scales. Except for semiconductor devices and integrated circuits, the production scale of most high-tech industries is small. For example, the industrial production scale of high-grade dye intermediates, pharmaceutical intermediates, some additives, catalysts, etc. in fine chemical products may be only a few tons or even a few kilograms per year. However, many users can be covered.
(3) The added value is high. Among the mechanical products, the prices of high-tech products and general electrical products are very different (Table 3). Of

[2] Li Longyun, A Preliminary Study of Regions with Advanced Technologies in Japan, 1986.

course, the general electrical products listed in the table also include parts of high-tech products.

(4) The consumption of resources and energy is small, and the production, transportation, and sales processes are time-sensitive. Due to the above characteristics, high-tech industries generally do not consume a large amount of raw materials in the production process. The utilization rate of resources and energy is high, and the product is light in weight and small in size. Therefore, the large traffic volume is not generated, and it is not necessary to cooperate with the railway, water transportation and other means of transport that undertake large traffic volume. However, the products of most high-tech industries are frequently updated and have a short life cycle. It is required to complete the preparation of raw materials, production and transportation of semi-finished products and finished products, storage and sales process as quickly as possible. All links are required to be closely connected with each other. Therefore, fast transport means are required for coordination.

The rapid development of high-tech industries has a great impact on the industrial structure and industrial distribution. Most importantly, it promotes the advancement of industrial structures. The proportion of processing and assembly industries is constantly increasing, while the proportion of basic raw material industries such as iron and steel, chemical industry and oil refining gradually declines. As the industrial structure changes, new changes are also taking place in the industrial distribution and even the city layout as well as the layout of infrastructure construction. In some

Table 3 Price per gram of mechanical products (JPY)

			1985.8
Light, thin, short and small products	Price	General electrical products	Price
35 mm fully automatic camera	140	Refrigerator	3.5"
Fast video recorder	120	Color TV (11" to 21")	8.0
8 mm compact camera	140	Complete set of recording accompaniment	5.5
Small secondary transistor phonograph	85	Electric vacuum cleaner	7.0
Stereo cassette recorder	130	Electric iron	9.0
Stereo FM/AM record player	180	Electric fan	5.0
Business-card like ultra-thin radio	650	Stereo phonograph (all-in-one)	9.5
LCD small color TV	190	Electrical appliance	6.0
Mini cassette tape recorder	240		
Portable electronic typewriter	50		

Source Iijima Zhenichi (Japan): "Selection of Airport-type Industrial Location and Regional Revitalization", "Geography" (Japan), October 1985, translated by Chen Huizeng countries and regions, the trend of building large-scale industries (energy and raw material industries) and port complexes near the sea or rivers has stopped. However, high-tech industries and corresponding constructions are undergoing rapid development in the areas around large cities with developed technology and economy and the areas with good scenery and pleasant climate. A new imbalance with different connotations is emerging.

2. **Location trend of high-tech industries**

As high-tech industries have many requirements that are different from general traditional industries in the process of production, storage, transportation and sales, they also have their own characteristics in terms of spatial distribution. That is to say, the location trend of high-tech industries is also slightly different from that of general traditional industries. First, high-tech industries shall be deployed in areas with a high level of the economy, science and technology, so as to obtain the required materials and equipment, a large number of technical workers and high-level engineering technicians, and gain the conditions for coordination. For example, the production of integrated circuit equipment and the material industry are the "two wings" of the integrated circuit industry. The development of integrated circuit equipment requires the cooperation of traditional optics, vacuum and precision mechanical technologies, and the materials of integrated circuits shall be supported by traditional metallurgical and chemical industry sectors. The production of high-tech industries also requires related industries to provide processing and analytical inspection equipment, various ultra-pure materials, ultra-pure gases, ultra-pure water and ultra-pure working conditions [4]. Secondly, it requires a favorable geographical location in a large region and even in the world, with a very convenient communication network to the outside world, and rapid transit that connects with external markets as well as financial and commercial centers. Most production in high-tech industries is carried out in clean air and even in closed containers. Therefore, their production location requires good climate conditions, and shall be far from major air and water pollution sources.

Table 4 can be obtained by comparing the location factors of high-tech industries and general traditional industries one by one (from meso and micro perspectives).

It is extremely important to consider the location of high-tech industries in a country or region from a macro perspective, especially in China, which has a vast territory, a low level of the economy, science and technology, and unbalanced regional development. To develop high-tech industries, it is required to perform uniform planning and select the regions with superior conditions for centralized development according to their own characteristics and requirements. In China, the areas with a developed economy, high level of science and technology as well as convenient transportation and communication are mainly located in the lower reaches of the Yangtze River (especially the Yangtze River Delta region), the Beijing-Tianjin region, the Pearl River Delta region, Liaoning in the northeast, Wuhan in the middle reaches of the Yangtze River, Chengdu in Sichuan, Hefei (a new technology center), etc. There are rich intellectual resources that reflect the scientific and technological level as well as the degree of economic development in these areas and cities, which are key areas for developing high-tech industries. According to the census data in 1983,

Table 4 Comparison of effects of location factors of high-tech industries and general industries[a]

Location factors	General raw material industries	General processing and manufacturing industries	Commerce and financial industries	High-tech industries
Fuel and power	+++	++	–	–
Water	+++	+	–	+
Land and topography	++	+	+	+
Geographical location	+	++	++	++
Economic foundation	+	++	+++	++
Scientific and technological level	–	+	–	++
Traffic	+++	++	++	++
Communication	+	+	+++	+++
Cooperation and agglomeration economies	++	+++	++	+
Labor force[b]	–	+	–	++
Environmental conditions	–	+	-	++

[a]The number of "+" signs indicates the degree of impact of each factor on the layout of each industry, and the "–" sign indicates that there is basically no impact
[b]The number of the labor force in China can be assumed to be unlimited. Here, it refers to the impact of the labor quality factor

national intellectuals (including college students in schools) accounted for 0.6% of the total population. Among them, Beijing (4.9%), Shanghai (3.4%), Tianjin (2.4%) and Liaoning (1.1%) all greatly exceeded the national average level.

3. **Location selection of high-tech industrial zones**

High-tech industries have production characteristics and location trends that are different from general traditional industries. When developing high-tech industries, all countries will appropriately concentrate relevant enterprises and supporting facilities in space and arrange them in groups, so as to form high-tech industrial zones. According to the industry, science and technology management system and specific regional conditions of each country, there are the following types of high-tech industrial zones.

1) Science and technology industrial park

The science and technology industrial park is a novel science-industrial production complex that emerges with the development of science and technology in the contemporary world, such as the "Silicon Valley" in the United States, the "Silicon Island"

and Tsukuba Science City in Japan, the "Electronic Industry Center" in Britain, the "Science and Technology Park" in Singapore, etc. Despite different names, they have much in common, that is, focus on developing the latest industrial technologies and emerging products in the contemporary world in regions with suitable conditions (the area generally ranges from a few square kilometers to several hundred square kilometers); within this scope, a batch of senior talents engaged in scientific, technological and economic research are concentrated, together with relevant scientific research institutions, professional companies and manufacturers, colleges and universities at home and abroad, to carry out the research on basic science, technical science and applied science as well as the trial manufacture of new products, so as to convert scientific research achievements into productivity quickly, and form a system combining production, scientific research and teaching. Many countries build such complexes in areas with pleasant climates. The internal environment is beautiful, and the green land is vast, similar to a park. Therefore, they are called "Science and Technology Industrial Parks".

The science and technology industrial parks integrate the location trends of high-tech industries. By summarizing the actual practices of the United States, Japan, France and Taiwan Province in China, the following requirements are mainly considered for the selection of parks.[3]

First, most of the science and technology industrial parks are located in the suburbs of large cities with developed industries or satellite cities near large cities. These areas are rich in intellectual resources. Especially, the research on natural science and technical science is strong with a high level. For this reason, it is generally necessary to be close to famous universities and institutions of higher learning. There are a large number of skilled experts, engineers and postgraduates in the industrial parks of various countries, which have experience in first-class scientific research and enterprise production management.

Second, more attention is paid to physical and economic geographic locations and their advantages when selecting the geographic space. The layout section is not far from or too close to large cities but is the urbanized area within big cities. Land and water transportation, especially air transportation, is developed, which makes it convenient to connect with companies at home and abroad, and the ever-changing technology and market information can be mastered in time.

Third, when planning a scientific industrial park, the construction land used for building production and auxiliary workshops shall be controlled as much as possible, and there must be green belts, large grasslands and even gardens. The production space (site) with purification, air conditioning and dedusting is required for the workshops of various precision instruments and electronic computers, and the environment requirements are particularly strict. Meanwhile, scenic areas, green belts as

[3] Chen Hanxin, Construction and Layout of Science and Technology Industrial Park, Productivity Distribution and Territorial Planning, Volume 4, 1986.

well as beautiful gardens and grasslands in industrial areas are particularly important for attracting scientists, entrepreneurs and engineers to work here, and can play an important role in improving labor productivity and creative labor in terms of R&D and production. For example, the silicon island of Japan is located between 30° and 40° north latitude, where the natural conditions are excellent: sunny and warm, with little snow in winter, lush trees and fresh air. There is Kumamoto, known as the "Forest City", and Miyazaki and Kagoshima, known as the "Sunshine City", which have created favorable conditions for the development of cutting-edge industries with their beautiful natural environment. Silicon Valley in San Francisco, the United States also has a quiet and beautiful natural geographical environment.

Fourth, the science and technology industrial park is a complex of science, design and production testing of industrial products. The combined features and inherent connection are considered in the spatial distribution of the region. It is very important to combine similar industries and other supporting industries in the region to form a production system, which is particularly important for areas with rapid economic development. The planning of industrial parks shall be predictable and coordinated with the overall planning and layout of the city. The inside and outside of the park shall not be connected with some harmful industries and polluting industries, and special attention shall be paid to the relationship between wind direction, water flow and water quality.

Fifth, the degree of impact of the scientific research center on the adjacent areas must also be considered. When determining the scale, nature and product direction of the development of the science and technology industrial park in this area, it is necessary to investigate, analyze and compare the economic structure and market conditions of the nearby areas, and then confirm the development steps, implementation policies and regional layout.

In addition, the science and technology industrial parks also require sufficient power supply, stable voltage and cycle, abundant water source, good water quality, as well as unobstructed water supply and drainage system.

The above requirements for site selection can be indicated by the actual location conditions of several famous science and technology industrial parks at home and aboard.[4]

(1) "Silicon Valley" in San Francisco. "Silicon Valley" is located in South Santa Clara County, San Francisco, California in the western United States. It is called "Silicon Valley" because of the mountains in the east and west and the particularly developed semiconductor industry. It covers an area of about 1350 square miles. The climate here is pleasant, like spring all the year round (the average maximum temperature is 17 °C in January, 28 °C in July and 23 °C in October). There is basically no rain from April to October, with snow in winter and no

[4] Chen Hanxin, Construction and Layout of Science and Technology Industrial Park, Productivity Distribution and Territorial Planning, Volume 4, 1986.

sandstorm all year round. Due to the superior climate conditions, strict pollution control and a lot of work done in traffic, transportation and greening, the whole area is like a big park, which creates a good environment for developing cutting-edge technologies. The development of "Silicon Valley" started at Stanford University in Palo Alto. In 1930, Stanford University developed linear accelerators, nuclear magnetic resonance and microwave technology, and used computers for the first time. After World War II, Stanford University had land and talent, but lacked funds. Professor Thalman encouraged entrepreneurs to invest in the land to create an industry-based park. Therefore, companies such as General Electric and Ford were established. Some government departments also established research units here or nearby, such as NASA's research center, Moffett Naval Air Station, Lawrence Laboratory, etc. Some large companies required advanced technology and were willing to cooperate with schools and companies in the "Silicon Valley". For example, Lockheed Missile and Space Company moved to the "Silicon Valley" in 1965. Shockley, the Nobelist in 1956, came to Palo Alto from Bell Labs to set up the Shockley Transistor Company. The Fairchild Semiconductor Company, which was established later, trained many talented and knowledgeable engineers and technicians. More than 40 semiconductor companies in the "Silicon Valley" were founded by the technicians from Fairchild. The development of the semiconductor industry laid a foundation for microelectronics. With the research, development and production of computers, the advanced technologies centered on microcomputers rapidly developed in the "Silicon Valley" area. In a word, the formation and development of the "Silicon Valley" underwent the transition from general electronic technologies to technologies centered on semiconductors, integrated circuits and microcomputers. At present, "Silicon Valley" has a population of 1.4 million, including 760,000 labor force. In 1982, there were 1774 advanced technology enterprises. Among them, there are 897 manufacturing plants, 391 enterprises that research and develop computer services and 486 distributors. At present, the output of semiconductors in the "Silicon Valley" has accounted for one-third of that in the United States.

(2) Both sides of "Route 128" in Boston. In the metropolitan area of Boston in the United States, private companies engaged in research and production have sprung up along the Lincoln Experimental Center (it belongs to the comprehensive research base for senior scientists and engineers engaged in air defense and rocket systems from the Massachusetts Institute of Technology) and on both sides of "Route 128" in the west of Boston, such as American Radio Company, AJack Company, Bona Royd Company and Wang An Computer Company, etc., which are new enterprises that combine science and production and are engaged in scientific research and product manufacturing. The nearby universities have also become world-renowned universities and research bases, forming a corridor-style science and technology industrial park.

(3) Ile de France Science City in France. It is located in the south of Paris, 25 km away from the downtown, with an area of 300 ha. There are expressways and perfect communication and information facilities. This science city has a history of more than 30 years and was formed naturally. In the early 1950s, the French Atomic Energy Commission established a nuclear research center and related enterprises here, and then the University of Paris Sud (Science and Engineering) moved in. The National Scientific Research Center began to establish laboratories here, and Thomson Corporation also established a R&D center in the area, initially forming a concentration of scientific research, teaching and industry. The fields engaged in included energy, electronics, computing technology, biology, etc., and then six schools, including the Higher Optical School, the Comprehensive Engineering School, and the Higher Electrical School, as well as enterprises such as EDF and General Electric moved into this area. This area was called the "Route 128" of France by the mid-1970s. At present, an important base for high-level scientific research, teaching and emerging industries has been formed, and the most advanced technology research and development are being carried out, such as electronics, photonics, new materials, robots, office automation, energy, transportation, biotechnology, medicine, food, etc.
(4) University of Cambridge and Glasgow in Britain. There are more than 300 companies engaged in computer, medical equipment, laser, engineering, etc. in a small area near the University of Cambridge in southern Britain. There are more than 270 various electronic production companies in an area of more than 100 km long and 45 km wide between the industrial cities of Glasgow and Dantie in another part of Britain. Many new electronics manufacturers are springing up in old stone houses in Scotland. The measures of tax reduction and investment incentives are taken in this area to attract many internationally renowned electronics companies to set up factories here. At the same time, there are six universities in this area, with a large number of scientists and researchers, whose rich research findings can be used by various companies in the country.
(5) Kyushu in Japan. The Kyushu region had been established as the semiconductor industry base in Japan since the mid-1960s. The Japanese built a "Silicon Island" similar to Silicon Valley in the United States about 15 years later. The geographical conditions here are very superior, the four seasons are mild and pleasant, and there is no sandstorm all the year round and little snow in winter. The trees are lush, and the air is fresh. The transportation conditions are also very convenient. Except for Saga Prefecture, there are airports in the rest six prefectures in the Kyushu region. Among them, there are four international airports with direct flights to Hong Kong, Singapore, Beijing, Shanghai, Tokyo, etc. The small integrated circuits can be put on the market as soon as possible through air transport.
(6) Hsinchu in Taiwan, China. The "Hsinchu Science and Industrial Park" has been established since 1977. It is located in the southeast of Hsinchu. At that time, the planned total area was about 2000 ha, with a planned total investment of about 100 million US dollars. It was planned to be developed in three phases within 10 years, and the construction was completed by 1988; it was planned to

introduce foreign advanced technology for combination with its own scientific research, production and education, so as to promote the development of the locally advanced industries, cultivate talents, and create conditions for establishing a novel science and technology industrial park. At present, the electronic computer system industries that have been introduced into the Hsinchu Industrial Park include precision electronics, precision instruments, precision machinery, new materials, etc. There are 14 factories under construction in the park with a total investment of 34 million US dollars. Hsinchu Park determines the guiding ideology of planning and the principles for introducing projects. First, pay attention to the advanced industries that it lacks, which can cooperate with the development of technology-intensive industries in Taiwan; second, the experimental instruments to be introduced must be quite advanced and can be applied and operated by its own senior scientific and technological talents; third, pay attention to cultivating its own scientific and technological talents, so that new technologies can be transformed into new industries, new technologies and new products that can enter the international market.

(7) Shenzhen in Guangdong, China. It is the first science and technology industrial park in mainland China, and was jointly constructed by Shenzhen and the Chinese Academy of Sciences since the mid-1980s. The park is located in the central and western parts of Shenzhen Special Economic Zone, adjacent to Shenzhen University and China Experimental University and not far from Nantou and Shekou. The moderate geographical location provides better cooperation and supporting conditions for the development of the industrial park. Surrounded by mountains and facing the sea, it has certain potential land reserves. The east–west trunk road of Shenzhen runs through the middle of the park, which connects the park with the Luohu urban area, Shekou, Nantou industrial area and Guangzhou. There is a freight road in the north of the park. Guangzhou-Shenzhen Expressway planned to be constructed will pass through the place one or two kilometers in the north of the park. After completion, it will provide a fast and convenient external communication channel for the park. A large international airport is built in Yokota, which enables the park to connect with all countries in the world. The park is adjacent to Shenzhen Bay in the south. If a special wharf can be built, it can provide cheap water transportation conditions for building materials such as cement, steel, sand and gravel at the initial stage of park construction, and can be converted into recreational facilities in the future.

2) Industrial technology center

The old industrial base areas of some western developed countries have experienced the so-called "structural crisis". In response to this crisis, various countries have adopted various policies to promote structural transformation and economic development. One of them is the "middle-class policy", that is, to revitalize the economy by building and renovating a large number of small and medium-sized enterprises. Compared with large enterprises, small and medium-sized enterprises can easily improve their production processes and product directions to meet market demands and add more jobs. However, small and medium-sized enterprises generally lack the strength of scientific research and trial manufacture. In response to such problems, the Federal Republic of Germany has established numerous industrial technology centers since the late 1970s.[5] The main function of the industrial technology center is to link scientific research with the economy, so that scientific research achievements can be converted into productivity as quickly as possible. It provides conditions and places for new products, new process technologies and trial manufacture, and acts as a development base for new industrial technologies. The conditions for location selection are basically the same as those of the science and technology industrial parks, that is, they are generally located in the suburbs or edge areas of large and middle-sized cities, or near universities or scientific research units. There shall be a developed and convenient transportation network for connecting large and medium-sized cities, and the communication and infrastructure supply systems such as water, electricity and drainage are complete. The direction of scientific research and trial manufacture of the industrial technology center is generally consistent with the industrial direction of the city where it is located. There are many small and medium-sized enterprises nearby, which need to develop new processes and new products. The industrial technology center has workshops and necessary materials suitable for the development and experiment of new technologies, and the scope of attraction and service is smaller than that of the science and technology industrial park. In the early 1980s, eight industrial technology centers were built in the Ruhr region within several years (almost one for each medium-sized city and above), which played an important role in the structural adjustment and revitalization of the old industrial base of the Ruhr region. In these industrial technology centers, the main industrial directions of research, application and production are: electronic technology, material technology, development and application of software systems, laser technology, measuring instruments, medical devices, etc.

3) Economic and technological development zone

In the past ten years, in order to introduce foreign capital and advanced technology, some developing countries have developed foreign trade, promoted the revitalization of their own economies, and selected a certain section near the foreign trade port and port city that is convenient for economic and technological exchanges with the outside world as an economic and technological development zone. The state or local

[5] There are many names for the industrial technology centers, such as the Technology Development and Technology Consulting Association, Technology and Development Center, Enterprise Innovation and Future Development Promotion Association, etc.

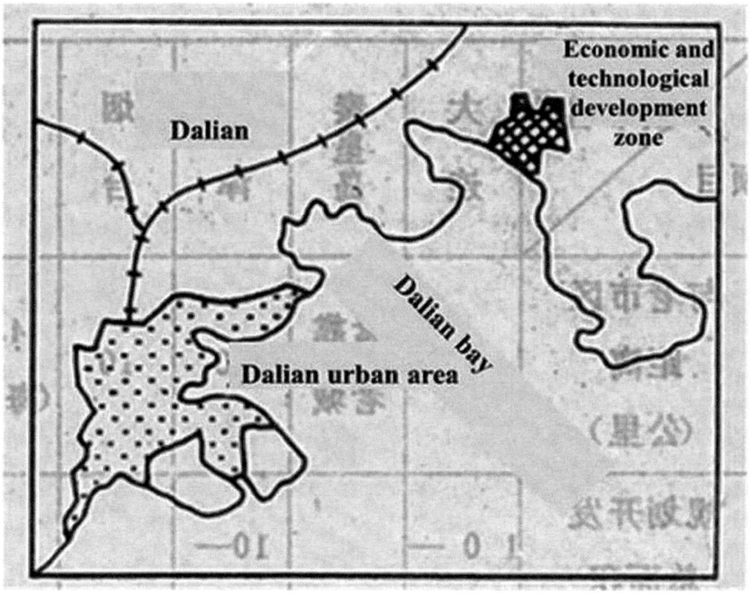

Fig. 8 Overall location of Dalian Economic and Technological Development Zone

governments have carried out unified planning, implemented "seven connections and one leveling", and constructed life service facilities, some standard workshops, etc., so as to attract foreign manufacturers to invest and set up factories. Most of the products are sold to the international market through foreign channels. The location of the economic and technological development zone does not need to be too far away from the original large and medium-sized cities. Otherwise, it will be inconvenient to connect with cities in all aspects. This is different from the special economic zone. Special policies are implemented in the special economic zone in terms of economy, management, wages and employment system, that is, the market system is adopted comprehensively, so that foreign ideas and cultures also enter the special zone. In order to prevent the intersection of different systems and avoid significant impacts on the country, it is generally required to have a clear geographical boundary with the region of the country. The special zone itself shall have complete and independent urban functions, and it is more favorable to be far away from large and middle-sized cities in terms of space. Dalian Economic and Technological Development Zone is located in Maqiaozi, more than 100 km away from downtown Dalian, because it was originally planned as a special zone (Fig. 8). Qinhuangdao Economic and Technological Development Zone is located in an independent section between the two rivers. It is very convenient to connect with cities, and the infrastructure construction is relatively compact and economical (Fig. 9 and Table 5).

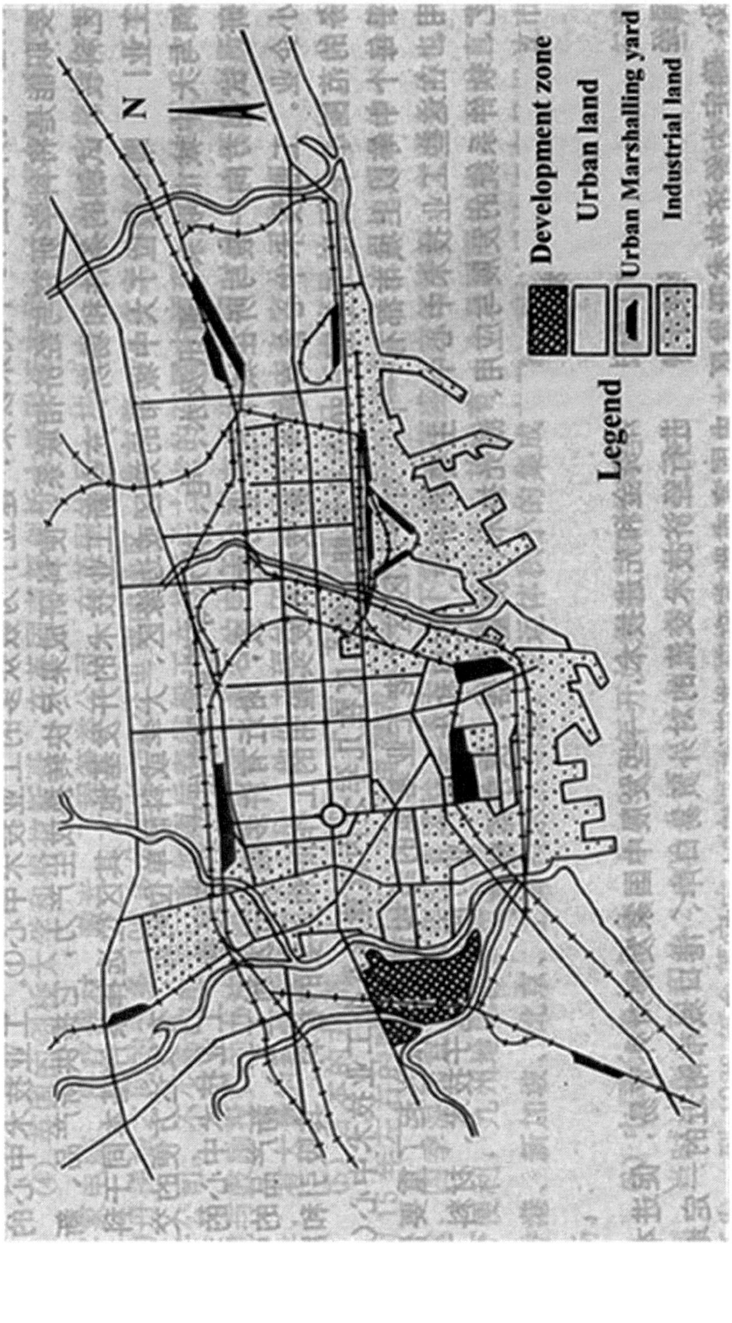

Fig. 9 Overall location of Qinhuangdao economic and technological development zone

Table 5 Scale of some economic and technological development zones in China and the spatial distance from the old city

Name project	Dalian	Qinghuangdao	Tianjin	Yantai	Qingdao		Lianyungang	Nantong
Distance from the old city (km)	150	Close to the old town	10	10	4.2 (Offshore)	110 (Onshore)		12
Total planned development area (km^2)	10–20		10–33			10		
Initial development area (km^2)	3	0.6	3	2	2	1.3		

Name project	Shanghai		Ningbo	Wenzhou	Zhanjiang	Fuzhou	Beihai
Distance from the old city (km)	6.5 (Hongqiao)	30 (Minhang)	23			24	
Total planned development area (km^2)			3.9–10	18	9.2	23	46
Initial development area (km^2)			1	2.2	2.2	1.4–4.4	1–2

Source Zhao Yanqing, Scale, Site Selection and Functions of Economic and Technological Development Zone, 1986

References

1. Y. Zhi, *Introduction to Industrial Economics* (China Renmin University Press, Beijing, 1985)
2. Institute of Economics of the Former Soviet Academy of Sciences, *Political Economy Textbook.* (People's Publishing House, New Delhi, 1955), p. 91
3. L. Dadao, *Location Theory and Regional Analysis Methods* (Science Press, Beijing, 1988), pp.15–16
4. W. Xinzhen, Issues on cutting-edge industrial distribution and technology-intensive urban construction. Econ. Geogr. 1 (1985)

Appendix A

On the Optimal Structure and Optimal Development of Regions

—Review and reanalysis since the proposal of the "point-axis system" and the T-shaped structure

Lu Dadao

Abstract
By elaborating on the relationship between various types of spatial structures and development and how to achieve optimal development through the optimal organization of the region, the objective process of socio-economic spatial organization and the formation of a "point-axis system" are demonstrated from the combination of theory and practice, indicating that the "point-axis system" theory can lead to the optimal development of a region or country. Therefore, the system is the most effective regional development model. Through the analysis of the practical effect of China's regional development for more than ten years, it demonstrates that the strategy of a T-shaped structure has played an important role in China's development.

When people discuss regional and national development, they always think of capital, market, resources, entrepreneurs, information, technology development capabilities, etc. Still, they do not consider the spatial aspect of socio-economic objects. Why do cities and villages vary in size? What is the law of size distribution? Are there any differences in the socio-economic development of urban residential areas of different scales and structures? Do various traffic lines, energy supply lines, and information lines affect regional development? How to organize them in space (region)? For the mundane situations in our life, we tend to ignore their connotations. One may consider how to obtain optimal development when considering regional or national development. However, less consideration is given to the optimal organization of regional or national socio-economic objects. Individuals especially do not consider the connection between optimal development and optimal organization.

D. Lu, *Regional Development and Its Spatial Structure*,
https://doi.org/10.1007/978-981-99-7683-6

It is discussed in this paper how to achieve optimal development through the optimal organization of the region, thus answering whether there is any optimal organization and any optimal spatial structure formed. The premise of the discussion is the development of the region under the condition of a homogeneous surface. That is, there is no difference in physical geographical elements within the region. At the same time, this paper further describes the practical effects in the past 15 years since the theory of the "point-axis system," and the T-shaped structure of China's territorial exploitation and economic distribution was proposed in 1984. The two issues are closely related. Practices have proved that there is obvious theoretical and practical significance to how the effects of the implementation of this spatial structure strategy and regional development direction directly serve the optimization of China's regional planning and regional development decision-making.

1. Development and spatial structure

The economic and social development of a country or region is the result of the interaction of socio-economic factors and natural factors. "Development" is inevitably manifested as the emergence of a socio-economic object. However, if several socio-economic objects emerge, a certain spatial organization will be generated within a certain range. To judge whether a spatial organization is inevitable for regional development and whether it can lead to the optimal development of the region, the criterion is whether the flow of people, material and energy in regional development is the most economical, whether the expenditure for the intermediate process of production and circulation is minimized, whether the connection between urban and rural residential areas is integrated, and whether the region changes from unbalanced development to more balanced development. In terms of long-term economic growth rates, the optimal development of the region is steady growth.

Location theorists and scholars of regional production complexes have long argued and proved that: any economic object must be connected with other objects for existence and operation in a homogeneous and undeveloped region. Due to individuals' need for social interaction and sharing of infrastructure (roads, mechanical power equipment, communication, energy and water supply, urban facilities, markets, etc.), socio-economic objects must be concentrated in a region or point, that is, agglomeration produces benefits, and association produces benefits. This basic principle is the same for the formation of villages in the early primitive society and the emergence of modern large cities [1].

With the development of economy and society, there will be an accumulation of socio-economic objects at more than one "point" or "region" inevitably, which produces a linear infrastructure that connects two or more "points" or "regions." "Points," "regions," and the linear infrastructure connecting "points" and "regions" constitute the most basic spatial pattern, which also forms the simplest spatial structure. In the process of regional socio-economic development, "points" gradually develop into cities [2]. However, the urban residential areas form a hierarchical system [3].

The spatial structure of social economy refers to the positional relationship and the agglomeration degree of socio-economic objects in space and the direction and intensity of interaction through linear infrastructure.

The spatial structure of regions and countries is the result of long-term socio-economic development. It is also the result of implementing related regional development policies according to the characteristics of nature, location, history, economy, and other factors of the regions. Individuals usually concentrate development activities at a few points and regions in the early stage of regional or national development, which can obtain better economic benefits than decentralized development. The reason why the spatial structure affects development is that agglomeration produces benefits. That is, there exist "agglomeration economies." However, excessive development of economic and social activities, or economic and social objects, at one "point" (city, region, etc.) will also lead to "agglomeration." Suppose the concentration scale of industry, transportation, population, power facilities, etc., is too large. In that case, there will inevitably be land and water shortage. The proposed infrastructure projects will be too complicated and costly (such as large-scale three-dimensional traffic projects, long-distance water diversion, environmental governance projects, etc.), the environment and ecological quality will deteriorate, and the construction costs of various projects will increase substantially. If the agglomeration continues, regional economic growth will slow down. Therefore, it is objectively required, and the economic strength at this time also allows us to adjust the direction of regional development gradually and focus on developing new regions to achieve more balanced development. Only in this way could we maintain a high economic growth rate.

For a long time, many countries and regions have considered balanced development between regions as an important development goal. However, the development practice of various countries has proved that: Balanced development is a long historical process; due to the differences in natural conditions, geographical location as well as an economic and historical basis between regions, balanced development is just relative; the greater the difference in stability factors between regions, the greater the relativity of such "balance," that is, the degree of difference.

The socio-economic spatial structure has different characteristics in different stages (levels) of socio-economic development. In other words, the spatial organization of socio-economic objects in the region has different development stages and patterns, and there is an optimal pattern. In the early stage, geographers and economists proposed various location theories, which were for a spatial organization with a single factor. The "spatial structure" discussed here has 3D characteristics. The "space" in the spatial structure concept of social economy is not equal to the "absolute space" in physics or the static "pure space" in geometry. Still, it refers to other types of objects such as agriculture, industry, urban residential areas, road and communication facilities, cultural and commercial supply facilities, etc., where the following phenomena are constantly taking place, such as commodity production, transportation of raw materials and finished products, information transmission, commodity sales, development of new districts, population flow (dispersion and agglomeration), town expansion, generation of new residential areas, diffusion of new technologies, etc. Each thing and object, as well as the movement between them, will form a spatial situation, and their combined relationship as a whole will create multiple spaces. It corresponds to a comprehensive and holistic space system, including a system of various elements that exist in space. The investigation and analysis of the space system emphasize the interconnectedness between various components in the system, that is, the subsystems, and the process of their connection and change, that is, the instant reflection of various dynamic sequences. At different moments of development, the relationships between various things mentioned above, including the positional relationship, will change to different degrees. Therefore, things and phenomena have two-dimensional meanings of length and width and are also given the vertical (time) concept. This three-dimensionality of space is an important characteristic of the "diseconomies of agglomeration" of the spatial structure. On the other hand, the three-dimensionality of space also refers to the positional relationship of related things and phenomena in space. To determine the locations of the above conceptual objects, three mutually perpendicular lines can be drawn from any point in space in terms of coordinates, like three lines of length, width, and height in geometry. In this way, focusing on the positional relationship and location structure in space, the three-dimensional term "spatial structure" is undoubtedly more appropriate than "territorial structure".

2. Diffusion, agglomeration and intrinsic mechanism for formation of "point-axis system"

To further answer the relationship between spatial structure and development, it is necessary to explore and reveal various types of spatial structures and their relationship with development.

From the preliminary proposal of the "point-axis system" model in 1984 [4] to the publication of the book *Location Theory and Regional Analysis Methods* in 1989, the elaboration of the internal mechanism of the formation of the "point-axis system" was still very simple [5]. There lack an "analysis of the intrinsic mechanism" for the theory of the "point-axis system" [6]. When completing the natural science foundation project "Regional Development and its Spatial Structure" from 1990 to 1994 and compiling the *Report on China's Regional Development* in recent years, the theory of the "point-axis system" was improved in terms of the development of a scientific basis and the analysis of intrinsic mechanism.

W. Christaller proposed the central place theory as early as the 1930s and deduced the formation of the urban hierarchy and its forming mechanism. Theoretical geographers such as T. Hagerstrand proved in the 1960s and 1970s that: Similar to the principle of spatial interaction of objects, socio-economic objects (especially industrial and tertiary industry facilities) have two tendencies of spatial diffusion and spatial agglomeration [7]. It could be seen from the growth pole theory proposed by the French economist F. Perroux in the 1950s that regional exploitation in a certain range often starts from the point occupied by one or a few enterprises. The growth pole theory is one of the bases of the unbalanced development theory. The theories in these aspects are the scientific basis for the author to propose the "point-axis system." Analysis and demonstration are carried out in the following four aspects based on these scientific principles:

(1) In the process of economic development, the "axis" formed by connecting the linear infrastructure is formed between "points" at the time when most socio-economic factors are concentrated on the "point." The linear infrastructure bundle that connects cities at various levels is called the "development axis," as it has the function of promoting the development of a region similar to a sector.

The "axis" has strong economic attraction and cohesion to nearby regions. Meanwhile, the "axis" is also the path (direction) for the outward diffusion of socio-economic factors at the "point". That is to say, socio-economic objects progressively diffuse in space in a "point-axis" pattern. The "point" here refers to central cities at all levels, and the "axis" refers to the industrial agglomeration belt connected by traffic and communication trunk lines as well as energy and water channels.

The "point-axis system" reflects the objective laws of socio-economic spatial organization and the formed spatial structure. That is to say, the socio-economic spatial organization of any region or country must be in the form of a "point-axis system" under normal development conditions. Of course, due to the differences in geographical bases and the levels of socio-economic development, the formation process of the "point-axis" spatial structure has different levels and scales. Especially the characteristics of the spatial structure are different at different stages of socio-economic development: There is a low-level stage of economic and social development before the "point-axis system" is formed. When the surface is a homogeneous space, that is, the stage of spatial equilibrium of development, the socioeconomic objects are distributed in an "orderly" state but an unorganized state. The unorganized state of the space is extremely inefficient. At the beginning of industrialization, points and axes emerge simultaneously, the local region begins to be organized, and the resource development and economy of the region enter a period of dynamic growth. In the middle stage of industrialization with rapid regional economic growth, the main point-axis system framework is formed, the social economy evolves rapidly, and the spatial structure changes greatly. The "point-axis" spatial structure system is formed, and the region enters a comprehensive and organized state. Its formation is the result of the long-term self-organization process of socio-economic factors, as well as the result of scientific regional development policies, plans, and planning. From the macroscopic perspective, the spatial structure is recovered in the "balanced" stage. In this stage, although social organizations and economic organizations are highly efficient, population growth and economic growth, as the signs of social development, are not at a high rate [2].

(2) With the further development of the regional social economy, the "point-axis" will inevitably develop into a "point-axis-aggregation area". The agglomeration area here is also a "point", a "point" with a larger scale and external force.
(3) The core of the theory of the "point-axis system" is about the "optimal structure and optimal development" of the region. That is to say: The "point-axis system"

is the optimal structure for regional development; it is inevitably required to organize socioeconomic objects in the model of the "point-axis system" to achieve the optimal development of the region.

(4) The socio-economic development of a region or country is driven by the development axis and the central place (city). The axis is the concentrated zone of regional economic facilities and social facilities. According to the principle of spatial interaction (the source of the theory shall be the principle of Newtonian mechanics), the belt has a "force field," that is, a stunning effect on the surrounding regions. The socio-economic objects near the axis generate a centripetal force, which does not point to a point (town) on the axis but to several points or a line. This indicates that the axis has an agglomeration or agglutination effect on the social economy of the nearby area, and drives regional development through the objects within the sphere of influence. The socio-economic facilities concentrated on the axis have a diffusion effect on nearby areas through products, information, technology, personnel, finance, etc. The diffused material and non-material elements act on the nearby area and are combined with the elements of the area to form new productivity and promote socio-economic development.

The "point-axis system" is the most effective mode of regional development as it reflects the objective process and laws of socio-economic spatial organization. This theory reveals the relationship between socioeconomic development and the spatial organization of socio-economic objects in the process of regional development and also answers the question of the relationship between geographical processes and geographical patterns in human geography. Within the range of human geography, we investigate the growth and layout of facilities (development objects) such as agriculture, industry, transportation and communication, tertiary industry and culture, regional economic and social development and its spatial changes, and discuss the role of factors such as capital, labor, technology, information, location, and market that affect development. Among them, the development trend at each moment is bound to have a corresponding spatial pattern. This pattern is dominated by the scale, direction, mode, and nature (level) of socioeconomic development. However, a significant impact will also be caused on the development by adjusting the pattern. This is also one of the important tasks of the research on regional issues.

The experience of the optimal development process of the region tells us that: the development of a country or a region is always from unbalanced to relatively balanced. An answer to the development mechanism could be obtained from the theory of the "point-axis system." That is to say, when the level of economic development is deficient in the early stage of regional development, the focus of capital investment and development activities is concentrated on a few centers and axes. Such key development can cause rapid growth of the regional economy. As development activities increase and the level of economic development increases, more lower-level central places and development axes appear, and the focus of development activities gradually shifts to more central places and development axes with a lower level. In this way, the rapid growth of the regional economy continues in another development stage.

It shall be admitted that the formation process of the spatial structure of national and regional development is related to the economic system implemented. However, by analyzing the formed structural "type" in the scientific sense, it is found that the spatial organization of production and social factors ("self-organized" or artificially organized) has objective laws regardless of the system. In other words, the spatial structure under different economic systems shall not differ.

3. The "point-axis system" can lead to the optimal development of a region or country

Several aspects of optimal development have been mentioned in the discussion above. For the "point-axis system," the spatial structure that configures and improves productivity and the spatial organization that implements the entire social economy according to the "point-axis" spatial structure system can produce the effects in the following aspects. As a result, the optimal development of the region can be achieved:

(1) Development and exploitation in the model of "point-axis system" can comply with the socio-economic development and meet the objective requirements that the objects must be agglomerated into points in space and exert the agglomeration effects. In different geographical scopes, when a reasonable structure of the industry, agriculture, and tertiary industry and the organic combination between economic activities and regional infrastructure is determined, a huge spatial agglomeration effect will be produced. Such effect can only be fully utilized through development in the form of a "point-axis system."
(2) The development of the model of the "point-axis system" can give full play to the role of central cities at all levels. A city, as the core and commander of the region, is generated from the regional economic development and develops relying on raw materials, energy, food, labor, etc., provided by the affected areas. A city supplies and commands its area with industrial products, information, technology, policies, etc. To develop a vast area, central cities at all levels must be taken as the emphasis of development at different levels.
(3) The development of the model of the "point-axis system" can realize the best spatial combination between the production distribution and the linear infrastructure so as to achieve the necessary consistency between key traffic arteries, energy and water source construction lines at all levels within the range and towns, industrial and mining areas under key construction, thus avoiding the mutual separation of the two in space and time that appeared in the previous practice.
(4) The development in the model of the "point-axis system" is favorable for the convenient connection between cities, regions, and between urban and rural areas. Therefore, it is objectively favorable for realizing the specialization and cooperation between regions and between cities, thus forming an organic

regional economic network. In the information-based and knowledge-based economy society, the role of such a network in promoting regional economic development is increasingly important.

(5) The determination of key development axes at all levels in China can better combine the national strategy with the regional strategy, so that all regions and departments in the area have a clear and unified regional development direction, which is favorable for improving the investment effect of capital construction and the level of management organization. The development axis with the intensive transportation industry belt as the connotation drives the development of the regional economy just like the rotation of the door axis drives the rotation of the entire door.

Since the founding of the People's Republic of China, China has carried out large-scale territorial exploitation and regional development. In terms of the allocation of productivity and the spatial organization of social development, it basically meets the requirements for the model of the spatial structure of the "point-axis system" except in the "third-line" construction period. In the 1950s, the Harbin-Dalian Railway was first recommended for the national key development axes, followed by the Beijing–Wuhan section of the Beijing–Guangzhou Railway and the Zhengzhou–Lanzhou section of the Longhai Railway. Among the 156 key projects (152 under actual construction), nearly 65 projects were located on the three key development axes, accounting for more than 40% of the total. In the 1960s, China carried out the "third line" construction on a large scale. The layout policy implemented at that time deviated from the requirements of the law of agglomeration economies. Most of the key development axes were located in the central and western regions of China, such as Baoji–Chengdu, Chengdu–Kunming, Xiangyang–Chongqing, Jiaozuo–Zhicheng, and Guiyang–Kunming Railways. The construction projects were not far from the linear infrastructure (most were newly constructed), but they were decentralized. The central towns subject to key development were not prominent. The "point" was separated from the axis development, which brought huge economic losses. In the first half of the 1970s, the "third line" construction continued, and the above-mentioned development axes were still the focus of national industrial and transportation construction. Small and medium-sized towns with industrial points as the main body began to form, and the benefits of the "point axis" began to appear partially. In the mid to late 1970s, most of the major foreign projects introduced by China were located in coastal zones and along the Yangtze River and were close to seawater and freshwater sources. Among them, some enterprises used water transportation to improve their investment and operation effects. These large enterprises were mainly from the new construction and expansion of thermal power plants, oil refineries, iron and steel works, and petrochemical works. Among the ten large oil refineries and four expanded oil refineries constructed and put into operation from 1972 to 1980, 8 were located along the Yangtze River. Large chemical fertilizer plants were also mainly constructed on two national first-level development axes. This was major progress in China's territorial exploitation and industrial distribution on the basis of summarizing past experience and lessons.

In the past 20 years since the reform and opening-up, China's territorial exploitation and economic distribution have been more and more operating in a larger space according to the economic laws, which is manifested as operation in the model of "point-axis system" more prominently. On a national scale, large-scale exploitation was carried out successively with the coastal zones and the areas along the Yangtze River as the first-level axes of national development in the late 1970s and late 1980s. The strategic transformation from "balanced growth" to "unbalanced growth" has been achieved with the two first-level axes. The two are the key zones for capital investment in China in the past 20 years and for the development of various economic and technological development zones, and also the "policy belts" with preferential policies for facilities. The main body of the coastal open area was within the range of dozens of kilometers along the coastal zone (260 cities and counties, about 32×10^4 km^2 in total), and there were 14 open cities and Shenzhen, Zhuhai, Xiamen, Shantou, and Hainan Island special economic zones. About half of China's major projects were concentrated in the coastal zone, which was one of the two largest development axes in China. In the Seventh 5-year Plan, the investment in the coastal zone axis accounted for about 30% of that in China, and the net increase in the total industrial and agricultural output value accounted for about 40% of that in China. From this, we could see the status of the coastal zone axis in China. The construction scale of the axis along the Yangtze River is second only to that of the coastal zone axis. Especially since the 1980s, Jiangsu, Anhui, Jiangxi, Hubei, Hunan, Sichuan and other provinces along the Yangtze River have all proposed essential points to develop the industrial belt along the Yangtze River, and the emphasis on the construction in the province was placed on the area along the Yangtze River (called "Wanjiang Economic Belt" in Anhui). A series of mega-projects such as the Yangtze 300,000 t ethylene project, the Wuhan 1.7 m rolling mill, and the Gezhouba Dam hydro-junction were constructed in succession. Correspondingly, the Yangtze River shipping and other water transportation facilities were constructed on a large scale. In terms of the second-level development axis, the industrial belts, such as the area along the Longhai, Jiaoji, Changda and Jiaozhi Railways and the coastal area of the upper and middle reaches of the Yellow River, were mainly constructed. In the 1990s, China decided to carry out the large-scale construction of Pudong, Shanghai, which brought a series of new development opportunities to the development of the Yangtze River Industrial Belt.

Technological innovation will change the socio-economic spatial structure. Today, we are at a great historical turning point in the development of a highly information-based society. The information communication networks such as information communication systems and public databases are being constructed and promoted on a large scale. The Internet will change all aspects of economic and social life. Information is becoming the new factor of production and location; and satellite communications and video conference have further shortened the spatial distance around the world. Industries and institutions that rely heavily on information tend to be concentrated in space, while production, management and living will be decentralized. It can be believed that the pattern of spatial organization of economic and social factors will mutate on the basis of the "point-axis system" model in the future.

4. Strategic significance of "T"-shaped spatial structure

The strategy of "T"-shaped spatial structure for China's territorial exploitation and economic distribution is proposed on the basis of the theory of industrial and traffic distribution and the theory of "point-axis system". The proposed practice basis is the author's summary of the location factors and distribution laws of national productivity development, especially industrial development.

4.1 From "point-axis system" to "T"-shaped macrostructure

According to the above discussion on the relationship between the theory of "point-axis system" and development, it can be concluded that: China's territorial exploitation, regional economic development and productivity distribution shall be carried out according to the scientific model of "point-axis system"; meanwhile, to determine the emphasis of regional development and territorial exploitation is to determine the key development axes at all levels of a country or region (including central cities of course).

According to this theoretical model and through analysis on the distribution of resources and economic potential in various regions in China and other factors, the eastern coastal zones and the coastal zones of the Yangtze River shall be the strategic emphasis of China's territorial exploitation and economic distribution. The two first-level axes form a "T" shape.

The strategy of "T"-shaped structure scientifically reflects the spatial distribution framework of China's economic development potential. This strategy achieves the optimal spatial combination of China's productivity distribution with transportation, water and soil resources, urban support and domestic and foreign markets (the long-term contradiction between China's productivity and regional development is that resources, water sources, transportation and other conditions are not coordinated in many areas). The construction of the two first-level axes can drive the development of the national economy.

4.2 There is a profound background of China's regional development in the 1980s for proposing the "T"-shaped structure

China faced these two extremely challenging problems in the 1980s: whether the coastal areas, which have just been started in terms of opening-up and development, can continue to be the strategic emphasis of national development and where is the emphasis of further expanding opening-up and development? There are different views and debates on China's regional development strategy: in the early and mid-1980s, the national leaders at that time visited the west for many times (January 5, 1983 and August 1984), proposed the "Western Development", and implemented the "strategic shift" to the west. "It is required to build the western region into a national modernization base within 5–10 years". In this case, the trend of "great strategic shift" to the west was very prevailing, and a variety of strategic shift theories that served this appeared in the academic circles; some scholars advocated the "gradient theory" of regional development, and claimed that the strategic emphasis of economic development should be shifted from the coastal regions to the central regions first, and then to the western regions; some scholars also proposed the "equilibrium theory", which advocated the balanced development of various zones and regions.

Faced with this situation, it is very important for China to implement the scientific regional development policies. If the opening-up and development of the coastal areas, which were the emphasis of national development at that time, could not be constantly strengthened, but the "strategic shift" was implemented, or if the key construction areas and non-key construction areas were simply divided, the harm would be very great. Since the founding of the People's Republic of China, the macro policies of China's territorial exploitation and economic distribution have undergone several major changes. Among them, two major changes have violated the geological foundation of China, thus causing great losses. In this context, the author participated in the compilation of the *Outline of Overall Planning for National Territory* from May 1985 to 1987, undertook the compilation of the "Overall Distribution of National Productivity", and proposed the "T"-shaped strategy which will take 2 first-level axes of zones along the coast and the Yangtze River as the basic framework for China's territorial exploitation and economic distribution in the next 15 years and even a longer period. This conclusion was clearly written in the "*Outline*". After several revisions, the "Outline" was sent to the whole country for trial implementation on March 25, 1987.

4.3 Role of proposing the strategy of "T"-shaped spatial structure

The concept of "point-axis system" and the "T"-shaped structural model of China's territorial exploitation and economic distribution proposed on this theoretical basis provided an operational theoretical structure for territorial development, product distribution and regional economic development, thus becoming the basis for China's land development and macroeconomic distribution since the mid-1980s. Since the mid-to-late 1980s, the strategy of "T"-shaped spatial structure has been implemented on a large scale throughout the country and in various regions. In reality, it has become the development goal and policy behavior of the country as well as many provinces, cities, and districts.

4.3.1 One of the important factors for the sustained and rapid growth of China's economy. Some people do not understand the background and connotation proposed in the strategy of "T"-shaped spatial structure, believing that this is an elementary recognition without any important scientific value and practical significance. In fact, the key development of zones along the coast and the Yangtze River has become one of the important factors for the sustained and rapid growth of China's economy for more than 10 years.

China has adopted the strategy of a "T-shaped structure". The practice of regional development for more than ten years has shown that this theory and viewpoint have played an immeasurable role in China's socioeconomic development. This macro-regional development strategy is in line with China's national conditions in the current stage, and a huge success has been achieved, thus greatly enhancing the comprehensive national strength. From 1978 to 1990, the annual average growth rate of national income was 8.5%, while that of coastal areas was about 11%; from 1981 to 1990, the fund for the construction of large and medium-sized projects in coastal areas accounted for about 47% of that of the whole country. However, the gross domestic product (GDP) provided by the coastal zones accounted for 58% of that of the whole country; from 1990 to 1995, the annual GDP growth of the whole country was about 11.5%, and that of the coastal areas was more than 15%. The growth of Anhui, Jiangxi, Hubei, Hunan, and other provinces along the Yangtze River also exceeded the national average level. For a long time, the foreign trade export value and the utilization of foreign capital in the coastal areas accounted for more than 85% of the whole country. In recent years, the adjustment of industrial structure and the integration of industrial development and international standards have achieved remarkable results. Several pillar industries, such as automobiles, communication equipment, steel, petrochemicals, biological pharmacy, computers, etc., have been strengthened and cultivated in some regions. Among them, the electronic industry has become a new growth point of the national economy. Electric appliances, integrated circuits, program-controlled switches, etc., have grown substantially, and more and more products have entered the world market. The rapid economic growth in the coastal zones and the improvement of internationalization are the foundation and guarantee for the sustainable development of China's national economy. Today, the GDP of coastal areas accounts for more than 59% of that of the whole country, and the import and export trade accounts for more than 85%. The main transportation and communication hubs, financial centers, residences of large domestic and foreign companies, etc., in China are centralized in coastal areas. Coastal areas are the main supporting force for China to become one of the world's economic powers.

While continuing to strengthen the economic development in coastal areas, relevant state departments and various provinces and cities along the Yangtze River have concentrated funds to develop and construct the Yangtze River Industrial Belt since the second half of the 1980s. In 1994, the State Council convened a meeting on the development planning for the Yangtze River Delta and areas along the Yangtze River. Many regions are publicizing and implementing the key development strategy of the "point-axis system" in the region and developing and constructing various types of intensive industrial belts in a centralized way.

Recent scientific research has shown that almost all rising provinces (referring to the increase in the proportion of economic aggregates in the whole country) in China are concentrated along the coast and the Yangtze River in the past 10 years, which proves the predictability of the huge development potential of the areas along the coast and the Yangtze River and the scientificalness of its theoretical viewpoints. In terms of GDP, the proportion of the 17 provinces and cities along the coast and the Yangtze River in the whole country increased from 73.57% in 1985 to 77.83% in 1994 (the coastal areas here are the 12 provinces and cities usually referred to, including Beijing, but excluding Taiwan for the time being; the areas along the Yangtze River include Shanghai, Jiangsu, Zhejiang, Jiangxi, Anhui, Hunan, Hubei and Sichuan, 8 provinces and cities in total).

4.3.2 If the "great strategic shift" were implemented in the 1980s, an obvious loss of national interests would be caused. If the "great strategic shift" had been implemented in the mid-1980s when the development and opening-up of the coastal areas had just started, the first "overall situation" proposed by Comrade Deng Xiaoping would be destroyed. The results would be as follows: The opening-up of the coastal areas and the vast economic development advantages brought will be suspended, and the comprehensive national strength of China will not be as powerful as it is now; the huge potential of the areas along the Yangtze River that connects the east, middle and west zones will not be exerted, which will greatly affect the enhancement of China's comprehensive national strength; since the comprehensive national strength is not as powerful as it is now, the conditions for implementing the strategy of West Development cannot be provided as it is now.

Since 1986, the "point-axis system" theory and the theory of the "point-axis system" have been widely cited and applied in domestic magazines, journals, and treatises. They not only have a significant impact on government departments but also are widely respected in academic circles. It is believed in Japan's "Manager" Monthly issued in October 1993 that the Yangtze River Basin is a high-speed growth area following the coastal areas, and China's economy has changed from the I-shaped strategy to the T-shaped strategy.

References

[1] E. V. Boventer, *Standortentscheidung und Raumstrukt ur, Hanover* (Hermane Schoedd Verlag K C, 1979), pp. 50–135
[2] L. Dadao, *Regional Development and Its Spatial Structure* (Beijing Sciences)
[3] W. Christaller, *Die Zentralen Orte in Sudldeutschland* (Custau Fischern Jena,1933)
[4] L. Dadao, A sdentific configuration of indust rial allocation in China in 2000. Gegrph. Sci. **6**(2), 159 (1986)
[5] L. Dadao, *Location Theory and the Methods of Regional Studies* (Beijing Science Press, Beijing, 1988), pp. 101–102
[6] Y. Wuyang, *Liang Inshe Advanced Eionomic Geography* (Being Peking Umiversity Press, 1997)
[7] T. Hegerstrand, *Aspekte der Raeum ichen Strukt ur von Sozialen Komm unikationsnetzn und der Ifomationsausbreitung, Kiepenheuer* (Witsch, Berlin, 1970), pp. 367–3 9

Appendix B

Analysis on the Formation Mechanism of "Point-Axis" Spatial Structure System

Lu Dadao

Abstract
The theory of "point-axis system" is established based on the "central place theory", but the theoretical content and application goals of the two are different. The theory of "point-axis system" has achieved huge success in the practice of China's territorial exploitation and regional development. On the basis of analyzing the formation mechanism of the spatial structure of the "point-axis system" caused by spatial aggregation and spatial diffusion, the relationship between the theory of the "point-axis system" and the growth pole theory and the network development model is elaborated.

On the basis of long-term research on industrial location factors and the laws on the industrial and traffic distribution, the author proposed the theoretical model of the "point-axis system" and the "T"-shaped strategy (that is, the strategic emphasis of China's territorial exploitation and economic distribution shall be placed in the zones along the coast the Yangtze River) for China's territorial exploitation and economic distribution in the report on the "Scientific Basis of Master Plan for China's Industrial Distribution in 2000" at the national symposium on economic geography and territorial planning convened in Urumqi in October 1984 according to the basic principles of location theory and "spatial structure" theory. The official text was published in Geographical Science in 1986. However, when the theoretical model was proposed, the scientific basis of the theoretical model and the mechanism of model formation were not deduced or elaborated on in depth. Over the past 10 years, the formation process of the "point-axis" spatial structure, the structure and type of the "development axis", "point-axis progressive diffusion", "point-axis-cluster area", etc. have been elaborated in the monograph ("Regional Development and its Spatial Structure". Science Press, 1995) and several papers; the basic characteristics of the spatial structure in different stages of socio-economic development, as well as the

D. Lu, *Regional Development and Its Spatial Structure*,
https://doi.org/10.1007/978-981-99-7683-6

impact of spatial accessibility and location differential rent on regional development, have been analyzed. Therefore, this theory basically forms a system. The formation mechanism of the spatial structure of the point-axis system is further elaborated in this paper only.

1. Key point of the theoretical model and its application in territorial exploitation and economic distribution

1.1 Basis and basic key points of the theory

The theory of "point-axis system" is established based on the "central place theory" of the German geographer, Walter Christaller. The "central place theory" is the doctrine about the spatial structure of cities and urban functions, sizes, and compositions within a certain region (country), that is, the city's "hierarchy-scale" doctrine or urban location theory, and a regular hexagon is used to vividly summarize the relationship between regional urban hierarchy and urban scale. This theory provides a methodological basis for urban planning and regional planning. The "central ground theory" is the theoretical basis for the theory of the "point-axis system." This is mainly reflected as follows: The "central place theory" elaborates the basic process of spatial aggregation and spatial diffusion of socio-economic objects, and the spatial structure of the "point-axis system" is also formed on the basis of the same basic process; the "central place theory" reveals the hierarchy-scale law of the formation of residential areas and cities, and the spatial structure of the "point-axis system" also has a hierarchical relationship. In addition, the two have the logics and stereotypes that are basically the same in terms of model thinking and derivation. However, the theory of the "point-axis system" and the "central place theory" are different in terms of the theoretical system and application goals. In a few words: The "central place theory" is a doctrine about the law of urban scale-hierarchy and a theoretical model of urban planning and urban construction; the theory of "point-axis system" is one of the theories about the socio-economic spatial structure (organization), and is a theoretical model for productivity distribution, territorial exploitation, and regional development.

The model of the "point-axis system" is a spatial structure formed by socio-economic objects in reality after long-term development, and it is the main content of the theory of "point-axis system" to elaborate the formation process, mechanism, characteristics, and application of this model. The basic key points of the theory are as follows.

Socio-economic objects are always interacting with each other in the category of region or space. This is also similar to the basic principle of spatial interaction of objects, and there are two tendencies of spatial agglomeration and spatial diffusion.

In the process of national and regional development, most of the socio-economic elements are agglomerated at "points" and linked by linear infrastructure, thus forming "axes". The "point" here refers to the residential areas and central cities at all levels, and the "axis" refers to the infrastructure bundle connected by traffic and communication trunk lines as well as energy and water channels; the "axis" has strong economic attraction and cohesion to nearby regions. The socio-economic facilities concentrated on the axis have a diffusion effect on nearby areas through products, information, technology, personnel, finance, etc. The diffused material and non-material elements act on the nearby area, and are combined with the productivity elements of the region, so as to form new productivity and promote socio-economic development. During the development of a country or region, the industrial agglomeration belt will definitely be formed on the "infrastructure bundle"; due to the differences in geographical bases and the characteristics of socio-economic development in different countries and regions, the formation process of the "point-axis" spatial structure has different inner dynamics, types, levels, and scales; the spatial structure formed by social economy also has different characteristics in different stages (levels) of socio-economic development. Such characteristic is reflected in the degree of agglomeration and dispersion and the interaction between socio-economic objects.

With the further development of the regional social economy, the "point-axis" will inevitably develop into a "point-axis-aggregation area." The agglomeration area here is also a "point", a "point" with a larger scale and external force. The "development axis" has different structures and types, and the "point-axis" spatial structure system also affects regional development through spatial accessibility, location differential rent, etc.

According to the above key points, the theoretical model summary can be obtained: The core of the theory of "point-axis system" is about the "optimal structure and optimal development" of the region. That is to say: The "point-axis system" is the optimal spatial structure for regional development; it is inevitably required to organize socio-economic objects in the model of "point-axis system" to achieve the optimal development of the region. This theory also answers the relationship between the development process and the geographic pattern in regional development: the development process will definitely form a certain spatial pattern, which will in return, affect the regional development process. The integration and coordination between the two means the optimal development of the region. The "point-axis system" reflects the objective laws of socio-economic spatial organization, and is the most effective spatial structure model for territorial exploitation and regional development.

1.2 Application and impact of "point-axis system" theory and model in China's territorial exploitation and regional development

The "point-axis" development can meet the objective requirements that socio-economic development must be agglomerated into points in space and exert the agglomeration effect. In the three different geographical scopes of industrial area, industrial city, and industrial zone, when a reasonable industrial structure and an organic combination between industry and regional infrastructure are determined, a huge spatial agglomeration effect will be produced. Such effect can only be fully utilized by developing in the form of a point-axis. The "point-axis" development can give full play to the role of central cities at all levels. A city, as the core of the region, is generated from the regional economic development, and develops relying on raw materials, energy, food, labor, etc., provided by the affected areas. A city supplies and commands its area with industrial products, information, technology, policies, etc. To develop a vast area, central cities at all levels must be taken as the emphasis of development at different levels. The "point-axis" development can realize the best spatial combination between the production distribution and the linear infrastructure, so as to achieve the necessary consistency between key traffic arteries, energy and water source construction lines at all levels within the range and towns, industrial and mining areas under key construction, The "point-axis" development is favorable for the convenient connection between cities, between regions, and between urban and rural areas. Therefore, it is objectively favorable for realizing the specialization and cooperation between regions and between cities, thus forming an organic regional economic network. The determination of key development axes at all levels in China can better combine the national strategy with the regional strategy, so that all regions and departments in the area have a clear and unified regional development direction, which is favorable for improving the investment effect of capital construction and the level of management organization.

The "T"-shaped macro strategy of China's territorial exploitation and economic distribution is proposed according to the theory of the "point-axis system" and on the basis of analyzing China's natural foundation, especially the characteristics of economic distribution and comprehensive national strength of China in the 1980s. What's the meaning?

The concept of the "point-axis system" and the proposal of taking the zones along the coast and the Yangtze River as the strategic emphasis of China's territorial exploitation and economic distribution ("T"-shaped structure) that is made on this theoretical basis were written into the *Overall Planning for National Territory*. The plan was sent to the whole country for trial implementation in 1987. This model had also been used for territory development planning in 23 provinces, cities, and districts in the whole country by the early 1990s. It provides an operational theoretical structure for territorial exploitation, industrial production distribution, and regional economic development. It has become the strategic goal and policy behavior of the country and many provinces, cities, and districts. Wei Houkai, a regional economist, wrote in a book that: The "T"-shaped point-axis development strategy adopted by the country "vigorously promotes the rapid growth of China's economy and the effective utilization of funds". While continuing to strengthen the economic development in coastal areas, relevant state departments and various provinces and cities along the Yangtze River have concentrated funds to develop and construct the Yangtze River Industrial Belt since the second half of the 1980s. Many regions are publicizing and implementing the key development strategy of the "point-axis system" in the region, and developing and constructing various types of intensive industrial belts in a centralized way. Recent research has shown that almost all rising provinces (referring to the increase in the proportion of economic aggregates in the whole country) in China are concentrated along the coast and the Yangtze River in the past 10 years.

It has always been widely recognized and quoted in China's territorial exploitation and planning departments as well as the fields of human geography and regional economics since 1986. The theory of the "point-axis system" and the "T"-shaped macro-structural strategy have been quoted and applied in a large number of magazines,, journals and treatises. Famous geographers,, economists and leaders of government departments, such as Gui Shiyong, et al. (1987), Zuo Dakang and Li Wenyan (1990), Kitamura Yoshiyuki (1991), Sun Shangqing (1996), Liu Jiang (1999), Wang Mengkui (1999), Li Boxi (1996), Chen Dongsheng (1999), Zhang Dunfu (1998), Lu Yuqi (1998), Bai Hejin, et al. (2000), Wang Yiming (2000), Wei Houkai (2001), etc. discussed the scientific value of the theory of "point-axis system" as well as the necessity and significance of China's "T"-shaped development strategy in their published works[1].

There is a profound background of China's regional development in the 1980s for proposing the "point-axis system" and the "T"-shaped structure. China faced these two extremely challenging problems in the 1980s: whether the coastal areas, which have just been started in terms of opening-up and development, can continue to be the strategic emphasis of national development and where is the emphasis of further expanding opening-up and development? Due to the loosening of the

---

[1] Gui Shilian and Wei Liqun, Several Ideas on Economic and Social Development Strategies by the End of this Century, "Research on Planned Economy", 1987, No.7, P2.

Zuo Dakang, Li Wenyan, Xu Yuexian and Wang Ping: *Developing Geographical Sciences Towards Economic Construction*, published on the *Progress in the Research on Geographical Science* edited by Zuo Dakang, P7, Science Press, 1990.

Kitamura Yoshiyuki (Toyo University in Japan): "Comment on *China's Industrial Geography* and *Theory and Practice of China's Industrial Distribution*, published on the *Journal of Geographical Sciences*, 1991, 46(3).

Edited by Sun Shangqing, Yangtze River Development and Opening, P 1 ~ 7. China Development Press, 1996.

Edited by Liu Jiang, Review and Prospect of Regional Development in China, Volume 32, comprehensive volume, P37, 1999.

Edited by Wang Mengkui. Review and Prospect of China's Economic Development, P85, Chinese Financial & Economic Publishing House, 1999.

Wang Mengkui, Li Shantong, et al. Research on Imbalance of Socio-economic Development in China. P224 ~ 225, Commercial Press, 2000.

Research Group of the Development Research Center of the State Council, China's Regional Coordinated Development Strategy, P5.P13.P26, China Economic Publishing House, 1994.

Edited by Chen Dongsheng, China's Regional Development Across the Century, P56–57, Economy and Management Publishing House, 1999.

Edited by Xu Guodi, Research on Coordinated Development of Regional Economy in China, Reform Press, P107, 1997.

Edited by Zhang Dunfu, Research on Regional Economic Development, P45–47, China Light Industry Press, 1998 (on the point-axis system).

Bai Hejin, et al., China's Economic and Social Development Strategy in the Early 21st Century. P333–334. China Planning Press, 2000.

Edited by Wang Yiming, Research on China's Regional Economic Policies, P100, China Planning Press, 2000.

Lu Yulin, Research on Spatial Structure in Regional Development, P226–227, 253, etc. Nanjing Normal University Press, 1998.

Macro Research Institute of State Planning Commission, Territorial Exploitation and Regional Economic Research Institute, 1991–1997 Research Report (1997, unpublished).

Wei Houkai, Towards Sustainable and Coordinated Development, P37–39, Guangdong Economy Publishing House, 2001.

planning system and leaders' frequent visits to the west at that time, the trend of "great strategic shift" to the west was very prevailing, and a variety of strategic shift theories that served this appeared in the academic circles; there was also the "gradient theory" of regional development, which claimed that the strategic emphasis of economic development should be shifted from the coastal regions to the central regions; there was also the "equilibrium theory", which advocated the balanced development of various zones and regions. This situation made us realize that: it is very important for China to implement the scientific regional development strategy! If the key construction areas and non-key construction areas are simply divided, the harm would be very great. Since the founding of the People's Republic of China, the macro policies of China's territorial exploitation and economic distribution have undergone several major changes. Among them, two major changes have violated the geological foundation of China, thus causing great losses.

If the "great strategic shift" was implemented in the mid-1980s, the opening-up of the coastal areas and the huge economic development advantages brought would be suspended, and there would be no powerful national strength for large-scale development of the west in the current stage. The Yangtze River connects the east, middle and west zones, and the areas along the Yangtze River and the Yangtze River Basin have huge development potential. If the huge potential of the areas along the Yangtze River were not exerted over the past ten years, it would greatly affect the enhancement of China's comprehensive national strength.

The western region of China is vast, with a large area of land unsuitable for economic development and human habitation. The spatial model of western development is a major issue related to the implementation of the western development strategy. It was proposed in the *Notice of the State Council on Implementing Several Policies and Measures for the West Development* (People's Daily, December 28, 2000) to "connect points with lines and drive surfaces with points" for the West Development and build an economic belt. This is the specific application of the "point-axis system" in development. As early as 1991, the famous Japanese geographer, Kitamura Yoshiyuki, commented that the "point-axis system" is a "spatial diffusion theory for development from point to line and from line to the surface."

2. How to form the "point-axis" spatial structure system

Why can the spatial structure of "point-axis system" reflect the actual organization form of socio-economic objects in space, and can it be applied to the socio-economic development planning? Among them, the main reason is the organization form formed due to the interaction of socio-economic objects in space, which is a scientific reflection of objective laws.

2.1 Agglomeration and diffusion are two trends of the spatial movement of socio-economic objects

Since the emergence of modern geography, some human geographers have been studying the agglomeration phenomenon and agglomeration process of socio-economic objects as well as the spatial pattern formed. It is concluded that: The process of forming the spatial pattern can be divided into two trends, that is, spatial agglomeration and spatial diffusion. The research on the processes and phenomena of spatial agglomeration and spatial diffusion is one of the foundations of many geographic and economic theories.

According to the principles of physics, all kinds of objects have their own potential energy in space, and they are always transporting and diffusing their own potential energy to the surrounding environment. In the process of regional development, the diffusion of such potential energy is manifested as product flow, fund flow, people flow, technology flow, information flow, policy flow, etc. These "flows" flow from the center point (area) to the surroundings and are re-aggregated in the directions and at distances different from the center and combined with the original local natural and socio-economic elements, thus forming a new gathering point. When elaborating on the formation mechanism of the central place, W. Christaller emphasized that: the accumulation of matter to the core is the basic phenomenon of objects. That is, the objects in space originate from the center and are diffused outwards; the central place of the region, as the core of the region, is the commander of a specific region, that is, the town.

Any economic object must be connected with other objects for existence and operation in a homogeneous and undeveloped region; Due to people's need for social interaction and sharing of infrastructure (roads, mechanical power equipment, communication, energy and water supply, urban facilities, markets, etc.), socio-economic objects must be concentrated in a region or point. That is, agglomeration produces benefits. This basic principle is the same for the formation of villages in the early primitive society and the emergence of modern large cities. However, the excessive agglomeration of socio-economic objects at one point will inevitably lead to a series of side effects, such as traffic congestion, the serious shortage of water and soil resources and the resulting high cost, deterioration of the ecological environment, and social management problems brought therefrom. In this case, it is required to implement a certain degree of decentralized or balanced development. This is the objective requirement of the laws on spatial agglomeration and spatial diffusion of socio-economic objects.

2.2 Progressive diffusion leads to the formation of a "point-axis system"

Socio-economic objects are diffused in space in a progressive manner. We can imagine that: productivity is evenly distributed in a stage with low productivity level and extremely slow socio-economic development; in the early stage of industrialization, industrial and mining residential areas or towns first appear at the intersection of two resource origins or two traffic facilities with the development of mineral resources and the development of commodity economy, and traffic lines are constructed between these two points to meet the needs of social and economic connections; resources and economic facilities continue to be concentrated at these two points due to the agglomeration effect factors. Several large enterprises are established at these two points. The traffic lines become linear infrastructure bundles consisting of communication lines, energy supply lines, and telecommunication lines. Some economic facilities are established along the line. At the same time, new agglomerations begin to appear at several points nearby, and the traffic lines are extended accordingly. In this way, such infrastructure bundles become the (development) "axes" with further agglomeration and diffusion functions. The reason why such axes are called "development axes" is that they can drive regional economic development, just like the rotation of the door axis drives the rotation of the entire door. With further development of the such a model, the infrastructure bundles with low level are formed between some secondary points. These new and lower-level infrastructure bundles gradually become development axes with good development conditions, high benef, it levels and concentration of population, economy and technology. Greater agglomeration is formed on the initial infrastructure bundles, and a large number of population and economic units are concentrated along the line, forming a large dense industrial belt. The third-level and fourth-level development axes will be formed if such development continues. Accordingly, the territorial organization of productivity will be further improved, forming a spatial structure system marked by a "point-axis".

The above process indicates that: Socioeconomic objects originate from one or several diffusion sources in the early stages of regional development. The socio-economic "flows" are progressively diffused along several linear infrastructures (bundles) (also called diffusion channels), forming new agglomerations in various degrees at positions with different distances from the center. Due to the law that the diffusive force attenuates as the distance increases, the scale of new agglomerations also decreases as the distance increases. As a result of the diffusion of diffusion sources in adjacent areas, the diffusion channels are connected to each other, becoming the development axis. With the further development of the social economy, the development axis is further extended, and the new gathering points and development axes with relatively small scales are continuously formed. This is the basic content of progressive diffusion.

The "point-axis progressive" diffusion can realize the transition from unbalanced development to the relatively balanced development of the region. The exploitation and development levels and strengths are always unbalanced in a country or a region. As the exploitation process extends (the extension of axes and "points"), the entire country or region is gradually covered with development axes and "points", the differences in levels of "points" and axes become smaller, and a relatively balanced state begins to form, that is, a state of spatial exploitation and development from "points" to "lines" to "surfaces" is formed. It can be seen that the development process of this region is the result of "point-axis progressive diffusion" through microscopic investigation.

This is: the socio-economic objects are diffused in the space in a "point-axis progressive" manner, and such diffusion will inevitably lead to the formation of the "point-axis" spatial structure.

2.3 Main stages of the formation of the "point-axis" spatial structure system

According to the analysis of the process model of the formation of economic and social spatial organizations, in reality, it could be seen that there are four main stages in the formation of the socio-economic "point-axis" spatial structure system. The socio-economic spatial structures of these four stages reflect the general laws of main countries and regions and are also consistent with the level of socio-economic development and the stage differences in structural characteristics.

At Stage I, that is, the equilibrium stage before the point-axis is formed, the surface is a homogeneous space. Although being distributed in an "orderly" state, the socio-economic objects established in the agricultural society (residential areas with towns as the mainstay) are in an unorganized state. This unorganized spatial state is extremely inefficient.

At Stage II, socio-economic objects begin to agglomerate, points and axes begin to be formed at the same time, the local region begins to be organized, and the resource development and economy of the region enter a period of rapid growth. If measured according to the stage of socio-economic development, the characteristics of the spatial structure belong to the early stage of industrialization.

At Stage III, the main framework of the "point-axis system" is formed, the social economy evolves rapidly, and the spatial structure changes greatly. It belongs to the characteristics of the spatial structure of the middle stage of industrialization.

At Stage IV, the "point-axis" spatial structure system is formed. The region enters a comprehensive and organized state. Its formation is the result of the long-term self-organization process of socio-economic factors, as well as the result of scientific regional development policies, plans, and planning. From the macroscopic perspective, the spatial structure is recovered in the "balanced" stage. In this stage, although social and economic organizations are highly efficient, population growth and economic growth, as the signs of social development, are not at a high speed.

2.4 "Point-axis system" and optimal development of the region

The "point-axis system" reflects the objective process and laws of socio-economic spatial organization, and is the most effective mode of regional development. If socio-economic development is organized according to the development model of the "point-axis system", the relationship between concentration and decentralization, between fairness and efficiency, and between unbalanced development and more balanced development can be scientifically handled.

It is a problem difficult to solve in the process of exploitation and development in many countries and regions to handle the relationship between centralization and decentralization, between fairness and efficiency, and between unbalanced development and more balanced development. This problem is particularly prominent in China. For the division of coastal and inland areas in the 1950s, the division of inland areas in the 1960s, the division of first-, second- and third-tier areas in the 1970s, and the "strategic shift" (from the coast to the west) that prevailed in the mid-1980s, it is hoped to well handle the relationship between centralization and decentralization, between fairness and efficiency, and between unbalanced development and more balanced development. The regional development strategies at that time all reflected the objective needs of territorial exploitation and economic development to a great extent. However, the practice of dividing key areas and non-key areas by region cannot provide a concrete framework that can be used for practical operation, which affected the benefits of China's territorial exploitation and economical construction.

After the "point-axis" spatial structure system was basically formed, the regions entered a comprehensive and organized state. Its formation is the result of the long-term self-organization process of socio-economic factors, as well as the result of scientific regional development policies, plans, and planning. From the macroscopic perspective, the spatial structure is recovered in the "balanced" stage. In this stage, although social and economic organizations are highly efficient, population growth and economic growth, as the signs of social development, are not at a high speed.

If the infrastructure and economic distribution are implemented according to the "point-axis" spatial structure system, scientific accessibility can be achieved for the region, so that the flow of goods, personnel, and information is the most reasonable. Therefore, the optimal development of the region can be achieved.

3. Similarities and differences between growth pole, point-axis system and network development model

3.1 Theory of "point-axis system" and growth pole theory

The growth pole theory was first proposed by the French economist, F. Perroux, in the 1950s. There have been many introductions to the growth pole theory in China. The theory is established based on the 'central place theory' from the perspective of its scientific basis. The key points are as follows:

1. In the context of high industrialization, the agglomeration of social production is first realized at the sites with the momentum for economic development. As the internal motion of the magnetic field is the strongest at the magnetic pole, such regional "polarization" of economic development is called the "Growth pole."
2. The industry that serves as the basis for the development and role of the growth pole is called a key industry. It is characterized by a large production scale, a powerful driving force for growth and extensive linkages with other industries. When the critical industry starts to grow, other industries in the region where the enterprise (or sector) is located also begin to grow. The momentum for economic growth will gradually penetrate and eventually spread to the entire region.
3. The growth pole acts on surrounding areas in two directions. The first is the "polarization process", that is, the agglomeration process. That is, the growth pole attracts the natural and socio-economic potential of the surrounding area, such as mineral resources, raw materials, labor force, investment, local industries, or enterprises, with its strong economic and technical strength and superior conditions. The second is the diffusion process, also known as the "leakage process", that is, the growth pole invests in the surrounding areas and provides other economic and technical assistance to form affiliated enterprises or subsidiaries, so as to provide markets for primary products in the surrounding areas and absorb the agricultural surplus labor force.
4. At the middle and late stages of the development and evolution of the growth pole, the "leakage" (diffusion) effect is the mainstay. At this stage, giving is more than absorption, and the regional development level tends to be balanced.

The common points between the "growth pole" and "point-axis system" models are as follows: The models are formed due to spatial agglomeration and spatial diffusion (spatial agglomeration before spatial diffusion). Another common point is the spatial form of socio-economic objects. However, there are differences between them: "growth pole" is a theoretical model that clarifies the role of central cities at all levels in regional economic growth, while the "point-axis system" model is an overall model of regional economic development; the "growth pole" theory emphasizes the economic process and economic connection formed by the spatial structhe ture, and is a theoretical model in the context of high industrialization. However, the theory of "point-axis system" emphasizes more on the spatial process, and is a theoretical model applicable to all development stages.

3.2 "Point-axis system" development model and network development model

The network development model was proposed by Chinese scholars in the 1990s. According to the narration of Wei Houkai, the network development model is a spatial structure under the condition that regional economic development has reached a high level. The key point is as follows: The development of the "point-axis system" has led to excessive agglomeration in the middle and later stages of industrialization, and regional economy requires the relatively balanced development, which is also possible to be implemented. With the development of traffic, communication and network technologies, the regional economy tends to be decentralized, thus forming network conditions.

The network development model is actually further development of the "point-axis system" model, and is a manifestation of the theoretical model. There is no essential difference between the two. Because the "point-axis system" development model emphasizes the following: within the range of a country (or a region), it is necessary to determine several linear infrastructure axes between regions, provinces and prefecture-level cities and with favorable development conditions, focus on the development of the axis zone, especially several points (city and urban area-development centers), and develop the resources located on the axis and within the direct attraction range of the axis in priority. With the continuous enhancement of economic strength, more and more attentions of economic development shall be placed on the development axis and development center at the lower level. In other words, the development axis is gradually extended to the less developed areas (including the area within the developed areas but far from the development axis and the development center). Points that were not considered as development centers previously are determined as development centers at lower levels, and new development centers are specified. In the same way, there are tertiary and quaternary centers around the secondary center. The functions of central cities at different levels are manifested in different production scales, technical levels, and product sales ranges in terms of industrial development distribution. In this way, the growth centers and development axes at different levels form a spatial network of social economy. It is the repeated coverage of central cities at different levels and development axis networks on the (regional) surface.

References

[1] L. Dadao, The scientific basis of the general map of China's industrial productivity distribution in 2000. Geogr. Sci. **6**(2), 110–118 (1986)
[2] W. Christaller. *Die Zentraler Orte in Sudteutschand* (Jena, 1933)
[3] W. Houkai, *Towards Sustainable and Coordinated Development* (Guangdong Economy Publishing House, Guangzhou, 2001)
[4] L. Dadao, *Theory and Practice of China's Industrial Distribution* (Science Press, Beijing, 1990)
[5] K. Yoshiyuki. Comment on China's industrial geography and theory and practice of China's industrial distribution. *J. Geograph. Sci.* **46**(3), 383–384 (1991)
[6] W. Christaller, Die Hierarchie der Staedie, in *The IGU Symposium in Urban Geography* (Lund, 1960)
[7] D. Sibao, "Growth Pole" model and economic development in underdeveloped regions. Econ. Geogr. **9**(4), 457 (1989)
[8] L. Dadao, *Regional Development and its Spatial Structure* (Science Press, Beijing, 1995)